D1243535

ELECTRICAL MACHINES

ELECTRICAL MACHINES

Turan Gönen

California State University, Sacramento

Power International Press

Carmichael, California

ELECTRICAL MACHINERY

Power International Press

P.O. Box 1294, Carmichael, CA 95609-1294

This book was set in Century Schoolbook by Modern Technical Typographers.
The editor was Joan Banna.
The assistant editor was Valerie Robbins.
The design and layout was done by Jacob Connell.
The production supervisor was Nenog Narut.
The art director was Amy Anderson.
The cover design and illustrations were done by Jacob Connell.

Printed in the United States of America

10   9   8   7   6   5   4   3   2   1

ISBN 0-9658944-2-8

Library of Congress Cataloging-in-Publication Data:

Gönen, Turan.
    Electrical machines / Turan Gönen.
       p.   cm.
    Includes bibliographical references and index.
    ISBN 0-9658944-2-8
    1. Electric machinery.   I. Title.
TK2000.G66   1998   97-27376
621.31'042--dc21   CIP

ABOUT THE AUTHOR

Turan Gönen is Professor of Electrical Engineering and Director of the Electrical Power Educational Institute at California State University, Sacramento. Previously, Dr. Gönen was Professor of Electrical Engineering and Director of the Energy Systems and Resources Program at the University of Missouri-Columbia. Professor Gönen also held teaching positions at the University of Missouri-Rolla, the University of Oklahoma, Iowa State University, Florida International University and Ankara Technical College. He has taught electrical machines and electric power engineering for over twenty years.

Dr. Gönen also has a strong background in power industry; for eight years he worked as a design engineer in numerous companies both in the United States and abroad. He has served as a consultant for the United Nations Industrial Development Organization (UNIDO), Aramco, Black & Veatch Consultant Engineers, and the public utility industry. Professor Gönen has written over 90 technical papers as well as four other books: *Electric Power Distribution System Engineering, Modern Power System Analysis, Electric Power Transmission System Engineering: Analysis and Design*, and *Engineering Economy for Engineering Managers*.

Turan Gönen is a senior member of the Institute of Electrical and Electronics Engineers and the Institute of Industrial Engineers. He serves on several Committees and Working Groups of the IEEE Power Engineering Society, and he is a member of numerous honor societies including Sigma Xi, Phi Kappa Phi, Eta Kappa Nu, and Tau Alpha Pi. Professor Gönen received the Outstanding Teacher Award at CSUS in 1997.

Dr. Gönen holds a BS and MS in Electrical Engineering from Istanbul Technical College (1964 and 1966, respectively), and a Ph.D. in Electrical Engineering from Iowa State University (1975). Dr. Gönen also received an MS in Industrial Engineering (1973) and a Ph.D. co-major in Industrial Engineering (1978) from Iowa State University, and a Master of Business Administration (MBA) degree from the University of Oklahoma (1980).

PREFACE

As electrical engineering programs have become overloaded with various new courses, many universities have started to offer only one course in electrical machinery. Therefore, the primary purpose in writing this book is to provide a meaningful and easily readable textbook for a three-semester-hour introductory-level electrical machinery course. Clearly, the purpose of this book is not an introduction to the design of electrical machinery. But is intended for students in electrical and other engineering disciplines. It is assumed that the students have already completed an electrical circuits analysis course and are familiar with electromagnetic fields.

This book has evolved from the content of courses given by the author at California State University, Sacramento; the University of Missouri at Columbia; the University of Oklahoma; Florida International University; and Iowa State University. It has been written for junior-level undergraduate students, as well as practicing engineers in the industry. The book is suitable for both electrical and non-electrical engineering students and has been particularly written for students or practicing engineers who may want to teach themselves.

Basic material has been explained carefully and in detail with numerous examples. Special features of the book include ample numerical examples and problems designed to use the information presented in each chapter. Each new term is clearly defined when it is first introduced in the text and a special effort has been made to familiarize the reader with the vocabulary and symbols used by the industry. Also, consistent with modern practice, the *International System* (SI) of units has been used throughout the book.

It is important for every electrical engineering student, regardless of his or her particular specialization area, to be familiar with the fundamental concepts involving three-phase circuits, power, and power measurement in ac circuits. However, it has been the experience and observation of the author throughout his long teaching career that such topics are often reviewed either inadequately or too quickly toward the end of a basic circuit analysis course, or not at all in some cases due to lack of time. Therefore, as a remedy, a brief review of these topics is included in the first two chapters of this book. Of course, instructors will decide for themselves whether or not to review them in detail, depending on the need.

Furthermore, it is a fact that most students who are taking a basic electrical machines course are not specializing in electrical power engineering. Certainly they will not be designing any electrical machines during their professional careers; probably not even those students specializing in electrical power engineering will need to do so. Therefore, based on such reasoning as well as the Power Engineering Society of the IEEE, a more general approach to electrical machines has been used throughout this book.

It is expected that a typical student has a working knowledge of complex algebra, sinusoidal analysis, phasor diagrams, phasor analysis, and other basic concepts. However, it is a good idea to review the first two chapters briefly to establish a common background before reading further chapters. It is recommended that the minimum amount of material covered in a one-semester course should include the chapters on magnetic circuits, transformers, induction machines, synchronous machines, and direct-current machines, depending upon the purpose of the course. A complete solutions manual is available for instructors from the publisher.

ACKNOWLEDGMENTS

The author wishes to express his appreciation to Dr. David D. Robb of D. D. Robb and Associates for his kind encouragement and invaluable suggestions and friendship over the years. The author also wishes to express his sincere appreciation to Dr. Paul M. Anderson of Power Math Associates for his continual encouragement and suggestions.

The author is most grateful to numerous colleagues, particularly Dr. Anjan Bose of Washington State University; Dr. Thomas H. Ortmeyer of Clarkson University; Dr. Juan C. Balda of University of Arkansas; Dr. Yahya Baghzouz of University of Nevada, Las Vegas; Dr. Alexander E. Emanuel of Worcester Polytechnic Institute; Dr. Adly A. Girgis and Dr. Elham B. Makram of Clemson University; Dr. Alvin Day of Iowa State University; Dr. G. T. Heydt of Arizona State University; Dr. Charles Slivinsky, Dr. Richard G. Hoft, Dr. Cyrus O. Harbourt, and Dr. James R. Tudor of the University of Missouri at Columbia; Professor John Pavlat and Dr. Kenneth C. Kruempel of Iowa State University; Dr. Max Anderson and Dr. Earl F. Richards of the University of Missouri at Rolla; Dr. James Story of Florida International University; Dr. Don Koval of the University of Alberta; and Dr. B. P. Lathi of California State University, Sacramento for their interest, encouragement, and invaluable suggestions. Also acknowledged are Dr. Salah Yousif, Dr. John Balachandra, Dr. Miroslav Markovic, and Dr. S. K. Ramesh of California State University, Sacramento, for their support and encouragement.

A special thank-you is extended to Gerhard W. Juette and Klaus Habur of Siemens A.G.; John R. Stoutland, Ron Stevens, and Darlene Heare of General Electric Company; Gary B. Lister of Canadian General Electric Company; Bill Petruska of ABB, Inc.; Andy Carpenter, Marilyn Muscenti of Reliance Electric, Inc.; Judy Chaves of North American Transformer, Inc.; and Robert Murray and Otto Stoll of MagneTex, Inc.

The author is also indebted to numerous students who studied portions of the book at California State University, Sacramento; the University of Missouri at Columbia; the University of Oklahoma, and Florida International University and made countless contributions and valuable suggestions for improvements.

TURAN GÖNEN

Sacramento, California
September, 1997

CONTENTS

ELECTRICAL MACHINES

CHAPTER 1

BASIC CONCEPTS

I can resist everything except temptation.
Oscar Wilde, *Lady Windermere's Fan*, 1892

I wouldn't belong to any club that would have me as a member.
Groucho Marx

If you are not puzzled yet, you are just not listening to me.
Author Unknown

1.1 INTRODUCTION

In the United States, the use of electrical energy increased quickly after 1882 and power plants mushroomed across the entire country.[1] The main reasons for such a rapid increase in demand for electrical energy are several: Electrical energy is, in many ways, the most convenient energy form. It can be sent by power lines over great distances to the consumption point and easily transformed into mechanical work, radiant energy, heat, light, or other forms. Electrical energy cannot be stored effectively, has also contributed to its growing use. Further, by generating electrical energy in very large power plants an *economy of scale* in such production can be achieved; i.e., the unit cost of electrical energy goes down with increasing plant size. In general, the use of electrical energy may include various kinds of conversion equipment in addition to transmission lines and control devices.

The structure of an electrical power or energy[2] system is very large and complicated. However, it can be represented basically by five main components. The *energy source* may be coal, natural gas, or oil burned in a furnace to heat

---

[1] As of 1991, the number of electrical motors in the 1-to-120 hp range was more than 125 million, according to a recent study by the U.S. Department of Energy. This study also stated that 53-58 percent of the electrical energy generated is consumed by electric motor-driven systems.

[2] The term *energy* is being increasingly used in the electrical power industry to replace the conventional term *power*, depending on the context. Here, the terms are used interchangeably.

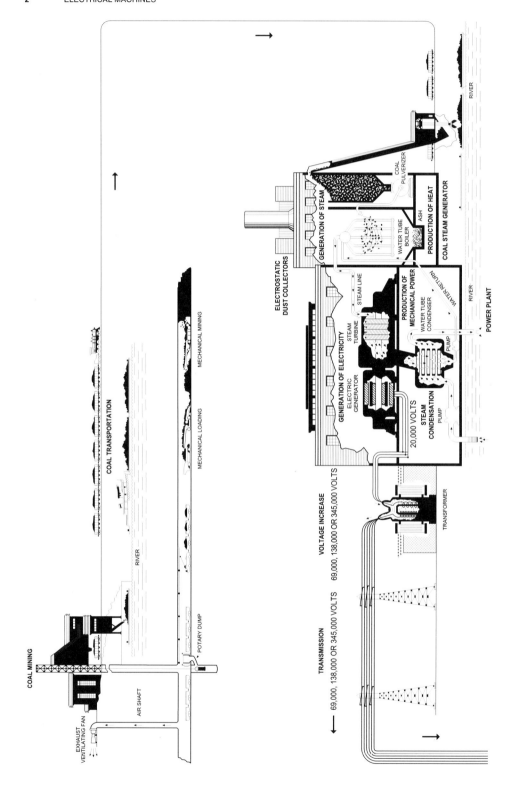

COAL MINING

EXHAUST VENTILATING FAN

AIR SHAFT

POTARY DUMP

COAL TRANSPORTATION

RIVER

MECHANICAL LOADING

MECHANICAL MINING

ELECTROSTATIC DUST COLLECTORS

GENERATION OF STEAM

COAL PULVERIZER

ASH

WATER TUBE BOILER

PRODUCTION OF HEAT

COAL STEAM GENERATOR

RIVER

GENERATION OF ELECTRICITY

STEAM LINE

STEAM TURBINE

ELECTRIC GENERATOR

PRODUCTION OF MECHANICAL POWER

WATER TUBE CONDENSER

WATER RETURN

PUMP

RIVER

POWER PLANT

20,000 VOLTS

STEAM CONDENSATION

PUMP

PUMP

TRANSFORMER

TRANSMISSION

VOLTAGE INCREASE

← 69,000, 138,000 OR 345,000 VOLTS

69,000, 138,000 OR 345,000 VOLTS →

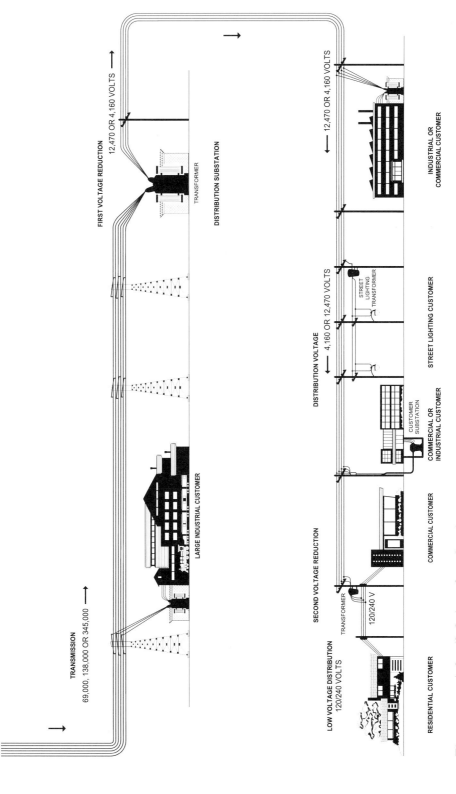

Figure 1.1 A detailed view of an electric power system.

water and generate steam in a boiler; it may be water in a dam; it may be oil or gas burned in a combustion turbine; or it may be fissionable material, which in a nuclear reactor will heat water to produce steam. The *generation system* converts the energy source into electrical energy. The *transmission system* transports this bulk electrical energy from the generation system to principal load centers where it is distributed through (usually extra) high-voltage lines. The *distribution system* distributes such energy to consumers by using lower voltage networks. Finally, the last component, that is *load*, utilizes the energy by converting it to a required form for lights, motors, heaters, or other equipment, alone or in combination.

Figure 1.1 shows a detailed view of an *electrical power system* that delivers energy from the source to the load connected to it. Note that the first transformer in the system (the one next to the power plant) is called a *step-up transformer*, and the second transformer (the one at the end of the transmission line) is called a *step-down transformer*.

According to the *energy conservation principle* of thermodynamics, energy is never used up; it is simply converted to different forms. Presently available energy conversion methods can be categorized into four different groups. The first group includes the conventional methods that generate more than 99 percent of today's electrical energy. They convert thermal energy from fossil fuels or from nuclear fission energy to mechanical energy via thermal energy and then to electrical energy; or they convert hydro energy to electrical energy. The second group contains methods that are technically possible but have low-energy conversion efficiency, such as the internal combustion engine and the gas turbine. The third group covers the methods capable of supplying only very small amounts of energy, e.g., photovoltaic solar cells, fuel cells, and batteries. The last group includes methods that are not technologically feasible but appear to have great potential, e.g., fusion reactors, magneto-hydrodynamic (MHD) generators, and electrogasdynamic generators. Figure 1.2 shows an installed 1300 MW cross compound turbine generator.

1.2 A BRIEF OVERVIEW OF BASIC ELECTRICAL MACHINES

In general, an electrical machine can be defined as an apparatus that can be used either to convert electrical energy into mechanical energy or to convert mechanical energy into electrical energy. If such a machine is used to convert electrical energy into mechanical energy, it is called a *motor*; however, if it is used to convert mechanical energy into electrical energy, it is called a *generator*. Any given machine can convert energy in either direction and can therefore be used either as a motor or as a generator. Such conversion is facilitated through the action of a magnetic field. A generator has a rotary motion provided by a prime mover which supplies mechanical energy input. The relative motion between the conductors and the magnetic field of a generator produces an electrical energy output. On the other hand, a motor has electrical energy supplied to its windings and a magnetic

Figure 1.2   An installed 1300 MW cross compound turbine generator.

field that generates an electromagnetic interaction to produce mechanical energy or torque.

In general, each machine has a nonmoving (i.e., stationary) part and a moving (i.e., nonstationary) part. Depending on whether such a machine functions as a generator or motor, the moving part that is attached to a mechanical system receives mechanical input or provides mechanical output. The motion of such a moving part can be linear (e.g., as in *linear motors*); vibrating or reciprocating (e.g., as in various electrical razors); or rotating.

In this book, only the rotating electrical machines will be reviewed. They include (1) polyphase synchronous machines, (2) polyphase induction (also called *asynchronous*) machines, and (3) dc machines. However, there are other rotating and linear machines that will not be included here. They operate basically on the same principles. Examples include:

- **Reluctance machines**, which are synchronous machines without the dc excitation and are used in timers, electrical clocks, and recording applications.
- **Hysteresis machines**, which are similar to reluctance machines with a solid cylinder rotor made up of a permanent magnet material that needs only one electrical input. They are used in phonograph turntables and in other constant-speed applications, such as electrical clocks.
- **Rotating rectifiers**, which have the same performance as regular synchronous machines except that field excitation is provided by an ac auxiliary generator and by rectifiers that are stationed on the rotor.

- **Permanent magnet machines**, which are ordinary synchronous machines with the field excitation provided by a permanent magnet. They have a very high efficiency since there are no field losses.
- **Beckey-Robinson** and **Nadyne Rice machines**, which are brushless synchronous machines that operates based on rotor magnetic structure with a changing reluctance. They are mainly used in aerospace applications.
- **Lundell machines**, which are also brushless synchronous machines (but need slip rings to supply a dc field) that operate based on rotor magnetic structures with a changing reluctance. They are mainly used in automotive alternators.
- **Inductor and flux-switch machines**, which are inductor flux-switch configurations based on a changeable-reluctance principle similar to the reluctance machines, and a function of rotor position accomplished by the rotor design. They can be used as brushless synchronous motors and generators in aerospace and traction applications.

In addition to basic rotating electrical machines, transformers will also be discussed. Even though a transformer involves the interchange of ac electrical energy from one voltage level to another, some of its principles of operation constitute the foundation for the study of electromechanical energy conversion. Thus, many of the relevant equations and conclusions of transformer theory are equally applicable to electromechanical energy conversion theory.

Rotating electrical machines have an outside (i.e., *stationary*) part that is called the stator and an inner (i.e., *rotating*) part that is called the rotor. As shown in Figure 2.1(a), the rotor is centered within the stator, and the space that is located between the outside of the rotor and the inside of the stator is called the air gap. The figure shows that the rotor is supported by a steel rod that is called a shaft. In turn, the shaft is supported by bearings so that the rotor can turn freely. Both the rotor and stator of a rotating machine, as well as a transformer, have windings, The terminology that is commonly used to describe the windings of basic electrical machines and transformers is given in Table 1.1.

It is important to point out that in the study of any electromechanical apparatus there is a need to model its electric circuit and one should be very familiar with the ac circuit analysis applicable to power circuits. Each electric circuit concept is analogous to a corresponding magnetic circuit concept.[3] Thus, to understand electrical machines, one needs a good knowledge of the concepts of both magnetic circuits and electrical power circuits. Therefore, a brief review of phasor representation is included in Appendix A. Also, the concepts of real and reactive powers in single-phase ac circuits are briefly reviewed in the following section of this chapter. In Chapters 2, and 3, the concepts associated with three-phase circuits and magnetic circuits are reviewed. It is hoped that such reviews are sufficient to provide a common

---

[3] In Table 1.1 "ac current" is grammatically a redundant statement. Nevertheless, ac and dc, originally used as abbreviations, are now commonly used as adjectives in engineering vocabulary.

TABLE 1.1 Terminology Used to Describe the Windings of Basic Electrical Machines and Transformers

Apparatus	Name of Winding	Location of Winding	Function of Winding	Type of Current in Winding
Synchronous machine	Armature	Stator	Input/output	ac
	Field	Rotor	Magnetizing	dc
Induction machine	Stator	Stator	Input	ac
	Rotor	Rotor	Output	ac
dc machine	Armature	Rotor	Input/output	ac in winding
	Field	Stator	Magnetizing	dc at brushes
Transformer	Primary	---	Input	ac
	Secondary	---	Output	ac

base, in terms of notation and references, in order to be able to follow the subsequent chapters.

1.3 REAL AND REACTIVE POWERS IN SINGLE-PHASE AC CIRCUITS

If the sinusoidal voltage across the terminals of a single-phase ac circuit is used *as a reference* and designated by the phasor $|V|\angle 0°$, and the phasor of the alternating current in the circuit is $|I|\angle\theta$, then the *real power* (i.e., *average* or *active power*) can be expressed as

$$P = |V||I| \cos\theta \tag{1.1}$$

or

$$P = V_{rms} I_{rms} \cos\theta \tag{1.2}$$

$$Q = |V||I| \sin\theta \tag{1.3}$$

or

$$Q = V_{rms} I_{rms} \sin\theta \tag{1.4}$$

Since sin θ is dimensionless, Q has the dimension of volt-ampere. However, to help distinguish between real and reactive powers, Q is measured in var, which stands for volt-ampere-reactive. The relation of a voltage with respect to a lagging current can be observed in the phasor diagram shown in Figure 1.3(a). The term *power factor* is used for the factor cos θ and sometimes the term *reactive factor* is used for the factor sin θ.

Apparent power S is the product of the phasor voltage magnitude and current magnitude. Therefore, it can be expressed as

$$S = |\boldsymbol{V}||\boldsymbol{I}| \tag{1.5}$$

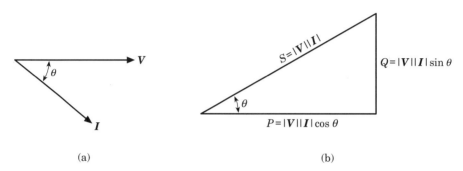

(a)   (b)

Figure 1.3   For a lagging current: (a) current and voltage phasor diagram; (b) power triangle.

The relationship between the real, reactive, and apparent powers can be represented by a triangle, known as the *power triangle*, as shown in Figure 1.3(b).

If the phasor voltage and phasor current are of a purely resistive network, the real power (i.e., average or active power) can be expressed as

$$P = |\boldsymbol{V}||\boldsymbol{I}| \cos \theta$$

$$= \frac{|V|^2}{R} \tag{1.6}$$

$$= |\boldsymbol{I}|^2 R$$

since cos $\theta = 1$.

Similarly, the reactive power for a purely capacitive network can be expressed as

$$Q = |\boldsymbol{V}||\boldsymbol{I}| \sin \theta$$

$$= \frac{|V|^2}{X}$$

$$= |I|^2 X \tag{1.7}$$

since $\sin \theta = 1$.

In power system computations, it has been customary to define a *phasor sum*[4] as,

$$S = P + jQ = VI^* \tag{1.8}$$

where S = complex (or phasor) power
I^* = conjugate of current phasor I

When the resultant S *complex* (or *phasor*) *power* is in rectangular form, then the related real and reactive power can be expressed as

$$P = \mathscr{R}e(VI^*) = VI \cos \theta \tag{1.9}$$

and

$$Q = \mathscr{I}m(VI^*) = VI \sin \theta \tag{1.10}$$

PROBLEMS

1.1 Assume that two load impedances Z_1 and Z_2 are connected in series with respect to each other but that loads Z_1, Z_2, and Z_3 require 5 kVA at 0.8 lagging power factor, 10 kVA at 0.9 lagging power factor, 5 kW at 0.6 lagging power factor, and 5 kVA at 0.8 lagging power factor, respectively. Determine the kW required by the total load, if the frequency is 60 Hz.

1.2 A 10-kW 220-V single-phase ac motor is operating at 0.7 lagging power factor. Find the value of the capacitor that needs to be connected in parallel with the motor, if the power factor is to be improved to 0.95 lagging.

1.3 Assume that a single-phase $2,400\angle 0°$ V bus is connected to a single phase 100-kW motor operating at a lagging power factor of 0.9, a lighting load of 100-kW operating at a unity power factor, and a static capacitor of 100-kvar. Determine the following:
   (a) The total complex power supplied by the bus to the three loads.
   (b) The total current supplied to the bus.
   (c) The power factor of the total load connected to the bus.

---

[4] The algebraic sign of reactive power has been a subject of debate for 25 years. Finally the convention defining S as VI^* was adopted by the American Institute of Electrical Engineers, approved by the American Standards Association, and published in *Electrical Engineering*[2]. Therefore, to obtain the proper sign for Q, it is necessary to calculate S as VI^*, rather than V^*I, which would reverse the sign for Q.

1.4   A single-phase $4,800\angle0°$ V bus is connected to a single-phase 100-kW motor operating at a lagging power factor of 0.8, a lighting load of 200-kW operating at a unity power factor, and static capacitors of 150-kvar. Determine the following:
(a) The total complex power supplied by the bus to the loads.
(b) The total current supplied to the bus.
(c) The power factor of the total load connected to the bus.

1.5   Consider Problem 1.3 and use the bus voltage of $2,400\angle0°$ V as the reference phasor and determine:
(a) The phasor current of the lighting load.
(b) The phasor current of the motor.
(c) The phasor current of the capacitor.
(d) The total phasor load current.
(e) Draw the phasor diagram of the voltage (all three load currents) and show how the three load currents combine to become the total load current in terms of phasor addition.

1.6   Consider Problem 1.4 and use the bus voltage of $4,800\angle0°$ V as the reference phasor. Determine the following:
(a) The phasor current of the lighting load.
(b) The phasor current of the motor.
(c) The phasor current of the capacitor.
(d) The total phasor load current.
(e) Draw the phasor diagram of the voltage, all three load currents, and show how the three load currents combine to become the total load current in terms of phasor addition.

CHAPTER 2

THREE-PHASE CIRCUITS

Imagination is more important than knowledge.
Albert Einstein

The great end of learning is nothing else but to seek for the lost mind.
Mencius, *Works*, 299 B.C.

Earn your ignorance! Learn something about everything
before you know nothing about anything.
Turan Gönen

2.1 INTRODUCTION

In a single-phase ac circuit, instantaneous power to a load is of a *pulsating* nature. Even at unity power factor (i.e., when the voltage and the current are in phase with respect to each other), the instantaneous power is less than unity (i.e., when the voltage and the current are not in phase). The instantaneous power is not only zero four times in each cycle but it is also negative twice in each cycle. Therefore, because of economy and performance, almost all electrical power is produced by *polyphase* sources (i.e., by those generating voltages with more than one phase[1]).

A polyphase generator has two or more single-phases connected so that they provide loads with voltages of equal magnitudes and equal phase differences.[2] For example, in a balanced *n-phase* system, there are *n* voltage sources

---

[1] A *phase* is one of three branches making up a three-phase circuit. For example, in a wye connection, a phase is made up of those circuit elements connected between one line and neutral; however, in a delta connection, a phase consists of those circuit elements connected between two lines.

[2] Therefore, a polyphase generator is somewhat analogues to a multicylinder automobile engine in that the power delivered is steadier. Consequently, there is less vibration in the rotating machinery, which, in turn, performs more efficiently.

connected together. Each phase voltage (or source) alternates sinusoidally, has the same magnitudes, and has a phase difference of $360/n$ degrees (where n is the number of phases) from its adjacent voltage phasors, except in the case of two-phase systems.[3] Generators of six-, twelve-, or even twenty-four-phases are sometimes used with *polyphase rectifiers* to supply power with low levels of ripples in voltage on the dc side in the range of kilowatts. Today, virtually all the power produced in the world is three-phase power with a frequency of 50 or 60 Hz. In the United States, 60 Hz is the standard frequency. Recently, six-phase power transmission lines have been proposed because of their ability to increase power transfer over existing lines and reduce electrical environmental impact.[4]

2.2 THREE-PHASE SYSTEMS

As previously stated, even though other polyphase systems are feasible, the power utility industry has adopted the use of *three-phase* systems as the standard. Consequently, most of the generation, transmission, distribution, and heavy-power utilization of electrical energy involves three-phase systems. A three-phase system is supplied by a three-phase generator (i.e., *alternator*) which consists essentially of three single-phase systems displaced in time phase from each other by one-third of a period, or 120 electrical degrees. The advantages of three-phase systems over single-phase systems are:

- Less conductor material is required in the three-phase transmission of power and therefore it is more economical.
- Constant rotor torque and therefore steady machine output can be achieved.
- Three-phase machines (generators or motors) have higher efficiencies.
- Three-phase generators may be connected in parallel to supply greater power more easily than single-phase generators.

Figure 2.1(a) shows the structure of an elementary three-phase and two-pole ac generator (also called an *alternator*). Its structure has basically two parts: the stationary outside part that is called the *stator* and the rotating inside part which is called the *rotor*. The field winding is located on the rotor and is excited by a direct current source through slip rings located on the common shaft. Thus, an alternator has a rotating electromagnetic field; however its stator windings are *stationary*. The elementary generator shown in Figure 2.1(a) has three identical stator coils $(aa', bb',$ and $cc')$, of one or more turns, displaced by 120° in space with respect to each other. If the rotor is driven counterclockwise at a constant speed, voltages will be generated in the three phases accord-

---

[3] In a *two-phase generator*, the two equal voltages differ in phase by 90°, but in the *three-phase generator* the three equal voltages have a phase-angle difference of 120°. However, the use of two-phase systems is very uncommon.
[4] In six-phase transmission lines the conductor potential gradients are lower, which reduces both audible noise and electrostatic effects without requiring additional insulation [36].

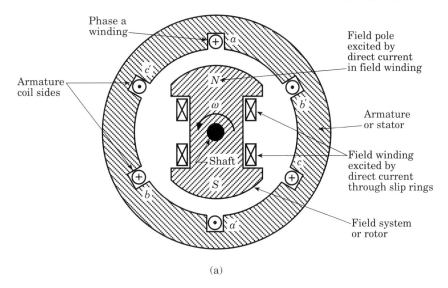

(a)

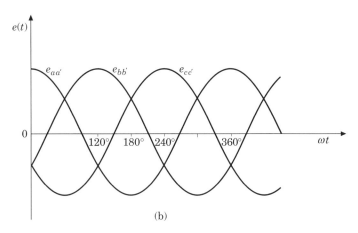

(b)

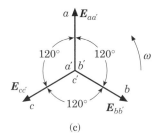

(c)

Figure 2.1   (a) Basic structure of an elementary three-phase two-pole ac generator; (b) generated voltage waveforms in time domain; (c) corresponding voltage phasors.

ing to Faraday's Law, as shown in Figure 2.1(b). Notice that the stator windings constitute the *armature* of the generator (unlike *dc* machines where the armature is the rotor). Thus the field rotates inside the armature. Each of the three stator coils makes up one phase in this single generator. Figure 2.1(b) shows the generated voltage waveforms in time domain, while Figure 2.1(c) shows the corresponding phasors of the three voltages. The stator phase windings can be connected in either wye or delta, as shown in Figure 2.2(a) and Figure 2.2(b), respectively. In wye configuration, if a neutral conductor is brought out, the system is defined as a *four-wire three-phase system*; otherwise, it is a *three-wire three-phase system*. In a delta-connection, no neutral exists and therefore it is a *three-wire three-phase system*.

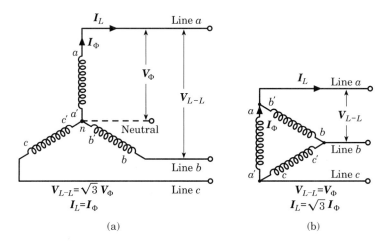

(a)   (b)

Figure 2.2   Stator phase windings connected in: (a) wye; (b) delta.

2.2.1 IDEAL THREE-PHASE POWER SOURCES

An ideal and balanced three-phase voltage source has three equal voltages that are 120° out of phase with respect to one another, as shown in Figure 2.3. Therefore, the balanced three-phase voltages given in the *abc phase sequence*[5] (or phase order) can be expressed as

$$V_a = V_\phi \angle 0° \tag{2.1}$$

$$V_b = V_\phi \angle 240° = V_\phi \angle -120° \tag{2.2}$$

$$V_c = V_\phi \angle 120° \tag{2.3}$$

---

[5] All phasors of a phasor diagram are assumed to rotate counterclockwise. A simple way of defining the phase sequence is to locate a point on any phasor in the system, e.g., V_a, and then move clockwise until the next two phasors are met, i.e., V_b and V_c. The phase sequence is then *abc*. In the United States almost all utility systems have the *abc* phase sequence.

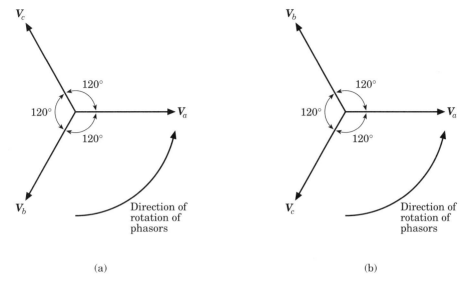

V_c

V_a

V_b

120°

120°

120°

Direction of
rotation of
phasors

V_b

V_a

V_c

120°

120°

120°

Direction of
rotation of
phasors

(a)   (b)

Figure 2.3   Phasor diagrams for balanced three-phase voltages, arranged in: (a) positive (or *abc*) phase sequence; (b) negative (or *acb*) phase sequence.

where V_ϕ is the magnitude of the phase voltage given in rms value. Similarly,
the balanced three-phase voltages given in the *acb phase sequence* can be expressed as

$$V_a = V_\phi \angle 0° \tag{2.4}$$

$$V_b = V_\phi \angle 120° \tag{2.5}$$

$$V_c = V_\phi \angle 240° = V_\phi \angle -120° \tag{2.6}$$

Furthermore, in the balanced three-phase systems, each phase has equal
impedance so that the resulting phase currents are equal and phase-displaced
from each other by 120°. The term **balanced** describes three-phase voltages or
currents, which are equal in magnitude and are 120° out of phase with respect
to each other, and form a symmetrical three-phase set. If the three-phase system is balanced, then equal real and reactive power flow in each phase. On the
other hand, if the three-phase system is not balanced, it may lack some or all of
the aforementioned characteristics.

Figures 2.4(a) and 2.4(b) show a wye-connected and a delta-connected[6] ideal three-phase source, respectively. The corresponding voltage and current phasor
diagrams are shown in Figure 2.5. The use of double-subscript notation greatly

---

[6] Since there is no neutral point in the delta-connected stator windings of a generator, this permits
currents of both fundamental and higher frequencies, especially of the third harmonic, to circulate
in the stator windings. This, in turn, causes greater heating of the stator windings. Therefore, a
wye-connected generator is usually preferable to a delta-connected generator.

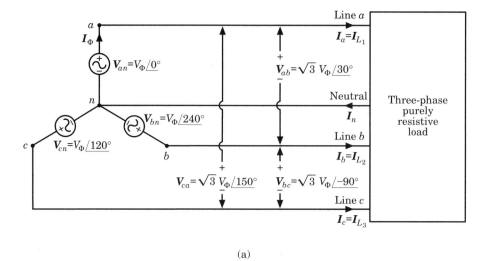

(a)

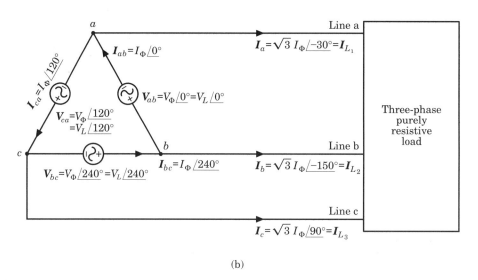

(b)

Figure 2.4   Ideal three-phase source connected in: (a) wye; (b) delta.

simplifies the phasor analysis. In the case of voltages, the subscripts indicate the points between which the voltage exists. Here, the first subscript is defined as positive with respect to the second. Therefore, the order of subscripts indicates the direction in which the voltage rise is defined. For example, $V_{an} = -V_{na}$. Hence, switching the order of the subscript causes a 180° phase shift in the variable. Similarly, in the case of currents, the subscript order indicates the *from-to* direction. The nodes a, b, and c are called the *terminals* or *lines* and the point n is called the *neutral*. The branches a-n, b-n, and c-n are defined as the *phases* of the source.

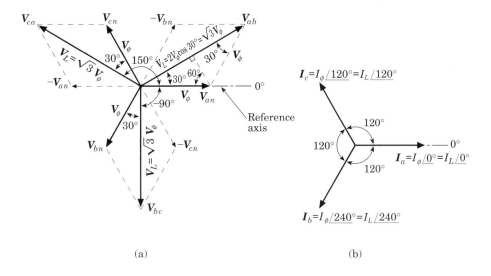

(a)

(b)

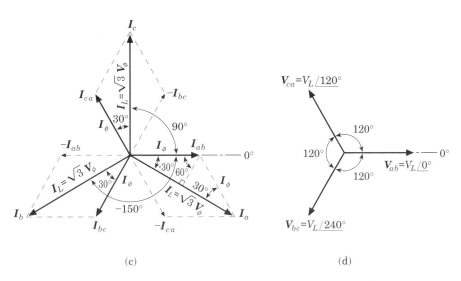

(c)

(d)

Figure 2.5   Phasor diagrams for three-phase sources: (a) phase and line voltages in a wye connection; (b) line currents in a wye connection; (c) phase and line currents in a delta connection; (d) line voltages in a delta connection.

The voltages V_{an}, V_{bn}, and V_{cn} are defined as the *line-to-neutral voltages* or *line-to-ground voltages* or simply *phase voltages*. The voltages V_{ab}, V_{bc}, and V_{cn} are defined as the *line-to-line voltages* or *phase-to-phase voltages* or simply *line voltages*. In general, whenever a three-phase voltage level is given, it is understood that it is a line voltage unless otherwise specified.

2.2.1.1 WYE-CONNECTED IDEAL THREE-PHASE SOURCE

Figure 2.5(a) illustrates how to determine the line voltages graphically from the given balanced phase voltages, if the source is connected in wye (or star). The line voltages can be determined mathematically as

$$V_{ab} = V_{an} + V_{nb}$$

$$= V_{an} - V_{bn}$$

$$= V_\phi \angle 0° - V_\phi \angle -120°$$

$$= V_\phi (1 + j0) - V_\phi \left(-\frac{1}{2} - j\frac{\sqrt{3}}{2} \right)$$

$$= \sqrt{3} V_\phi \angle 30°$$

$$= V_L \angle 30° \tag{2.7}$$

Similarly,

$$V_{bc} = V_{bn} - V_{cn}$$

$$= \sqrt{3} V_\phi \angle -90°$$

$$= V_L \angle 90° \tag{2.8}$$

and

$$V_{ca} = V_{cn} - V_{an}$$

$$= \sqrt{3} V_\phi \angle 150°$$

$$= V_L \angle 150° \tag{2.9}$$

where V_ϕ = the magnitude of the phase voltage,
V_L = the magnitude of the line voltage,

$$V_L = \sqrt{3} V_\phi \tag{2.10}$$

so that

$$V_\phi = |V_{an}| = |V_{bn}| = |V_{cn}| \tag{2.11}$$

and

$$V_L = |V_{ab}| = |V_{bc}| = |V_{ca}| = \sqrt{3} V_\phi \tag{2.12}$$

The line voltages are also $120°$ out of phase with respect to each other and form a symmetrical three-phase set. Figure 2.5(b) shows that each current lags its phase voltage by an equal phase angle. However, the phase and line currents are the same in a wye-connection. Figures 2.6(a) and 2.6(b) show alternativeaaaa ways of drawing the phasors given in Figures 2.5(a) and 2.5(b), respectively. As can be seen in those figures, the sum of the line voltages is zero for a balanced system, that is

$$V_a + V_b + V_c = 0 \tag{2.13}$$

and similarly,

$$I_a + I_b + I_c = 0 \tag{2.14}$$

Therefore, the neutral conductor does not carry any current (i.e., $I_n = 0$) if the source and load are both balanced. Otherwise,[7]

$$I_n = -(I_a + I_b + I_c) \tag{2.15}$$

2.2.1.2 DELTA-CONNECTED IDEAL THREE-PHASE SOURCE

Figure 2.5(c) illustrates how to determine the line currents graphically from the given balanced phase currents, if the source is connected in delta (or

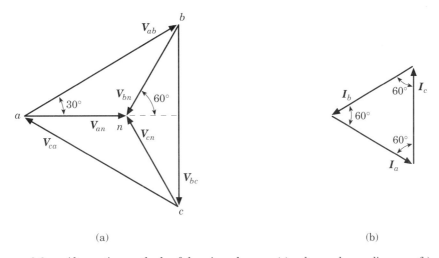

(a)   (b)

Figure 2.6   Alternative methods of drawing phasors: (a) voltage phasor diagram; (b) current phasor diagram.

---

[7] Most commercial generators are wye-connected, with their neutral grounded through a resistor. Such a grounding resistor (typically, $700\ \Omega$) limits ground-fault (i.e., short circuit) currents and therefore substantially reduces the amount of possible damage to apparatus in the event of a ground fault.

mesh). The line currents can also be determined mathematically. For example, if the balanced phase currents are given in *abc phase sequence* as

$$\boldsymbol{I}_{ab} = I_\phi \angle 0° \tag{2.16}$$

$$\boldsymbol{I}_{bc} = I_\phi \angle 240° = \boldsymbol{I}_\phi \angle -120° \tag{2.17}$$

$$\boldsymbol{I}_{ca} = I_\phi \angle 120° \tag{2.18}$$

as shown in Figure 2.5(c). The corresponding line currents can be found from the KCL as

$$\boldsymbol{I}_a = \boldsymbol{I}_{ab} - \boldsymbol{I}_{ca}$$

$$= I_\phi \angle 0° - I_\phi \angle 120° \tag{2.19}$$

$$\boldsymbol{I}_b = \boldsymbol{I}_{bc} - \boldsymbol{I}_{ab} \tag{2.20}$$

$$\boldsymbol{I}_c = \boldsymbol{I}_{ca} - \boldsymbol{I}_{bc} \tag{2.21}$$

where I_ϕ = the magnitude of the phase current
I_L = the magnitude of the line current

Since

$$\boldsymbol{I}_L = \sqrt{3}\boldsymbol{I}_\phi \tag{2.22}$$

then

$$|\boldsymbol{I}_a| = |\boldsymbol{I}_b| = |\boldsymbol{I}_c| = I_L = \sqrt{3}\,I_\phi \tag{2.23}$$

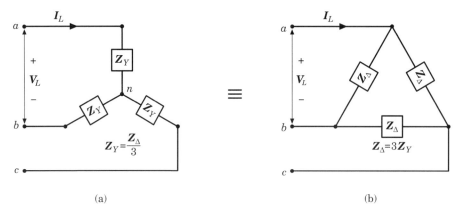

Figure 2.7   Balanced three-phase loads: (a) wye-connected load; (b) delta-connected load.

The line currents are also 120° out of phase with respect to each other and form a symmetrical three-phase set. Figure 2.5(c) shows that the phase and line currents are not in phase with each other. The phase and line voltages are the same in a delta connection, as shown in Figure 2.5(d).

Furthermore, for easier calculation, it is possible to replace any balanced three-phase delta-connected ideal source with an equivalent three-phase wye-connected ideal source. In this case, the magnitudes of the wye voltages are $1/\sqrt{3}$ times the magnitudes of the delta voltages. The wye voltages are out of phase with the corresponding delta voltages by 30°. Thus, if the phase sequence is abc, the wye voltages lag the delta voltages by 30°. Otherwise, the wye voltages lead the delta voltages by 30°.

2.2.2 BALANCED THREE-PHASE LOADS

Three-phase loads may be connected either in wye connections or delta connections. As shown in Figures 2.7(a) and 2.7(b), respectively. In a wye connection, the line voltages are 30° ahead of the corresponding phase voltages. However, the line currents and the corresponding phase currents are the same, as shown in Figure 2.5(b). The magnitudes of line voltages are $\sqrt{3}$ times those for phase voltages. In a delta connection, the line currents are 30° behind the corresponding phase currents, as shown in Figure 2.5(c). Here, the magnitudes of line currents lag the line-to-neutral voltages. The line currents also lag the line-to-neutral voltages by the phase-impedance angle, regardless of whether the circuit is wye or delta.

When the impedances in all three phases are identical the load is said *to be balanced.* If a balanced three-phase source is connected to a balanced load over an inherently balanced transmission or distribution line, then the total system is balanced. If the balanced three-phase load is wye-connected,

$$Z_{an} = Z_{bn} = Z_{cn} = Z_{\mathrm{Y}} \qquad (2.24)$$

then

$$
\begin{aligned}
I_{an} &= \frac{V_{an}}{Z_{an}} \\
&= \frac{V}{|Z_{\mathrm{Y}}|} \angle -\theta
\end{aligned} \qquad (2.25)
$$

$$
\begin{aligned}
I_{bn} &= \frac{V_{bn}}{Z_{bn}} \\
&= \frac{V}{|Z_{\mathrm{Y}}|} \angle 240° - \theta
\end{aligned} \qquad (2.26)
$$

$$I_{cn} = \frac{V_{cn}}{Z_{cn}}$$

$$= \frac{V}{|Z_Y|} \angle 120° - \theta \tag{2.27}$$

where

$$Z_Y = Z_Y \angle \theta \tag{2.28}$$

By applying the KCL at the point n,

$$I_n = I_{an} + I_{bn} + I_{cn} = 0 \tag{2.29}$$

and therefore the neutral conductor does not exist (from a theoretical point of view) and has no effect on the system. Since

$$V_\phi = V_{an} = V_{bn} = V_{cn} \tag{2.30}$$

and

$$I_\phi = I_{an} = I_{bn} = I_{cn} \tag{2.31}$$

then the total three-phase power of the load can be expressed as

$$P_{3\phi} = 3V_\phi V_\phi \cos \theta \tag{2.32}$$

where

$$V_\phi = \frac{V_L}{\sqrt{3}} \tag{2.33}$$

and

$$I_\phi = I_L \tag{2.34}$$

Therefore,

$$P_{3\phi} = \sqrt{3} V_L I_L \cos\theta \tag{2.35}$$

Similarly, the total three-phase power of the load can be expressed as

$$Q_{3\phi} = 3V_\phi I_\phi \sin \theta \tag{2.36}$$

or

$$Q_{3\phi} = \sqrt{3} V_L I_L \sin \theta \tag{2.37}$$

Therefore, the total apparent power of the load can be found as

$$S_{3\phi} = \left(P_{3\phi}^2 + Q_{3\phi}^2 \right)^{1/2} \tag{2.38}$$

or

$$S_{3\phi} = 3S_{1\phi} \tag{2.39}$$

or

$$S_{3\phi} = 3V_\phi I_\phi \tag{2.40}$$

Substituting equations (2.33) and (2.34) into equation (2.40),

$$\boldsymbol{S}_{3\phi} = \sqrt{3}V_L I_L \angle\theta \tag{2.41}$$

where V_L = the magnitude of the line voltage
$\quad\quad I_L$ = the magnitude of the line current
$\quad\quad \theta$ = the power factor angle by which the line current lags or leads the
$\quad\quad\quad$ line voltage (or the angle of the impedance in each phase)

The power factor of the three-phase load is still cos θ. If the balanced three-phase load is delta-connected,

$$\boldsymbol{Z}_{ab} = \boldsymbol{Z}_{bc} = \boldsymbol{Z}_{ca} = \boldsymbol{Z}_\Delta = Z_\Delta \angle\theta \tag{2.42}$$

then

$$\boldsymbol{I}_{ab} = \frac{\boldsymbol{V}_{ab}}{\boldsymbol{Z}_{ab}} = \frac{V_L}{Z_\Delta} \angle -\theta \tag{2.43}$$

$$\boldsymbol{I}_{bc} = \frac{\boldsymbol{V}_{bc}}{\boldsymbol{Z}_{bc}} = \frac{V_L}{Z_\Delta} \angle 240° -\theta \tag{2.44}$$

$$\boldsymbol{I}_{bc} = \frac{\boldsymbol{V}_{bc}}{\boldsymbol{Z}_{bc}} = \frac{V_L}{Z_\Delta} \angle 240° -\theta \tag{2.45}$$

Therefore, the line currents can be found from

$$\boldsymbol{I}_a = \boldsymbol{I}_{ab} - \boldsymbol{I}_{ca} \tag{2.46}$$

$$\boldsymbol{I}_b = \boldsymbol{I}_{bc} - \boldsymbol{I}_{ab} \tag{2.47}$$

$$\boldsymbol{I}_c = \boldsymbol{I}_{ca} - \boldsymbol{I}_{bc} \tag{2.48}$$

Since

$$I_L = I_a = I_b = I_c = \sqrt{3} I_\phi \tag{2.49}$$

$$V_L = V_\phi \tag{2.50}$$

then the total three-phase power of the load can be expressed as

$$P_{3\phi} = 3 V_\phi I_\phi \cos \theta \tag{2.51}$$

or

$$P_{3\phi} = \sqrt{3}\, V_L I_L \cos \theta \tag{2.52}$$

which is identical to equation (2.35). Similarly, the total three-phase reactive power of the load connected in delta can be found from equations (2.36) or (2.37). The total apparent power of the load can be found from equations (2.40) or (2.41). The power expressions are independent of the type (i.e., wye or delta) of load connection, as long as

$$Z_\Delta = 3 Z_Y \tag{2.53}$$

and since

$$\mathbf{Z} = Z \angle \theta$$

$$\theta = \arg |\mathbf{Z}_\Delta| = \arg |\mathbf{Z}_Y| \tag{2.54}$$

The real and reactive powers per phase can be found directly from the complex power as

$$\mathbf{S}_{1\phi} = P_{1\phi} + j Q_{1\phi} = \mathbf{V}_\phi \mathbf{I}_\phi^* \tag{2.55}$$

or

$$\mathbf{S}_{1\phi} = P_{1\phi} + j Q_{1\phi} = \mathbf{Z}^* |I_\phi|^2 \tag{2.56}$$

or

$$\mathbf{S}_{1\phi} = P_{1\phi} + j Q_{1\phi} = \mathbf{Y}_\phi^* |V_\phi|^2 \tag{2.57}$$

Thus, the three-phase complex, real, and reactive powers ~~be~~ can found from

$$\mathbf{S}_{3\phi} = 3 \mathbf{S}_{1\phi} = 3 P_{1\phi} + j 3 Q_{1\phi} \tag{2.58}$$

or

$$\mathbf{S}_{3\phi} = P_{3\phi} + j Q_{3\phi} = \sqrt{3} V_L \mathbf{I}_L^* \tag{2.59}$$

Table 2.1 provides a summary comparison of the basic variables of delta- and wye-connected balanced three-phase loads. Notice that the connection type does not affect the power calculations.

TABLE 2.1 Comparison of Balanced Three-Phase Loads Connected in Wye and Delta

Three-Phase Load	Δ-Connected Load	Y-Connected Load
Load impedance	$Z_\Delta = 3Z_Y$	$Z_Y = Z_\Delta/3$
Line current	$I_L = \sqrt{3}I_\phi$	$I_L = I_\phi$
Line-to-line voltage	$V_L = V_\phi$	$V_L = \sqrt{3}\,V_\phi$
Three-phase real power	$P_{3\phi} = \sqrt{3}\,V_L I_L \cos\theta$	$P_{3\phi} = \sqrt{3}\,V_L I_L \cos\theta$
Three-phase reactive power	$Q_{3\phi} = \sqrt{3}\,V_L I_L \sin\theta$	$Q_{3\phi} = \sqrt{3}\,V_L I_L \sin\theta$
Three-phase apparent power	$S_{3\phi} = \sqrt{3}\,V_L I_L$	$S_{3\phi} = \sqrt{3}\,V_L I_L$

The virtue of working with balanced systems is that they can be analyzed on a single-phase basis. The current in any phase is always the phase-to-neutral voltage divided by the per-phase load impedance. Therefore, it is not necessary to calculate the currents in the remaining phase separately. Calculating the current, voltage, and power in only one phase is sufficient in an analysis because of the complete symmetry that exists between the three phases. The knowledge of these variables in one phase which is referred to as the *reference phase*, directly provides information about all phases. This type of analysis is called *per-phase analysis*. This characteristic of balanced three-phase systems is the basis for the use of *one-line diagrams*. In these diagrams a circuit composed of three or more conductors is pictorially represented by a single line with standard symbols for transformers, switchgear, and other system components.

Example 2.1

Assume that the phase voltages of a wye-connected source (at its terminals) are given as $V_{an} = 277.13\angle 0°$ V, $V_{bn} = 277.13\angle 240°$ V, and $V_{cn} = 277.13\angle 120°$ V. Determine the following:
(a) The line voltages of V_{ab}, V_{bc}, and V_{ca}
(b) If a balanced wye-connected three-phase load of $Z_{an} = Z_{bn} = Z_{cn} = 10\angle 30°$W is connected to the source, find all the phase and line currents.

SOLUTION:
(a) The line voltages are found as

$$V_{ab} = V_{an} - V_{bn} = 277.13\angle 0° - 277.13\angle 240° = 480\angle 30° \text{ V}$$

$$V_{bc} = V_{bn} - V_{cn} = 277.13\angle 240° - 277.13\angle 120° = 480\angle -90°\text{V}$$

$$V_{ca} = V_{cn} - V_{an} = 277.13\angle 120° - 277.13\angle 0° = 480\angle 150° \text{ V}$$

(b) Since in a wye-connected three-phase load the phase and line currents are the same,

$$I_a = I_{an} = \frac{V_{an}}{Z_{an}} = \frac{277.13\angle 0° \text{ V}}{10\angle 30° \,\Omega} = 27.713\angle -30°\text{A}$$

$$I_b = I_{bn} = \frac{V_{bn}}{Z_{bn}} = \frac{277.13\angle 240° \text{ V}}{10\angle 30° \,\Omega} = 27.713\angle 210°\text{A}$$

$$I_c = I_{cn} = \frac{V_{cn}}{Z_{cn}} = \frac{277.13\angle 120° \text{ V}}{10\angle 30° \,\Omega} = 27.713\angle 90°\text{A}$$

Example 2.2

A balanced delta-connected three-phase load withdraws 200 A per phase with a leading power factor of 0.85 from a 12.47 kV line. Determine the following:
(a) The line current of the load.
(b) The phase voltage of the load.
(c) The three-phase apparent power.
(d) The three-phase real power.
(e) The three-phase reactive power.
(f) The three-phase complex power.

SOLUTION:
(a) Since the load is connected in delta, the line current of the load is

$$I_L = \sqrt{3}I_\phi = \sqrt{3} \times 200 = 346.41 \text{ A}$$

(b) The phase voltage of the load is

$$V_\phi = V_L = 12{,}470 \text{ V}$$

(c) The three-phase (or total) apparent power is

$$S_{3\phi} = \sqrt{3}V_L I_L = \sqrt{3}(12{,}470 \text{ V})(346.41 \text{ A}) = 7{,}482 \text{ kVA}$$

or

$$S_{3\phi} = 3V_\phi I_\phi = 3(12{,}470 \text{ V})(200\,\text{A}) = 7{,}482 \text{ kVA}$$

(d) The three-phase (or total) real power is

$$P_{3\phi} = \sqrt{3}V_L I_L \cos\theta = \sqrt{3}(12,470\text{V})(346.41\text{ A})0.85$$

$$= 6,359.7\text{ kW}$$

or

$$P_{3\phi} = S_{3\phi}\cos\theta = (7,482\text{ kVA})0.85 = 6,359.7\text{ kW}$$

(e) The three-phase (or total) reactive power is

$$Q_{3\phi} = \sqrt{3}V_L I_L \sin\theta = \sqrt{3}(12,470\text{ V})(346.41\text{ A})0.5268$$

$$= 3,941.52\text{ kvar}$$

or

$$Q_{3\phi} = S_{3\phi}\sin\theta = (7,482\text{ kVA})0.5268 = 3,941.52\text{ kvar}$$

(f) The three-phase (or total) complex power is

$$\boldsymbol{S}_{3\phi} = \sqrt{3}\boldsymbol{V}_L \boldsymbol{I}_L^* = P_{3\phi} - jQ_{3\phi} = (6,359.7\text{ kW}) - j(3,941.52\text{ kvar})$$

$$= 7,482.01\angle - 31.79°\text{ kVA}$$

Example 2.3

A balanced three-phase load of 8,000 kW with a lagging power factor of 0.90 is supplied by a three-phase 34.5 kV[8] line. If the line resistance and inductive reactance are given as 5 and 7 per phase (i.e., line-to-neutral), determine the following:

(a) The line current of the load.
(b) The power factor angle of the load.
(c) The line-to-neutral voltage of the line at the receiving end (i.e., the load side).
(d) The voltage drop due to line impedance.
(e) The line-to-line voltage of the line at the sending end.
(f) The power loss due to line impedance.

SOLUTION:
(a) From equation (2.52), the line current of the load is

$$I_L = \frac{P_{3\phi}}{\sqrt{3}V_L \cos\theta} = \frac{8,000\text{ kW}}{\sqrt{3}(34.5\text{ kV})0.90} = 148.75\text{ A}$$

---

[8] Unless otherwise specified, it is customary to assume a phase-to-phase voltage or line-to-line (i.e., *line*) voltage.

(b) The power factor of the load is

$$\theta = \cos^{-1} 0.90 = 25.8°$$

(c) The line-to-neutral voltage of the line at the receiving end is

$$V_{R(LN)} = \frac{V_L}{\sqrt{3}} = \frac{34,500 \text{ V}}{\sqrt{3}} = 19,918.6 \text{ V}$$

(d) The voltage drop in the line due to line impedance is

$$I_L Z = [148.75(\cos 25.8° - j\sin 25.8°)](5 + j7)$$

$$= 1,279.3\angle - 28.6° \text{ V}$$

(e) The line-to-neutral voltage at the sending end is

$$V_{S(LN)} = V_{R(LN)} + I_L Z$$

$$= 19,918.6\angle 0° + 1,279.3\angle - 28.6$$

$$= 21,050.7\angle - 1.7° \text{ V}$$

Therefore, the line-to-line voltage is

$$V_{S(L)} = \sqrt{3} V_{S(LN)} = \sqrt{3}(21,050.7 \text{ V}) = 36,460.9 \text{ V}$$

(f) The power loss due to line resistance is

$$P_{3\phi} = 3I_L^2 R = 3(148.75)^2(5) = 331.9 \text{ kW}$$

Example 2.4

A balanced three-phase delta-connected load is supplied by a balanced wye-connected source over a balanced three-phase line. The source voltage data are given in abc phase sequence in which V_{an} is $7.62\angle 0°$ kV and the line impedance is $1 + j7\ \Omega$. If the balanced load consists of three equal impedances of $15 + j10\ \Omega$, determine the following:
(a) The line currents I_a, I_b, and I_c.
(b) The phase voltages V_{ab}, V_{bc}, and V_{ca} of the delta-connected load.
(c) The phase currents I_{ab}, I_{bc}, and I_{ca} of the delta-connected load.
(d) The phasor diagram of the phasors found in parts (a), (b) and (c).

SOLUTION:
(a) Converting the given delta-connected load to its equivalent wye-connected form,

$$Z_Y = \frac{Z_\Delta}{3} = \frac{15 + j10}{3} = 5 + j3.33 \ \Omega \text{ per phase}$$

Therefore,

$$I_a = \frac{7,620\angle 0°}{6 + j10.33} = 637.7\angle -59.9° \text{ A}$$

$$I_b = \frac{7,620\angle -120°}{6 + j10.33} = 637.7\angle -179.9° \text{ A}$$

$$I_c = \frac{7,620\angle 120°}{6 + j10.33} = 637.7\angle 60.1° \text{ A}$$

(b) The line-to-neutral voltages at the wye-connected load can be found as

$$V_{an} = I_a Z_a = (637.7\angle -59.9°)(6\angle 33.7°) = 3,830.9\angle -26.2° \text{ V}$$

$$V_{bn} = I_b Z_b = (637.7\angle -179.9°)(6\angle 33.7°) = 3,830\angle -146.2° \text{ V}$$

$$V_{cn} = I_c Z_c = (637.7\angle 601°)(+6\angle 33.7°) = 3,830.9\angle 93.8° \text{ V}$$

Therefore,

$$V_{ab} = V_{an} - V_{bn} = 3,830.9\angle -26.2° - 3,830.9\angle -146.2° = 6,635.3\angle 3.8° \text{ V}$$

$$V_{bc} = V_{bn} - V_{cn} = V_{ab}\, e^{-j120°} = 6,635.3\angle -116.2° \text{ V}$$

$$V_{ca} = V_{cn} - V_{an} = V_{ab}\, e^{j120°} = 6,635.3\angle 123.8° \text{ V}$$

Alternatively,

$$V_{ab} = \sqrt{3} V_{an} \angle \theta_{an} + 30° = \sqrt{3}(3,830.9\angle -26.2° + 30°\,)\text{V}$$

$$= 6,635.3\angle 3.8° \text{ V}$$

$$V_{bc} = \sqrt{3} V_{bn} \angle \theta_{bn} + 30° = \sqrt{3}(3,830.9\angle -146.2° + 30°\,)\text{V}$$

$$= 6,635.3\angle -116.2° \text{ V}$$

$$V_{ca} = \sqrt{3} V_{cn} \angle \theta_{cn} + 30° = \sqrt{3}(3,830.9\angle 93.8° + 30°\,)\text{V}$$

$$= 6,635.3\angle 123.8° \text{ V}$$

(c) Thus,

$$I_{ab} = \frac{V_{ab}}{Z_\Delta} = \frac{6{,}635.3\angle 3.8^\circ \text{ V}}{18.03\angle 33.7^\circ \ \Omega} = 368\angle -29.9^\circ \text{ A}$$

$$I_{bc} = \frac{V_{bc}}{Z_\Delta} = \frac{6{,}635.3\angle -116.2^\circ \text{ V}}{18.03\angle 33.7^\circ \ \Omega} = 368\angle -149.9^\circ \text{ A}$$

$$I_{ca} = \frac{V_{ca}}{Z_\Delta} = \frac{6{,}635.3\angle 123.8^\circ \text{ V}}{18.03\angle 33.7^\circ \ \Omega} = 368\angle 90.1^\circ \text{ A}$$

(d) The phasor diagram is shown in Figure 2.8.

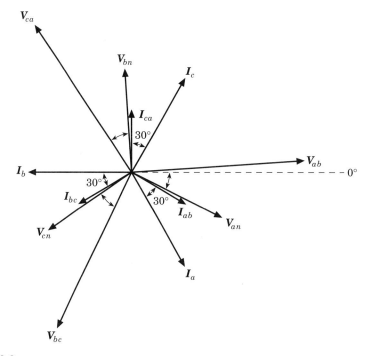

Figure 2.8

2.3 UNBALANCED THREE-PHASE LOADS

If an unbalanced three-phase load connected in delta or wye is present in an otherwise balanced three-phase system, the method of *symmetrical components*[9] is normally used to analyze the system. However, in simple situations the direct application of conventional circuit theory can be used without much difficulty, as in the following example.

---

[9] For the theory and applications of symmetrical components, see *Electric Power Transmission System Engineering* and *Modern Power System Analysis* of Gönen [34-35].

Example 2.5

An unbalanced three-phase delta-connected load is supplied by a balanced three-phase source through a power line. The load impedances Z_{ab}, Z_{bc}, and Z_{ca} are given as $5 + j5\ \Omega$, $5 - j5\ \Omega$, $5 + j0\ \Omega$, respectively. The power line impedance is given as $2 + j2\ \Omega$ per phase. In the event that the line-to-line voltages $V_{a'b'}$, $V_{b'c'}$, and $V_{c'a'}$ are $110\angle0°$ V, $110\angle240°$ V, and $110\angle120°$ V, respectively, determine the following:

(a) The line currents $I_{a'a}$, $I_{bb'}$, and $I_{c'c}$.

(b) The line-to-line voltages V_{ab}, V_{bc}, and V_{ca}.

SOLUTION:

(a) First convert the delta-connected load to its equivalent wye-connected form, as shown in Figure 2.9(b),

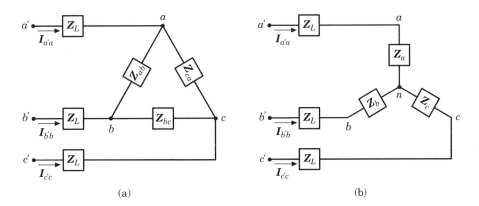

(a)   (b)

Figure 2.9

$$Z_a = \frac{Z_{ab}Z_{ca}}{Z_{ab} + Z_{bc} + Z_{ca}}$$

$$= \frac{(5 + j5)(5 + j0)}{(5 + j5) + (5 - j5) + (5 + j0)}$$

$$= \frac{5(5 + j5)}{15} = 1.67 + j1.67\ \Omega$$

and

$$Z_b = \frac{Z_{ab}Z_{bc}}{Z_{ab} + Z_{bc} + Z_{ca}} = \frac{(5 + j5)(5 - j5)}{15} = 3.33 + j0\ \Omega$$

$$Z_c = \frac{Z_{bc}Z_{ca}}{Z_{ab} + Z_{bc} + Z_{ca}} = \frac{(5 - j5)(5 + j0)}{15} = 1.67 - j1.67\ \Omega$$

By using KVL,

$$V_{a'b'} = V_{a'a} + V_{an} + V_{nb} + V_{bb'} = 110\angle 0°$$

$$V_{b'c'} = V_{b'b} + V_{bn} + V_{nc} + V_{cc'} = 110\angle 240°$$

where

$$V_{a'a} + V_{an} = I_{a'a}(Z_L + Z_a)$$

$$V_{b'b} + V_{bn} = I_{b'b}(Z_L + Z_b)$$

$$V_{c'c} + V_{cn} = I_{c'c}(Z_L + Z_c)$$

but

$$I_{c'c} = -(I_{a'a} + I_{b'b})$$

Therefore, $I_{c'c}$ can be eliminated so that

$$\begin{bmatrix} Z_L + Z_a & -Z_L - Z_b \\ Z_L + Z_c & 2Z_L + Z_b + Z_c \end{bmatrix} \begin{bmatrix} I_{a'a} \\ I_{b'b} \end{bmatrix} = \begin{bmatrix} 110\angle 0° \\ 110\angle 240° \end{bmatrix}$$

or

$$\begin{bmatrix} 3.67 + j3.67 & -5.33 - j2 \\ 3.67 + j0.33 & 9 - j2.33 \end{bmatrix} \begin{bmatrix} I_{a'a} \\ I_{b'b} \end{bmatrix} = \begin{bmatrix} 110\angle 0° \\ 110\angle 240° \end{bmatrix}$$

By using determinants and *Cramer's rule*,

$$I_{a'a} = \frac{\Delta_1}{\Delta} = \frac{\begin{vmatrix} 110\angle 0° & -5.33 - j2 \\ 110\angle 240° & 9 - j2.33 \end{vmatrix}}{\begin{vmatrix} 3.67 + j3.67 & -5.33 - j2 \\ 3.67 + j0.33 & 9 - j2.33 \end{vmatrix}}$$

$$= \frac{110\angle 0°(9.2967\angle -14.51°) - (5.6929\angle 200.57°)110\angle 240°}{5.1902\angle 45°(9.2967\angle -14.51°) - (5.6929\angle 200.57°)3.6848\angle 5.14°)}$$

$$= 13.85\angle -6.87° \text{ A}$$

Similarly,

$$I_{b'b} = \frac{\Delta_2}{\Delta} = \frac{\begin{vmatrix} 3.67 + j3.67 & 110\angle 0° \\ 3.67 + j0.33 & 110\angle 240° \end{vmatrix}}{69.178\angle 29.04°} = 9.27\angle 217.48° \text{ A}$$

Therefore,

$$I_{c'c} = -(13.85\angle - 6.87° + 9.27\angle 217.48°) = 9.70\angle 131.23° \text{ A}$$

As a check,

$$I_{a'a} + I_{b'b} + I_{c'c} = 13.85\angle - 6.87° + 9.27\angle 217.48 + 9.70\angle 131.23° = 0$$

(b) The line-to-neutral voltages can be found as

$$V_{an} = I_{a'a} Z_a = (13.85\angle - 6.87° \text{ A})(2.36\angle 45°) = 32.69\angle 38.13° \text{ V}$$

$$V_{bn} = I_{b'b} Z_b = (9.27\angle 217.48° \text{ A})(3.33\angle 0°) = 30.87\angle 217.48° \text{ V}$$

$$V_{cn} = I_{c'c} Z_c = (9.70\angle 131.23° \text{ A})(2.36\angle - 45°) = 22.89\angle 86.23° \text{ V}$$

Therefore, the line-to-line voltages can be found as

$$V_{ab} = V_{an} - V_{bn} = 32.69\angle 38.13° - 30.87\angle 217.48° = 63.56\angle 37.82° \text{ V}$$

$$V_{bc} = V_{bn} - V_{cn} = 30.87\angle 217.48° - 22.89\angle 86.23° = 49.08\angle 238° \text{ V}$$

$$V_{ca} = V_{cn} - V_{an} = 22.89\angle 86.23° - 32.69\angle 38.13° = 24.35\angle 173.74° \text{ V}$$

Notice that the unbalanced loads destroy the symmetry between the phasors, and cause the resulting currents and voltages not to have the simplicity that is characteristic of a balanced three-phase system.

Example 2.6

An unbalanced three-phase, four-wire wye-connected load, as shown in Figure 2.10, is connected to a balanced three-phase four-wire source. The load impedances Z_a, Z_b and Z_c are given as $100\angle 50°$, $150\angle - 140°$, and $50\angle - 100°$ Ω per phase, respectively. If the line voltage is 13.8 kV, determine the following:
(a) The line and neutral currents.
(b) The total power delivered to the loads.

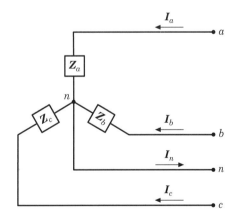

Figure 2.10

SOLUTION:
(a) Taking the line-to-line neutral voltages of phase-a voltage as the reference,

13,800

$$V_\phi = \frac{V_L}{\sqrt{3}} = \frac{\cancel{13,000}\text{ V}}{\sqrt{3}} = 7,967.4 \text{ V}$$

so that

$$V_{an} = 7,967.4\angle 0° \text{ V}$$

$$V_{bn} = 7,967.4\angle -120° \text{ V}$$

$$V_{cn} = 7,967.4\angle 120° \text{ V}$$

Therefore, the line currents can be found as

$$I_a = \frac{V_{an}}{Z_a} = \frac{7,967.4\angle 0° \text{ V}}{100\angle 50° \text{ }\Omega} = 79.7\angle -50° \text{ A}$$

$$I_b = \frac{V_{bn}}{Z_b} = \frac{7,967.4\angle -120° \text{ V}}{150\angle -140° \text{ }\Omega} = 53.1\angle 20° \text{ A}$$

$$I_c = \frac{V_{cn}}{Z_c} = \frac{7,967.4\angle 120° \text{ V}}{50\angle -100° \text{ }\Omega} = 159.3\angle 220° \text{ A}$$

Thus,

$$I_n = -(I_a + I_b + I_c)$$

$$= -(79.7\angle -50° + 53.1\angle 20° + 19.3\angle 220°) = 146.8\angle 81.8° \text{ A}$$

(b) Therefore, the power delivered by each phase is

$$P_a = V_{an} I_a \cos \theta_a = (7,967.4 \text{ V})(79.7 \text{ A}) \cos 50° = 408,171.3 \text{ W}$$

$$P_b = V_{bn} I_b \cos \theta_b = (7,967.4 \text{ V})(53.1 \text{ A}) \cos 140° = 324,089.6 \text{ W}$$

$$P_c = V_{cn} I_c \cos \theta_c = (7,967.4 \text{ V})(159.3 \text{ A}) \cos 100° = 220,395.5 \text{ W}$$

Thus, the total power delivered is

$$P_{3\phi} = P_a + P_b + P_c = 952,656.3 \text{ W} \cong 952.6 \text{ kW}$$

2.4 MEASUREMENT OF AVERAGE POWER IN THREE-PHASE CIRCUITS

A *wattmeter* is a device that has a potential coil and a current coil, which are designed and connected in such a way that its pointer's deflection is proportional to $VI\cos \theta$. Here, V is the rms value of the voltage applied across the potential coil, I is the rms value of the current passing through the current coil, and θ is the angle between the voltage and the current phasors involved. The direction in which the pointer deflects depends on the instantaneous polarity of the current-coil current and the potential-coil voltage. Thus, each coil has one terminal with a polarity mark ±, as shown in Figure 2.11. The wattmeter deflects in the right direction when the polarity-marked terminal of the potential coil is connected to the phase in which the current coil has been inserted.

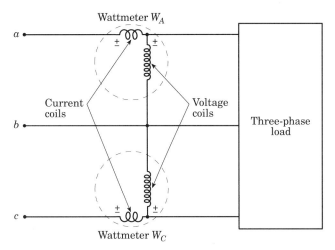

Figure 2.11   Connection diagram for the two-wattmeter method of measuring three-phase power.

If a separate wattmeter is used to measure the average (real) power in each phase, the total real power in a three-phase system can be found by adding the three wattmeter readings. If the load is delta-connected, each wattmeter has its current coil connected on one side of the delta and its potential coil connected line to line. If the load is wye-connected and the neutral wire does exist, the potential coil of each wattmeter is connected between each phase and the neutral wire. However, in actual practice, it may not be possible to have access to either the neutral of the wye connection or the individual phases of the delta connection in order to connect a wattmeter in each of the phases. In these cases, three wattmeters can be connected as shown in Figure P2.7. The common point 0 is a floating potential point. The total real power used by the load, whether it is delta- or wye-connected, balanced or unbalanced, is the sum of the three wattmeter readings. Therefore,

$$P_{3\phi} = W_a + W_b + W_c \tag{2.75}$$

It is also possible in delta- and wye-connected loads to use only two wattmeters to measure the total real power, as shown in Figure 2.11. This method is known as the *two-wattmeter method* of measuring three-phase power. The algebraic sum of the readings of the two wattmeters[10] will give the total average power consumed by the three-phase load. Thus,

$$P_{3\phi} = W_a + W_c \tag{2.76}$$

where the wattmeter connected on phase-a provides the reading of

$$W_a = V_{ab} I_a \cos \theta_a \tag{2.77}$$

where θ_a is the angle between phasors V_{ab} and I_a. Similarly,

$$W_c = V_{cb} I_c \cos \theta_c \tag{2.78}$$

where θ_c is the angle between phasors V_{cb} and I_c. Notice that the reading of the wattmeter W_c is determined by V_{ab} and I_c. Even though the sum of the two readings depends only on the total power of the load, the individual readings depend on the phase sequence. Now assume that the phase sequence is abc and that the voltage V_{ab} is the reference phasor, as shown in Figure 2.12(a). Also assume that the load is either balanced wye- or delta-connected with a lagging power factor angle of θ. From Figure 2.12(a), it can be observed that the angle between phasors V_{ab} and I_a is $(30° + \theta)$ and that between phasors V_{cb} and I_c is a $(30° - \theta)$. As mentioned before, the angle θ is the *load power-factor angle* or the angle associated with the load impedance. Therefore, equation (2.77) and (2.78) can be expressed, respectively,[11] as

$$W_a = V_L I_L \cos (30° + \theta) \tag{2.79}$$

and

$$W_c = V_L I_L \cos (30° - \theta) \tag{2.80}$$

where

$$\theta_a = 30° + \theta \tag{2.81}$$

$$\theta_c = 30° - \theta \tag{2.82}$$

---

[10] If one meter reads backwards, reverse its current coil and subtract its reading from the other wattmeter.

[11] If the phase sequence is acb,

$$W_a = V_L I_L \cos (30° - \theta)$$

and

$$W_c = V_L I_L \cos (30° + \theta)$$

so that

$$P_{3\phi} = W_a + W_c$$

and that the total reactive power is

$$Q_{3\phi} = \sqrt{3} (W_a - W_c)$$

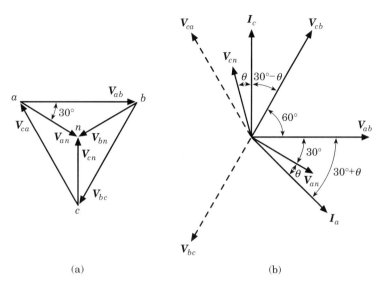

Figure 2.12   Phasor diagrams for the two-wattmeter method.

and where V_L and I_L are magnitudes of the line-to-line voltage and line current, respectively. Thus, the total average power can be determined as

$$P_{3\phi} = W_a + W_c$$

$$= \sqrt{3}\, V_L I_L [\cos (30° + \theta) + \cos (30° - \theta)]$$

$$= \sqrt{3}\, V_L I_L \cos \theta \qquad (2.83)$$

and the total reactive power can be determined as

$$Q_{\phi} = \sqrt{3}\,(W_a - W_c)$$

$$= \sqrt{3}\, V_L I_L [\cos (30° - \theta) - \cos (30° + \theta)]$$

$$= \sqrt{3}\, V_L I_L \sin \theta \qquad (2.84)$$

By observing equations (2.79) and (2.80), the following conclusions can be reached for the two-wattmeter method of measuring three-phase power in a balanced circuit.

- If the load power-factor equals 0.5, one of the wattmeters shows zero.
- If the load power-factor is less than 0.5, one of the wattmeters shows a negative value.
- If the load power-factor is greater than 0.5, both wattmeters show a positive value.
- Reversing the phase sequence will interchange the readings on the wattmeters.

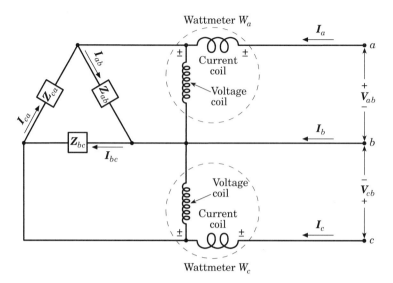

Figure 2.13

Example 2.7

An unbalanced three-phase delta-connected load, as shown in Figure 2.13, is supplied by a balanced three-phase system given in abc phase sequence in which V_{ab} is $220\angle 0°$ V. The load impedances Z_{ab}, Z_{bc} and Z_{ca} are given as $10\angle 0°$, $5\angle 60°$ and $15\angle -20°$ W, respectively. Determine the following:
(a) The phase currents I_{ab}, I_{bc} and I_{ca}.
(b) The line currents I_a, I_b and I_c.
(c) The powers absorbed by the individual impedances of the load.
(d) The total power absorbed by the load.
(e) The power recorded on each wattmeter.

SOLUTION:
(a) The phase currents can be found as

$$I_{ab} = \frac{V_{ab}}{Z_{ab}} = \frac{220\angle 0° \text{ V}}{10\angle 0° \ \Omega} = 22\angle 0° \text{ A}$$

$$I_{bc} = \frac{V_{bc}}{Z_{bc}} = \frac{220\angle -120° \text{ V}}{5\angle 60° \ \Omega} = 44\angle -180° \text{ A}$$

$$I_{ca} = \frac{V_{ca}}{Z_{ca}} = \frac{220\angle 120° \text{ V}}{15\angle -20° \ \Omega} = 14.67\angle 140° \text{ A}$$

(b) Therefore, the line currents are

$$I_a = I_{ab} - I_{ca} = 22\angle 0° - 14.67\angle 140° = 34.5\angle -15.8° \text{ A}$$

$$I_b = I_{bc} - I_{ab} = 44\angle - 180° - 22\angle 0° = 66\angle 180° \text{ A}$$

$$I_c = I_{ca} - I_{bc} = 14.67\angle 140° - 44\angle - 180° = 34.1\angle 16° \text{ A}$$

(c) The powers absorbed by the individual impedances of the load can be found as

$$P_{ab} = \mathscr{R}e(V_{ab} I^*_{ab}) = \mathscr{R}e[(220\angle 0°)(22\angle 0°)] = 4.84 \text{ kW}$$

$$P_{bc} = \mathscr{R}e(V_{ab} I^*_{bc}) = \mathscr{R}e\{(220\angle - 120°)(44\angle 180°)] = 4.84 \text{ kW}$$

$$P_{ca} = \mathscr{R}e(V_{ca} I^*_{ca}) = \mathscr{R}e[(220\angle 120°)(14.67\angle - 140°)] = 3.02 \text{ kW}$$

(d) The total power absorbed by the load is

$$P = P_{ab} + P_{bc} + P_{ca} = 12.7 \text{ kW}$$

(e) The power recorded by the wattmeter a is

$$W_a = |V_{ab}| \, |I_a| \cos \theta_a$$

where θ_a = the angle between V_{ab} and I_a

Therefore,

$$W_a = |220\angle 0°| \, |34.5\angle 15.8°| \cos 15.8° = 7.3 \text{ kW}$$

or alternatively,

$$W_a = \mathscr{R}e(V_{ab} I^*_a) = \mathscr{R}e[(220\angle 0°)(34.5\angle 15.8°) = 7.3 \text{ kW}$$

Similarly,

$$W_c = \mathscr{R}e(V_{bc} I^*_c) = \mathscr{R}e[(220\angle - 120°)(34.1\angle - 16°)] = 5.4 \text{ kW}$$

Therefore, the total power read by the wattmeters is 12.7 kW.

2.5 POWER FACTOR CORRECTION

In general, loads on electric utility systems have two components: active power (measured in kilowatts) and reactive power (measured in kilovars). Active power has to be generated at power plants whereas reactive power can be supplied by either power plants or capacitors. If reactive power is supplied only by power plants, each system component, including generators, transformers, and transmission and distribution lines, has to be increased in size accordingly. However, by using capacitors, the reactive power demand as well as line currents are reduced from the capacitor locations all the way back to power plants.[12] As a result, losses and loadings are reduced in distribution lines, substation transformers, and transmission lines. The power-factor correction generates savings in capital expenditures and fuel expenses through a release of power capacity and a decrease in power losses in all the equipment between the point of installation of the capacitors and the source power plants.

The *economic power factor* is the power factor at which the economic benefits of using capacitors equals the cost of capacitors. However, the correction of power factor to unity becomes more expensive with respect to the marginal cost of the capacitors installed. It has been found in practice that the economic power factor is about 0.95. In distribution systems, including industrial applications, *shunt* capacitors are used and are connected in delta or wye. However, in transmission systems, the capacitors are connected in *series* with the line involved.

Example 2.8

Assume that a 2.4 kV single-phase circuit supplies a load of 294 kW at lagging power factor and that the load current is 175 A. To improve the power factor, determine the following:
(a) The uncorrected power factor and reactive load.
(b) The new corrected power factor after installing a shunt capacitor unit with a rating of 200 kvar.

SOLUTION:
(a) Before the power-factor correction,

$$S_{old} = V I = (2.4 \text{ kV})(175 \text{ A}) = 420 \text{ kVA}$$

Therefore, the *uncorrected* power-factor can be found as

$$PF_{old} = \cos \theta_{old} = \frac{P}{S_{old}} = \frac{294 \text{ kW}}{420 \text{ kVA}} = 0.7$$

and the reactive load is

$$Q_{old} = S_{old} \times \sin(\cos^{-1} PF_{old}) = (420 \text{ kVA})(0.7141) = 300 \text{ kvar}$$

(b) After the installation of the 200 kvar capacitors,

$$Q_{new} = Q_{old} - Q_{cap} = (300 \text{ kvar}) - (200 \text{ kvar}) = 100 \text{ kvar}$$

---

[12] For further information, see Chapter 8 of *Electric Power Distribution System Engineering* of Gönen [33].

and therefore the *new* (or *corrected*) power factor is

$$PF_{new} = \cos\theta_{new} = \frac{P}{(P^2 + Q_{new}^2)^{1/2}}$$

$$= \frac{294 \text{ kW}}{[(294 \text{ kW})^2 + (100 \text{ kvar})^2]^{1/2}} = 0.95 \text{ or } 95 \text{ percent}$$

Example 2.9

A three-phase 400 hp 60 Hz 4.16 kV wye-connected induction motor has a full-load efficiency of 86 percent, and a lagging power factor of 0.8. If it is necessary to correct the power factor of the load to a lagging power factor of 0.95 by connecting three capacitors at the load, find the following:
(a) The rating of such a capacitor bank in kvar.
(b) The capacitance of each single-phase unit, if the capacitors are connected in delta in μF.
(c) The capacitance of each single-phase unit, if the capacitors are connected in wye in μF.

SOLUTION:
(a) The input power of the induction motor is

$$P = \frac{(400 \text{ hp})(0.7457 \text{ kW/hp})}{0.86} = 346.84 \text{ kW}$$

The reactive power of the motor at the *uncorrected* power factor is

$$Q_{old} = P \times \tan\theta_{old}$$

$$= (346.84 \text{ kW}) \times \tan(\cos^{-1}0.8)$$

$$= (346.84 \text{ kW}) \times 0.75$$

$$= 260.13 \text{ kvar}$$

The reactive power of the motor at the *corrected* power factor is

$$Q_{new} = P \times \tan\theta_{new}$$

$$= (346.84 \text{ kW}) \times \tan(\cos^{-1}0.95)$$

$$= (346.84 \text{ kW}) \times 0.3287$$

$$= 114 \text{ kvar}$$

Thus, the reactive power provided by the capacitor bank is

$$Q_{cap} = Q_{old} - Q_{new}$$

$$= (260.13 \text{ kvar}) - (114 \text{ kvar})$$

$$= 146.13 \text{ kvar}$$

Therefore, the rating of the capacitor bank is 146.13 kvar.
(b) If the capacitors are connected in delta, the line current is

$$I_L = \frac{Q_{cap}}{\sqrt{3}V_L} = \frac{146.13 \text{ kvar}}{\sqrt{3}(4.16 \text{ kV})} = 20.28 \text{ A}$$

and thus the current in each capacitance of the delta-connected capacitor bank is

$$I_{cap} = \frac{I_L}{\sqrt{3}} = \frac{20.28 \text{ A}}{\sqrt{3}} = 11.71 \text{ A}$$

Therefore, the reactance of each capacitor is

$$X_{cap} = \frac{V_L}{I_{cap}} = \frac{4,160 \text{ V}}{11.71 \text{ A}} = 355.25 \text{ }\Omega$$

and the capacitance of each unit is

$$C = \frac{10^6}{\omega X_{cap}} = \frac{10^6}{(2\pi f)X_{cap}}$$

$$= \frac{10^6}{2\pi(60 \text{ Hz})(355.55 \text{ }\Omega)} = 7.47 \text{ }\mu F$$

(c) If the capacitors are connected in wye in the capacitor bank,

$$I_{cap} = I_L = 20.28 \text{ A}$$

and therefore,

$$X_{cap} = \frac{V_{LN}}{I_{cap}} = \frac{4,160 \text{ V}}{\sqrt{3}(20.28 \text{ A})} = 118.43 \text{ }\Omega$$

Therefore, the capacitance of each unit is

$$C = \frac{10^6}{2\pi f X_{cap}} = \frac{10^6}{2\pi(60 \text{ Hz})(118.43 \text{ }\Omega)} = 22.4 \text{ }\mu F$$

PROBLEMS

2.1   A three-phase wye-connected induction motor is supplied by a three-phase and four-wire system with line-to-line voltage of 220 V and the impedance of the motor is $6.3 + j3.05$ Ω per phase. Determine the following:
(a)   The magnitude of the line current.
(b)   The power factor of the motor.
(c)   The three-phase average power consumed by the motor.
(d)   The current in the neutral wire.

2.2   A balanced three-phase load of 15 MVA with a lagging load factor of 0.85 is supplied by a 115 kV subtransmission line. If the line impedance is $50 + j100$ Ω per phase, determine the following:
(a)   The line current of the load.
(b)   The power factor angle of the load.
(c)   The line-to-neutral voltage of the line at the receiving end.
(d)   The voltage drop due to the line impedance.
(e)   The line-to-line voltage of the line at the sending end.
(f)   The power loss due to the line impedance.

2.3   A balanced three-phase delta-connected load is supplied by a balanced three-phase wye-connected source over a balanced three phase line. The source voltages are in abc phase sequence in which V_{an} is 19.49∠0° kV and the line impedance is $10 + j80$ Ω per phase. If the balanced load consists of three equal impedances of $60 + j30$ Ω, determine the following:
(a)   The line currents I_a, I_b, and I_c.
(b)   The phase voltages V_{ab}, V_{bc}, and V_{ca} of the load.
(c)   The phase currents I_{ab}, I_{bc}, and I_{ca} of the load.

2.4   Assume that the impedance of a power line connecting buses 1 and 2 is 50∠70° Ω and that the bus voltages are 70,000∠25° and 65,000∠20° V per phase, respectively. Determine the following:
(a)   The real power per phase that is being transmitted from bus 1 to bus 2.
(b)   The reactive power per phase that is being transmitted from bus 1 to bus 2.
(c)   The complex power per phase that is being transmitted.

2.5   Solve Problem 2.4 assuming that the line impedance is 50∠26° per phase.

2.6   An unbalanced three-phase, three-wire, wye-connected load is connected to a balanced three-phase, three-wire, wye-connected source, as shown in Figure P2.6. If the line-to-neutral source voltages V_a, V_b and V_c are 220∠30° V, 220∠270° V and 220∠150° V, respectively, and the load impedances Z_a, Z_b, and Z_c are 4∠0° Ω, 5∠90° Ω, and 8∠30° Ω per phase, respectively. Determine the following:

(a) The mesh currents I_1 and I_2 using determinants and *Cramer's rule*.
(b) The line currents I_a, I_b, and I_c.
(c) The potential difference between the source neutral N_S and the common node of the load, that is, N_L.
(d) Whether or not a neutral wire connecting the neutral point N_S and N_L is required.

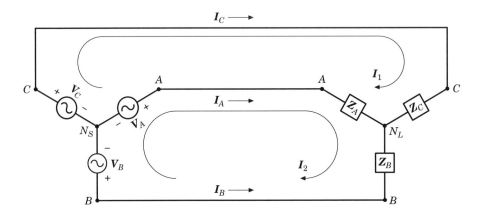

Figure P2.6

2.7   Consider Problem 2.6 and assume that three wattmeters are connected to measure the total power received by the unbalanced three-phase load, as shown in Figure P2.7. Ignore the small impedance of the current coils in the wattmeters and determine the following:
(a) The power recorded on each wattmeter.
(b) The total power absorbed by the load.

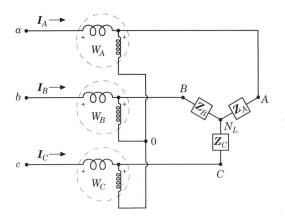

Figure P2.7

2.8 Consider Problem 2.7 and assume that only two wattmeters are used and connected as shown in Figure P2.8. Determine the following:
(a) The power recorded on each wattmeter.
(b) The total power absorbed by the load.

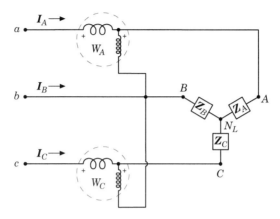

Figure P2.8

2.9 If the impedance of a transmission line connecting buses 1 and 2 is $50\angle 80°$ Ω and the bus voltages are $70\angle15°$ kV and $68\angle0°$ kV per phase, respectively, determine the following:
(a) The complex power per phase that is being transmitted from bus 1 to bus 2.
(b) The active power per phase that is being transmitted.
(c) The reactive power per phase that is being transmitted.

2.10 A three-phase motor is connected to a three-phase line that has an abc phase sequence and is supplied by 15 A current at 0.85 lagging power factor. If a single-phase motor withdrawing 5 A current at 0.707 lagging power factor is connected across lines a and b of the three-phase power line, determine the total current in each line.

2.11 Three loads are connected to a 208-V three-phase power source that has an abc phase sequence. The first load is a wye-connected three-phase motor withdrawing a line current of 20 A at a 0.8 lagging power factor. The second load is a single-phase load between lines a and b and withdraws a 10 A current at a 0.8 leading power factor. The third load is also a single-phase motor connected between lines b and c and withdraws a 7 A current at a 0.707 lagging power factor. Use the voltage V_{ab} as the reference phasor and determine:
(a) All line and phase voltages.
(b) All line currents.
(c) The total input power in watts.

2.12 A three-phase 60-Hz wye-connected synchronous generator is providing power to two balanced three-phase loads. The first load is delta-connected and made up of three $12\angle45°$ Ω impedances, while the second load is wye-connected and made up of three $5\angle60°$ Ω impedances. Determine the following:

(a) Total (i.e., equivalent) load impedance per-phase (i.e, line-to-neutral).

(b) The line current I_a at the generator terminal. Use $V_a = (208/\sqrt{3})\angle0° \cong 120\angle0°$ V.

(c) The total complex power provided by the generator.

2.13 A three-phase 60-Hz wye-connected synchronous generator has balanced line-to-line voltages of 480 V at its terminals. The generator is supplying power to two balanced and delta-connected three-phase loads. The first load is made up of three $15\angle-30°$ Ω impedances, while the second load is made up of three $18\angle50°$ Ω impedances. Determine the following:

(a) Total (i.e, equivalent) load impedance per phase (i.e, line-to-neutral).

(b) The line current I_a at the generator terminal. Use a phase voltage of $V_a = 277.1281\angle0°$ V (since $480/\sqrt{3}$ V = 277.1281 V.)

(c) The total complex power provided by the generator.

2.14 If a balanced, three-phase, 15 MW total load is fed by a 138-kV power line at a 0.85 lagging power factor, determine the following:

(a) The line current.

(b) The value of the capacitor in μF per phase, if a wye-connected capacitor bank is used to correct the power factor to a 0.95 lagging power factor.

2.15 If a balanced, three-phase, 20 MW total load is fed by a 138-kV (line-to-line) power source at a 0.8 lagging power factor, determine the following:

(a) The line current.

(b) The value of the capacitor in μF per phase, if a wye-connected capacitor bank is used to correct the power factor to a 0.9 lagging power factor.

(c) If a delta-connected capacitor bank is used in Part (b), find the value of each capacitor in such a bank. [Hint: Use the results of Part (b).]

2.16 Consider the balanced and delta-connected three-phase load that is shown in Figure P2.16. Assume that each impedance is $9 + j9$ Ω and that the line-to-line voltage, with a magnitude of 4,160 V, is supplied by a three-phase source where $V_{bc} = V_{bc}\angle0°$. In other words, assume that the voltage source phasors are aligned in the same fashion as the circuit symbols between nodes a, b, and c. Determine the following:

(a) The line currents I_a, I_b, and I_c.

(b) The total (i.e, three-phase) average power in watts.

2.17 Consider the balanced and delta-connected three-phase load that is shown in Figure P2.16. Assume that each impedance is $3 + j6\ \Omega$ and that the load is supplied by a balanced three-phase source where $V_{ab} = 12{,}470\angle 30°$ V so that $V_a = V_{an} \cong 7{,}200\angle 0°$ V. (In other words, line voltages lead phase voltages by 30°.) Determine the following:

(a) The line currents I_a, I_b, and I_c.

(b) The total (i.e, three-phase) average power in watts.

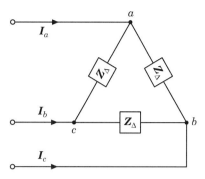

Figure P2.16

2.18 Consider the delta-connected three-phase load that is shown in Figure P2.18. Assume that the delta-connected unbalanced load is supplied by the same three-phase source that is given in Problem 2.16.

Determine the following:

(a) The line currents I_a, I_b, and I_c.

(b) Find the voltage V_{de}.

(c) The total (i.e, three-phase) average power in watts.

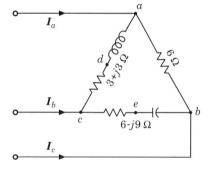

Figure P2.18

2.19 Assume that the delta-connected load is an unbalanced three-phase load and is supplied by the same balanced three-phase power source that is given in Problem 2.17. If $Z_{ab} = 5 + j5\ \Omega$, $Z_{bc} = 5 - j5\ \Omega$, $Z_{ca} = 3 + j6\ \Omega$, and $V_{ab} = 12{,}470\angle 30°$ V, determine the following:

(a) The line currents I_a, I_b, and I_c.

(b) The total (i.e, three-phase) average power in watts.

2.20 A three-phase 60-Hz 5-MVA synchronous generator has balanced line-to-line voltages of 8,320 V at its terminals. If the generator is supplying power to a wye-connected balanced load of 5 MVA at a 0.8 lagging power factor, determine the following:

(a) The total complex power absorbed by the wye-connected load.

(b) The load impedance per phase.

2.21 Solve Problem 2.20 but assume that the 5 MVA balanced three-phase load is delta-connected rather than wye-connected.

2.22 A three-phase 60-Hz 3,000-kVA wye-connected synchronous generator has balanced line-to-line voltages of 4,160 V at its terminals. The inter-

nal impedance of the generator is $Z_s = j3.5 \ \Omega$ per phase. A delta-connected balanced three-phase load of $15 + j30 \ \Omega$ per phase is connected to the generator over an S switch. The line to-line voltage at the switch before it is closed is 4,160 V. Use phase-a voltage as the *reference phasor* and determine the following:
(a) The percent voltage drop of no-load voltage of the terminal voltage at the switch when the load is connected.
(b) The total complex power delivered to the load.

2.23 Assume that a balanced three-phase wye-connected source is connected to a balanced three-phase wye-connected load and that a neutral conductor connects the neutral points of the source and the load. Assume *abc* phase sequence and $V_a = 230 \angle 0° \ V$ at the load terminal. If the load impedances are $Z_a = Z_b = Z_c = 2.3 \angle 30° \ \Omega$, determine the following:
(a) The line currents I_a, I_b, and I_c.
(b) The current in the neutral conductor, that is, I_n.

2.24 A balanced three-phase delta-connected load is connected to a balanced three-phase source of abc sequence. If $V_{ab} = 460 \angle 30° \ V$ and the load impedances $Z_{ab} = Z_{bc} = Z_{ca} = 4.6 \angle 5° \ \Omega$, determine the following:
(a) The line voltages of V_{bc} and V_{ca}.
(b) The currents I_{ab}, I_{bc}, and I_{ca} of the load.
(c) The line currents I_a, I_b, and I_c of the load.

2.25 Solve Problem 2.22 but assume that the generator rating is 5 MVA and that its internal impedance is $Z_s = j2 \ \Omega$ per phase. The delta-connected and balanced three-phase load is made up of $3 + j5 \ \Omega$ per phase impedances. Use the phase-a voltage of $2,400 \angle 0° \ V$.

2.26 Solve Problem 2.13 but assume that the first load is made up of three $12 \angle 40° \ \Omega$ impedances, and that the second load is made up of three $18 \angle 80° \ \Omega$ impedances.

2.27 Solve Problem 2.12 but assume that the first load is wye-connected and made up of three $6 \angle 45° \ \Omega$ impedances, and that the second load is delta-connected and made up of three $9 \angle 75° \ \Omega$ impedances. Use $V_a = (480 / \sqrt{3}) \angle 0° = 277.1281 \angle 0° \ V$ for the voltage at the generator terminal.

2.28 Two three-phase generators are supplying the same load bus, as shown in Figure P2.28. Both generators produce balanced voltages of *abc* phase sequence. Use $V_a = 120 \angle 0° \ V$ and $V_a' = 115 \angle 0°$ and V as the reference voltages for the left and right generators, respectively. If a balanced three-phase load is connected in the middle of the bus, as shown in the figure, determine the following:
(a) The phasor currents I_a, I_a', and I_a''. (Hint: First convert the delta-connected load into its equivalent wye-connected load and then apply the *nodal analysis method* using the *line-to-neutral approach*

since both the generators and the load are balanced.)

(b) The total complex power supplied to the load.

(c) The line-to-line voltage at the terminals of the load.

(d) The complex power supplied by the left generator to the bus.

(e) The complex power supplied by the right generator to the bus.

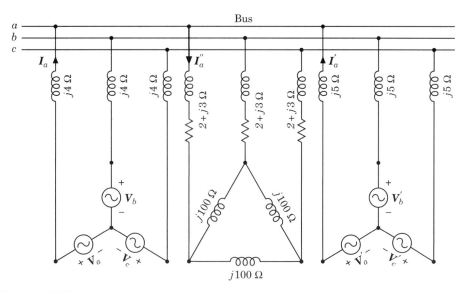

Figure P2.28

2.29 Solve Example 2.5 but use the following data. Let $\mathbf{Z}_{ab} = 6 - j6\ \Omega$, $\mathbf{Z}_{bc} = 0 + j6\ \Omega$, $\mathbf{Z}_{ca} = 6 + j6\ \Omega$, and use a line impedance $\mathbf{Z}_L = 3 + j3\ \Omega$ per phase. Use $\mathbf{V}_{a'b'} = 120\angle 0°$ V, $\mathbf{V}_{b'c'} = 120\angle 240°$ V and $\mathbf{V}_{c'a'} = 120\angle 120°$ V as the balanced source voltages.

2.30 Consider Figure P2.30 and assume that the balanced delta-connected load is made up of three impedances of $\mathbf{Z}_\phi = 27.71\angle -40°\ \Omega$ per phase

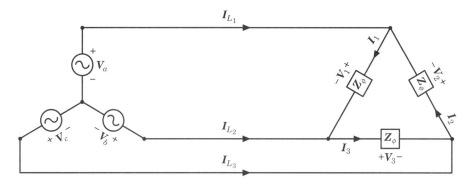

Figure P2.30

and that source voltages V_a, V_b, and V_c are 277.1∠0° V, 277.1∠−120° V, and 277.1∠120° V, respectively. Determine the following:
(a) The load voltages V_1, V_2, and V_3.
(b) The load (phase) currents I_1, I_2, and I_3.
(c) The line currents I_{L1}, I_{L2}, and I_{L3}.

2.31 Solve Example 2.7 but use the following data. Assume that $V_{ab} = 480∠0°$ V and that the load impedances Z_{ab}, Z_{bc}, and Z_{ca} are given as 20∠30° V, 10∠−60° Ω, and 15∠45° Ω, respectively.

2.32 A balanced and delta-connected three-phase voltage source is supplying power to a balanced and delta-connected three-phase load, as shown in Figure P2.32. The source voltages V_{ab}, V_{bc}, V_{ca} are 208∠30° V, 208∠270° V and 208∠150° V, respectively. The load impedances are $Z_{ab} = Z_{bc} = Z_{ca} = 2 + j3$ Ω. Determine the following:
(a) The load (phase) currents I_{ab}, I_{bc}, and I_{ca}.
(b) The line currents I_a, I_b, and I_c.
(c) The total real and reactive power supplied to the load.
(d) The total complex power supplied to the load.

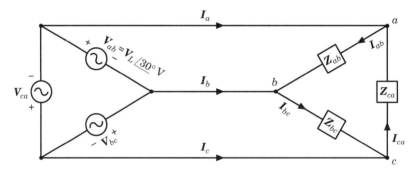

Figure P2.32

2.33 An unbalanced, three-phase, three-wire, wye-connected load is connected to a balanced, three-phase, wye-connected source and only two wattmeters are used and connected as shown in Figure P2.8. The line-to-neutral source voltages V_a, V_b, and V_c are 220∠30° V, 220∠270° V, and 220∠150° V, respectively, and the line currents I_a, I_b, and I_c are 71.62∠−11° A, 61.28∠16.2° A, and 13.26∠133.6°, respectively. Determine the following:
(a) The power recorded on each wattmeter.
(b) The total power absorbed by the load.

2.34 Consider Figure P2.32 and assume that a delta-connected source is supplying power to a delta-connected balanced load. If the generator has $V_{ab} = 480∠30°$ V, $V_{bc} = 480∠270°$ V, and $V_{ca} = 480∠150°$ V and load impedances of $\boldsymbol{Z}_{ab} = \boldsymbol{Z}_{bc} = \boldsymbol{Z}_{ca} = 5 + j5\ \Omega$, determine the following:

(a) The load (phase) currents $\boldsymbol{I}_{ab}$, $\boldsymbol{I}_{bc}$, and $\boldsymbol{I}_{ca}$.

(b) The line currents $\boldsymbol{I}_a$, $\boldsymbol{I}_b$, and $\boldsymbol{I}_c$.

(c) The total real and reactive power supplied to the load.

(d) The total complex power supplied to the load.

(e) Draw the current and voltage diagrams.

2.35   Solve Problem 2.34 but assume that $\boldsymbol{V}_{ab} = 380\angle 30°$ V, $\boldsymbol{V}_{bc} = 380\angle 270°$ V, and $\boldsymbol{V}_{ca} = 380\angle 150°$ V and that $\boldsymbol{Z}_{ab} = \boldsymbol{Z}_{bc} = \boldsymbol{Z}_{ca} = 2 + j3\ \Omega$.

2.36   Assume that a three-phase 480 V power line is supplying a delta-connected load and that $\boldsymbol{V}_{ab} = 480\angle 30°$ V, $\boldsymbol{V}_{bc} = 480\angle 270°$ V, and $\boldsymbol{V}_{ca} = 480\angle 150°$ V. Also assume that the current in each phase of the delta-connected load is 10 A at a 0.85 lagging power factor. If $\boldsymbol{V}_{an}$ is used as a *reference phasor* and the phase sequence is *abc,* determine the following:

(a) Draw the circuit, as given.

(b) The line-to-line voltages of $\boldsymbol{V}_{ab}$, $\boldsymbol{V}_{bc}$, and $\boldsymbol{V}_{ca}$.

(c) The line-to-neutral voltages of $\boldsymbol{V}_{an}$, $\boldsymbol{V}_{bn}$, and $\boldsymbol{V}_{cn}$.

(d) All phase and line currents.

(e) Draw the phasor diagram for the phase and line voltages.

(f) Draw the phasor diagram for the phase and line currents.

2.37   Solve Problem 2.36 but assume that $\boldsymbol{V}_{ab} = 380\angle 30°$ V, $\boldsymbol{V}_{bc} = 380\angle 270°$ V, and $\boldsymbol{V}_{ca} = 380\angle 150°$ V and that the current in each phase of the delta-connected load is 5 A at a 0. 75 lagging power factor.

2.38   A 2.4 kV single-phase circuit supplies a load of 250 kW at a lagging power factor and the load current is 160 A. If it is necessary to improve the power factor, determine the following:

(a) The uncorrected power factor and reactive load.

(b) The new corrected power factor after installing a shunt capacitor unit with a rating of 250 kvar.

2.39   A three-phase 50-hp 60-Hz 480-V wye-connected induction motor has a full-load efficiency of 0.85 percent, and a lagging power factor of 0.75. If it is required to correct the power factor of the load to a lagging power factor of 0.95 by connecting three capacitors, find the following:

(a) The rating of such a capacitor bank, in kV.

(b) The capacitance of each single-phase unit, if the capacitors are connected in delta, in μF.

(c) The capacitance of each single-phase unit, if the capacitors are connected in wye, in μF.

MAGNETIC CIRCUITS

All is flux, nothing stays still.
Heraclitus, 500 B.C.

Time crumbles things; everything grows old
under the power of Time
and is forgotten through the lapse of Time.
Aristotle, 352 B.C.

Lord, why is it that wisdom comes so late and life is so short?
Turan Gönen

3.1 INTRODUCTION

Today, the phenomenon of *magnetism*[1] is used in the operation of a great number of electrical apparatus including generators, motors, transformers, measuring instruments, televisions, radios, telephones, tape recorders, computer memories, computer magnetic tapes, car ignition tapes, refrigerators, air conditioners, heating equipment, and power tools. A material that has the ability to attract iron and steel is called a *magnet*. Magnets can be categorized as being *permanent* or *temporary,* based on their ability to retain magnetism. Figure 3.1(a) shows a permanent (bar) magnet and its magnetic field. Notice that the magnetic *flux* (Φ) lines (i.e., the magnetic *lines of force* [2]) are continuous, and come from the north pole and go toward the south pole. The direction of this field can be established using a compass (which is simply a freely sus-

---

[1] The phenomenon of magnetism has been recognized since 600 B.C. (by the ancient Greeks). However, the first experimental work was performed in the sixteenth century by the English physician, Gilbert, who discovered the existence of a magnetic field around earth. Also, Oersted recognized that current-carrying conductors could have magnetic effects. Further studies, done by Ampère, on the magnetic field around current-carrying loops led to the theory of magnetism on an atomic scale. This and other experiments that were performed by Henry and Faraday established the foundation for the development of modern electrical machinery.

[2] It is interesting to note that the lines of force in reality do not exist, but the concept is sometimes beneficial in describing the properties of magnetic fields.

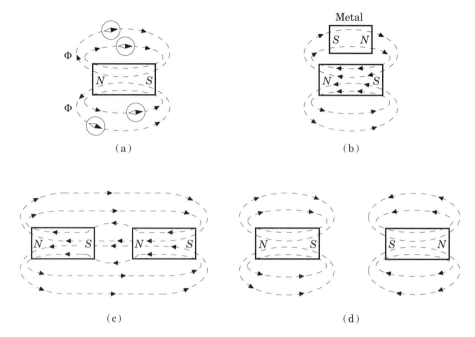

Figure 3.1 Magnetic field of: (a) single permanent magnet; (b) metal in the vicinity
of a permanent magnet; (c) two permanent magnets with unlike poles;
(d) two permanent magnets with like poles.

pended magnetized steel needle) since the marked end[3] of a compass needle
always points to the earth's magnetic north pole. As shown in Figure 3.1(b),
when a permanent magnet is placed near a metal, the magnetic lines go through
the metal and magnetize it. If two permanent magnets are located close together as shown in Figure 3.1(c), the magnets are attracted toward each other
since the direction of the magnetic lines of force of both magnets is the same.
However, if the two magnets are located in the opposite direction as shown in
Figure 3.1(d), the two magnets are repelled and forced apart since the magnetic
lines of force go from north to south and are opposing.

3.2 MAGNETIC FIELD OF CURRENT-CARRYING
CONDUCTORS

As illustrated in Figure 3.2(a), when a conductor carries an electric current I, a magnetic field is created around the conductor.[4] The direction of magnetic lines of force (or field) is determined using Ampère's **right-hand rule**
which is illustrated in Figure 3.2(c). It shows that *if the conductor is held in the*

---

[3] Since, according to the rule of magnetic attraction and repulsion, unlike magnetic poles attract
and like poles repel, the marked end of the compass needle is really a south pole.
[4] Oersted discovered a definite relationship that exists between electricity and magnetism in 1819.

right hand with the thumb pointing in the direction of the current flow, the fingers then point in the direction of the magnetic field around the conductor. Thus, the conversion of energy between mechanical and electrical forms is achieved through magnetic fields.

Figure 3.2(b) shows the magnetic fields around a conductor carrying current toward the reader and away from the reader, respectively. Note that, in the figure, the symbol ⊙ denotes a cross-sectional view of a conductor carrying current toward the reader, while the symbol ⊕ denotes the current flowing away from the reader. Figure 3.2(d) illustrates the magnetic field around a coil made up of two parallel conductors. Similarly, Figures 3.2(e) and 3.2(f) show the magnetic fields around a coil made up of four conductors all carrying current away from the reader and toward the reader, respectively.

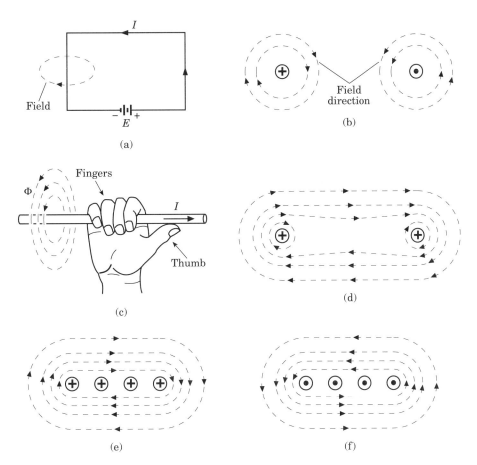

Figure 3.2 Magnetic lines of field: (a) around a current-carrying conductor; (b) around a current-carrying conductor toward the reader and away from the reader; (c) determined by using Ampère's right hand rule; (d) around two parallel conductors; (e) around four conductors all carrying current away from the reader; (f) around four conductors all carrying current toward the reader.

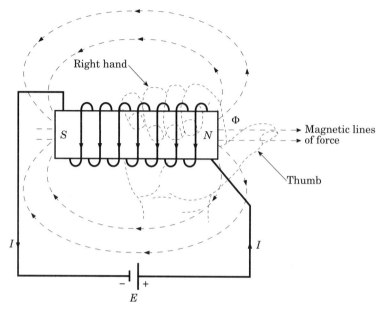

Figure 3.3   Magnetic field of a current-carrying coil.

Figure 3.3 shows a current-carrying coil that is formed by wrapping a conductor (or wire) on a hollow cardboard or fiber cylinder. The magnetic lines of force (i.e., flux) are concentrated within the cylinder. Each turn of the wire develops a magnetic field in the same direction. Because the direction of current flow is the same in all turns of the wire, the resultant magnetic field generated inside the coil is all in the same direction. The polarity of such a coil can also be found by using the right-hand rule, as illustrated in Figure 3.3. *If the coil is held in the right hand with the fingers pointing in the direction of the current in the coil, the thumb then points toward the north pole of the coil.* In Figure 3.3, the end of the coil where the flux comes out is the *north pole* of the coil. Figure 3.4 shows the magnetic circuit of a two-pole dc generator. A required strong magnetic field is produced by the two field coils wound around the iron pole cores. As the armature (which is located on the rotor) is rotating through the magnetic field, an ***electromotive* force** (emf) is generated in the armature conductors. The measure of a coil's ability to produce flux is called ***magnetomotive* force** (mmf). The mmf of a magnetic circuit corresponds to the emf in an electric circuit. The mmf of a coil depends on the amount of the current flowing in the coil and the number of turns in the coil. The product of the current in amperes and the number of turns is called the **ampere-turns** of the coil.

3.3 AMPÈRE'S MAGNETIC CIRCUITAL LAW

Figure 3.5(a) shows a ring-shaped coil of N turns supplied by a current I. The dotted circular line represents an arbitrary closed path that has the same *magnetic field intensity* value, H, over an elementary length, $d\ell$ at any location

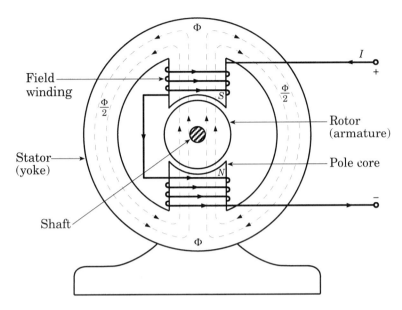

Figure 3.4   The magnetic circuit of a two-pole generator.

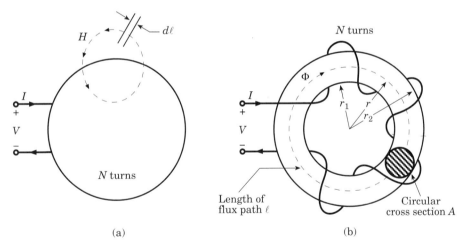

Figure 3.5   Illustration of a magnetic field of: (a) a current-carrying coil of N turns; (b) toroidal coil.

on the path. In contrast, Figure 3.5(b) shows a magnetic circuit with a ring-shaped magnetic core called a *toroid* with a coil wound around the entire core. If a current I flows through a coil of N turns, a field is created in the toroid.

According to **Ampère's Circuital law**, the magnetic potential drop around a closed path is balanced by the mmf, giving rise to the field (i.e., the mmf encircled by the closed path). For the average path at a mean radius r, the magnetic field intensity H is related to its source NI by Ampère's Circuital law,

$$\oint H \cdot d\ell = \mathscr{F} = NI \tag{3.1}$$

The line integral can be solved easily if the closed path of integration is selected along the direction of H, and as previously noted, H has a constant value along the mean path ℓ. Thus,

$$H\ell = NI \tag{3.2}$$

or

$$H2\pi r = NI \tag{3.3}$$

from which

$$H = \frac{NI}{2\pi r} \tag{3.4}$$

The quantity NI is called the *magnetomotive force* (**mmf**), and its unit is *ampere-turn* (A·turn) or *ampere* (A) since N has no dimensions.

$$\mathscr{F} = H\ell = NI \quad \text{A·turns} \tag{3.5}$$

from which

$$H = \frac{F}{\ell} \quad \text{A·turns/m} \tag{3.6}$$

or

$$H = \frac{NI}{\ell} \quad \text{A·turns/m} \tag{3.7}$$

The magnetic field intensity H describes the field produced by the mmf, and its unit is ampere-turn per meter. Also, the magnetic field intensity H produces a **magnetic flux density** B everywhere it exists. They are related to each other by

$$B = \mu H \quad \text{Wb/m}^2 \text{ or T} \tag{3.8}$$

The value of B depends not only on H (and thus the current), but also on the medium in which H is located. The SI unit[5] of B is weber/m^2 (Wb/m^2) or tesla (T). The effect of the medium is presented by its permeability[6] μ in henrys/m (H/m). Here, the μ represents the relative ease of establishing a magnetic field in a given material. The permeability of free space is called μ_0 and it has a value of

$$\mu_0 = 4\pi \times 10^{-7} \ \text{H/m} \tag{3.9}$$

which is approximately the same for air. The permeability of any other material with respect to the permeability of free space is called its **relative permeability** μ_r. Therefore,

$$\mu_r = \frac{\mu}{\mu_0} \tag{3.10}$$

The relative permeability μ_r is dimensionless[7] and equal to 1.0 for free space The permeability of any material can be expressed as

$$\mu = \mu_r \mu_0 = \frac{B}{H} \tag{3.11}$$

For materials used in electrical machines, the value of μ_r can be as high as several thousands. The larger the value of μ_r, the smaller the current that is needed to produce a given flux density B in the machine. By substituting equation (3.7) into equation (3.8), the magnitude of the flux density can be expressed as

$$B = \mu H = \frac{\mu NI}{\ell} \tag{3.12}$$

The total flux crossing of a given cross-sectional area A can be found from

$$\Phi = \int_A B \cdot dA \tag{3.13}$$

---

[5] Older units of magnetic flux density (i.e., the flux per unit area) that are still in use include lines/in^2, kilolines/in^2, and gausses (G). Note that 1 gauss = 1 maxwell/cm^2 and 1 T = 10 kG = 10^4 G. Therefore, if a flux density is given in lines/in^2, it must be multiplied by 1.55 x 10^{-5} to convert it to Wb/m^2 or T.

[6] Permeability, based on equation (3.8), can be defined as the ratio of change in magnetic flux density to the corresponding change in magnetic field intensity. Therefore, in a sense, permeability is not a constant parameter but depends on the flux density or on the applied mmf that is used to energize the magnetic circuit.

[7] From the relative permeability point of view, all materials can be classified into four distinct groups: (1) *diametric* ($\mu_r = 1.0^-$), (2) *nonmagnetic* ($\mu_r = 1.0$), (3) *paramagnetic* ($\mu_r = 1.0^+$), and (4) *ferromagnetic* ($1.0^+ < \mu_r < \infty$). Furthermore, a special type of diamagnetism has started to make the headlines recently. It is a case of *perfect diamagnetism* (known as the *Meissner effect*) that takes place in particular types of materials (which are called **superconductors**) at temperatures near absolute zero. They are increasingly used in certain types of electromagnetic devices, including various rotating machines and switching devices. In such superconductive materials, the flux density approaches zero and the relative permeability is basically zero. Therefore, a magnetic field cannot be developed in the superconducting material.

where dA is the differential unit of area. Therefore,

$$\Phi = BA \quad \text{Wb} \tag{3.14}$$

This equation is correct if the flux density vector is perpendicular to the place of area A, and if the flux density is constant at each location in the given area. For the toroid shown in Figure 3.5(b), the average flux density may correspond to the path at the mean radius of the toroid. Thus, the total flux[8] in the core is

$$\Phi = BA = \frac{\mu NIA}{\ell} \quad \text{Wb} \tag{3.15}$$

or

$$\Phi = B\pi r^2 = \frac{\mu NI\pi r^2}{\ell} \quad \text{Wb} \tag{3.16}$$

The product of the winding turns N and the flux Φ that links them is called the **flux linkage**. Flux linkage is usually denoted by the Greek letter λ (lambda) and expressed as

$$\lambda = N\Phi \quad \text{Wb} \tag{3.17}$$

3.4 MAGNETIC CIRCUITS

Consider the simple magnetic core shown in Figure 3.6(a), by substituting equation (3.5) into equation (3.15),

$$\Phi = \frac{\mu NIA}{\ell}$$

$$= \frac{NI}{\ell / \mu A}$$

$$= \frac{NI}{\mathcal{R}} \tag{3.18}$$

$$= \frac{\mathcal{F}}{\mathcal{R}} \tag{3.19}$$

from which

---

[8] The SI unit for magnetic flux is *webers* (Wb). The older unit of flux was the *line* or *maxwell*. Thus, 1 weber $= 10^8$ maxwells $= 10^8$ lines $= 10^5$ kilolines.

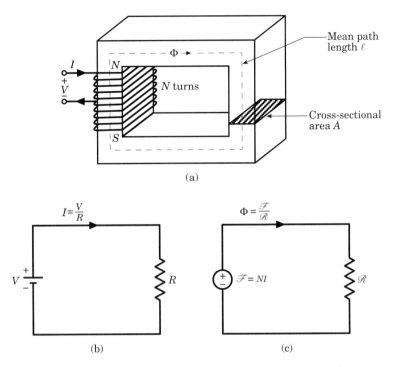

Figure 3.6   (a) A simple magnetic core; (b) its electric circuit; (c) its magnetic circuit analogue.

$$\mathscr{F} = \Phi \mathscr{R} \tag{3.20}$$

where $\mathscr{R}$ is the reluctance of the magnetic path and therefore

$$\mathscr{R} = \frac{\ell}{\mu A} \tag{3.21}$$

for uniform permeability μ, cross-sectional area A, and mean path length ℓ of the magnetic circuit. The reluctance[9] can also be expressed as

$$\mathscr{R} = \frac{\mathscr{F}}{\phi} \tag{3.22}$$

The reciprocal of the reluctance is called the **permeance** of the magnetic circuit and is expressed as

---

[9] In the SI system, no specific name is given to the dimension of reluctance except to refer to it as so many units of reluctance. One can observe from equation (3.22) that its real dimensions are *ampere · turns/weber*. In some old literature, the word *rels* has been used as the unit of reluctance.

$$\mathscr{P} = \frac{1}{\mathscr{R}}$$

3.23)

Therefore, the flux given by equation (3.19) can be expressed as

$$\Phi = \mathscr{F}\mathscr{P}$$

(3.24)

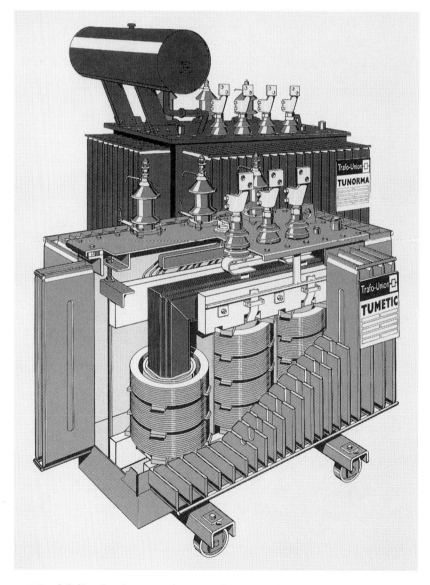

Figure 3.7 Oil distribution transformers: Cutaway of a TUMETIC transformer with an oil expansion tank shown in the background, and a TUNORMA with an oil expansion tank shown in the background. (Courtesy of Siemens AG.)

Figure 3.7 shows a cutaway view of an oil distribution transformer that reveals its iron core magnetic circuit and coil construction. Figure 3.8 shows a 630 kVA, 10/0.4 kV solid dielectric transformer.

In many aspects the electric and magnetic circuits are analogous. For example, notice the analogy between the electric circuit shown in Figure 3.6(b) and the magnetic circuit shown in Figure 3.6(c). Note that the flux in the magnetic circuit acts like the current in the electric circuit, the reluctance in the magnetic circuit can be treated like the resistance[10] in the electric circuit, and the magnetomotive force in the magnetic circuit can be treated like the electromotive force in the electric circuit. Equation (3.20) is often referred to as **Ohm's law of the magnetic circuit**. However, electric and magnetic circuits are not analogous in all respects. For instance, energy must be continuously provided when a direct current is flowing in an electric circuit, whereas in the case of a magnetic circuit, once the flux is established, then it remains constant. Similarly, there are no *magnetic insulators, only electric insulators*.

It is interesting that reluctances and permeances connected in series and parallel are treated in the same manner as resistances and conductances connected in series and parallel, respectively. For example, the equivalent reluctance of a number of reluctances connected in series with respect to each other is

$$\mathscr{R}_{eq} = \mathscr{R}_1 + \mathscr{R}_2 + \mathscr{R}_3 + \ldots \tag{3.25}$$

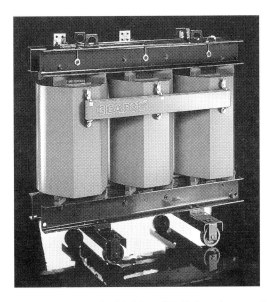

Figure 3.8   A 630-kVA, 10/0.4-kV GEAFOL solid dielectric transformer. (Courtesy of Siemens AG.)

---

[10] The resistance of a wire of length ℓ and cross-sectional area A is given by $R = \ell/\rho A$ where ρ is the conductivity of the material in S/m.

Similarly, the equivalent reluctance of a number of reluctances connected in parallel with respect to each other is

$$\frac{1}{\mathscr{R}_{eq}} = \frac{1}{\mathscr{R}_1} + \frac{1}{\mathscr{R}_2} + \frac{1}{\mathscr{R}_3} + \ldots \qquad 3.26)$$

Alternatively, first the equivalent permeance is found by using

$$\mathscr{L}_{eq} = \mathscr{L}_1 + \mathscr{L}_2 + \mathscr{L}_3 + \ldots \qquad (3.27)$$

and then the equivalent reluctance is determined as

$$\mathscr{R}_{eq} = \frac{1}{\mathscr{L}_{eq}} \qquad (3.28)$$

Also, equation (3.15) is substituted into equation (3.17)

$$\lambda = \frac{\mu N^2 I A}{\ell} \quad \text{Wb} \qquad (3.29)$$

so that flux linkage is directly proportional to the coil current.

The **inductance** L of a coil is defined as the flux linkage per ampere of current in the coil and measured in *henries* (H). Therefore,

$$L = \frac{\lambda}{I} \quad \text{H} \qquad (3.30)$$

From equations (3.21), (3.29), and (3.30), it can be shown that inductance can be related to reluctance as

$$L = \frac{N^2}{\mathscr{R}} \quad \text{H} \qquad (3.31)$$

Alternatively, inductance can be expressed in terms of permeance as

$$L = N^2 \mathscr{L} \quad \text{H} \qquad (3.32)$$

Example 3.1

Consider the toroid that is shown in Figure 3.5(b) with outside and inside radiuses of 5 and 4 cm, respectively. Assume that 500 turns are wound around the toroid of soft steel casting to produce a total flux of 168.5856×10^{-5} Wb in the core. If the magnetic field intensity in the core is 1000 A·turns/m, determine:

(a) The length of the average flux path in the toroid and the cross-sectional area perpendicular to the flux.
(b) The flux density in the core.
(c) The required magnetomotive force.
(d) The amount of current that must flow in the turns of the toroid.

SOLUTION:
(a) Since the mean radius r is

$$r = \frac{r_1 + r_2}{2} = \frac{4+5}{2} = 4.5 \text{ cm or } 4.5 \times 10^{-2} \text{ m}$$

the length of the average flux path is

$$\ell = 2\pi r = 2\pi(4.5 \times 10^{-2}) = 0.2827 \text{ m}$$

Thus, the cross-sectional area is

$$A = \pi r^2$$

$$A = \frac{\pi r^2}{4} = \frac{\pi\left(4.5 \times 10^{-2} \text{ m}\right)^2}{4} = 159.043 \times 10^{-5} \text{ m}^2$$

$$7.854 \times 10^{-5}$$

(b) The flux density in the core is

$$B = \frac{\Phi}{A} = \frac{168.5856 \times 10^{-5} \text{ Wb}}{159.043 \times 10^{-5} \text{ m}^2} = 1.06 \text{ Wb/m}^2 \text{ or } 1.06 \text{ T}$$

$$\gtrsim 1.46$$

(c) Since the magnetic field intensity is given as $1000 \text{ A} \cdot \text{turns/m}$, the required mmf is

$$\mathscr{F} = H\ell = (1000 \text{ A} \cdot \text{turns/m})(0.2827 \text{ m}) = 282.7 \text{ A} \cdot \text{turns}$$

(d) Since

$$\mathscr{F} = NI$$

therefore

$$I = \frac{\mathscr{F}}{N} = \frac{282.7 \text{ A} \cdot \text{turns}}{500 \text{ turns}} = 0.5654 \text{ A}$$

Example 3.2

Resolve Example 3.1 but assume that the toroid is made of plastic. The permeability of plastic is the same as that for free air. Assume that the total flux amount is the same as before but that the magnetic field intensity is un-known.

SOLUTION 1:
(a) As before,

$$r = 4.5 \times 10^{-2} \text{ m}, \quad \ell = 0.2828 \text{ m, and } A = 159.043 \times 10^{-5} \text{ m}^2$$

(b) The flux density in the core is still

$$B = 1.06 \text{ Wb/m}^2 \text{ or } 1.06 \text{ T}$$

(c) Since $B = \mu_0 H$, the magnetic field intensity in the plastic core is

$$H = \frac{B}{\mu_0} = \frac{1.06 \text{ Wb/m}^2}{4\pi \times 10^{-7} \text{H/m}} = 843{,}521.1983 \text{ A} \cdot \text{turns/m}$$

and since

$$\mathscr{F} = H\ell = NI$$

$$\mathscr{F} = H\ell = (843{,}521.1983 \text{ A} \cdot \text{turns/m}) \, 0.2827 \text{ m}$$

$$= 238{,}463.44 \text{ A} \cdot \text{turns}$$

(d) Therefore

$$I = \frac{\mathscr{F}}{N} = \frac{238{,}463.44 \text{ A} \cdot \text{turns}}{500 \text{ turns}} = 476.93 \text{ A}$$

Note that the current required to produce the same amount of flux has increased by 843.5 times from that which was required in Example 3.1, when the core was made of soft steel casting.

SOLUTION 2:
(a) The same as before.
(b) The same as before.
(c) First find the reluctance as

$$\mathscr{R} = \frac{\ell}{\mu_0 A} = \frac{0.2827 \text{ m}}{(4\pi \times 10^{-7} \text{ H/m})(159.043 \times 10^{-5} \text{ m}^2)}$$

$$= 141{,}449{,}489.8 \text{ A} \cdot \text{turns/Wb}$$

Since

$$\mathscr{F} = \Phi \mathscr{R}$$

$$= (168.5856 \times 10^{-5} \text{ Wb})(141{,}449{,}489.8 \text{ A} \cdot \text{turns/Wb})$$

$$= 238{,}463.47 \text{ A} \cdot \text{turns}$$

(d) Thus,

$$I = \frac{\mathscr{F}}{N} = \frac{238{,}463.47 \text{ A} \cdot \text{turns}}{500 \text{ turns}} = 476.93 \text{ A}$$

Example 3.3

Consider Example 3.2 and determine:
(a) The flux linkage in the core.
(b) The inductance of the coil by using equation (3.30).
(c) The inductance of the coil by using equation (3.31).

SOLUTION:
(a) From equation (3.17),

$$\lambda = N\Phi = (500 \text{ turns})(168.5856 \times 10^{-5} \text{ Wb}) = 0.8429 \text{ Wb}$$

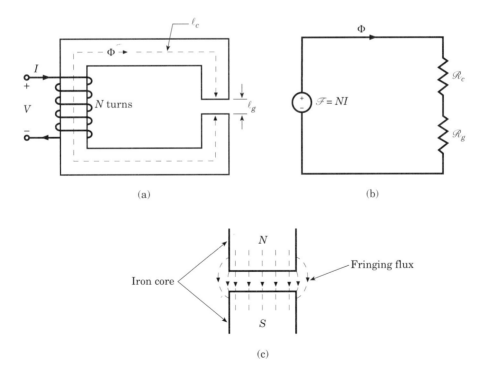

Figure 3.9 (a) A simple magnetic circuit with an air gap; (b) its magnetic circuit analogue; (c) the fringing effect of magnetic flux at an air gap.

(b)   From equation (3.30),

$$L = \frac{\lambda}{I} = \frac{0.8429 \text{ Wb}}{476.93 \text{ A}} = 0.0018 \text{ H}$$

(c)   From equation (3.31),

$$L = \frac{N^2}{\mathscr{R}} = \frac{500^2}{141{,}449{,}489.8} = 0.0018 \text{ H}$$

3.5 MAGNETIC CIRCUIT WITH AIR GAP

Air gaps are fundamental in many magnetic circuits currently in use. As shown in Figure 3.4, every electromechanical energy converter is made up of two parts, namely (1) the stator, and (2) the rotor embedded in the air gap of the stator.

As shown in Figure 3.9(a), essentially the same flux is present in the magnetic core and the air gap. To sustain the same flux density, the air gap must have much more mmf than the magnetic core. If the flux density is high, the magnetic core section of the magnetic circuit may show the saturation effect. However, the air gap section of the magnetic circuit will remain unsaturated due to the fact that the B & H curve for the air medium is linear, with a constant permeability.[11] In Figure 3.9(a), ℓ_c is the length of the magnetic core, while ℓ_g is the length of the air gap. Since there is more than one material involved, such a magnetic circuit is said to be made of a *composite structure*. Figure 3.9(b) shows the equivalent magnetic circuit that has the reluctance of the air gap $\mathscr{R}_g$ in series with the reluctance of the core $\mathscr{R}_c$. Applying Ampère's law, the required mmf can be found from

$$\mathscr{F} = H_c \ell_c + H_g \ell_g \tag{3.33}$$

or

$$NI = H_c \ell_c + H_g \ell_g \tag{3.34}$$

The resulting flux can be found from

$$\Phi = \frac{NI}{\mathscr{R}_c + \mathscr{R}_g} \tag{3.35}$$

where the reluctances for the core and air gap are

---

[11] The flux density in the air gap can be easily measured by the use of an instrument known as a *gausemeter*. The principle of the design of such an instrument is known as the *Hall effect*.

$$\mathscr{R}_c = \frac{\ell_c}{\mu_c A_c} = \frac{\ell_c}{\mu_0 \mu_r A_c} \qquad (3.36)$$

and

$$\mathscr{R}_g = \frac{\ell_g}{\mu_0 A_g} \qquad (3.37)$$

where ℓ_c = the mean length of the core

ℓ_g = the length of the air gap

A_c = the cross-sectional area of the core

A_g = the cross-sectional area of the air gap

The associated flux densities can be found by

$$B_c = \frac{\Phi_c}{A_c} = \frac{\mu_c}{\ell_c} \mathscr{F} \qquad (3.38)$$

and

$$B_g = \frac{\Phi_g}{A_g} = \frac{\mu_0}{\ell_g} \mathscr{F} \qquad (3.39)$$

Since the individual *mmf drops* in Figure 3.9(b) can be expressed as

$$H_c \ell_c = \Phi \mathscr{R}_c \qquad (3.40)$$

and

$$H_g \ell_g = \Phi \mathscr{R}_g \qquad (3.41)$$

by substituting equations (3.40) and (3.41) into equation (3.33)

$$\mathscr{F} = NI = \Phi(\mathscr{R}_c + \mathscr{R}_g) \qquad (3.42)$$

If there is an air gap in a magnetic circuit, there is a tendency for the flux to bulge outward (or spread out) along the edges of the air gap, as shown in Figure 3.9(c), rather than to flow straight through the air gap parallel to the edges of the core. This phenomenon is called **fringing** and is taken into account by assuming a larger (effective) air gap cross-sectional area. The common practice is to use an (effective) air-gap area by adding the air-gap length to each of the two dimensions which make up the cross-sectional area. Thus, the new (effective) air-gap area becomes

$$A_g = (a + \ell_g)(b + \ell_g) \qquad (3.43)$$

where a and b are the actual core dimensions of a given rectangular-shaped core. The corrected gap area slightly reduces the gap reluctance. The relative effect of fringing increases with the length of the air gap.

3.6 A BRIEF REVIEW OF FERROMAGNETISM

Magnetic materials which include certain forms of iron and its alloys in combination with cobalt, nickel, aluminum and tungsten are called *ferromagnetic materials*.[12] They are relatively easy to magnetize since they have a high value of relative permeability μ_r. These ferromagnetic materials are classified as *hard* or *soft* materials. Soft ferromagnetic materials include most of the soft steels, iron, nickel, cobalt, and one rare-earth element, as well as many alloys of the four elements. Hard ferromagnetic materials comprise the permanent magnetic materials such as alnico (which is iron alloyed with aluminum, nickel, and cobalt), the

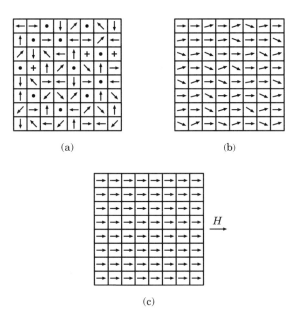

(a)   (b)

(c)

Figure 3.10   (a) Magnetic domains oriented randomly; (b) magnetic domains becoming magnetized; (c) magnetic domains fully magnetized (lined up) by the magnetic field H.

---

[12] There are also other magnetic materials that have been in use in recent years. They include (1) ferromagnetic materials, (2) superparamagnetic materials, and (3) ferrofluidic materials. The *ferromagnetic materials* are ferrites and therefore made up of iron oxides. They include permanent magnetic ferrites (for example, strontium or barium ferrites), manganese-zinc ferrites, and nickel-zinc ferrites. The *superparamagnetic materials* are made up of powdered iron (or other magnetic material) particles that are mixed in a nonferrous epoxy or other plastic material. Permalloy is an example of such a material and is made up of molybdenum-nickel-iron powder. Finally, the *ferrofluidic materials* are magnetic fluids that are made up of three components: a carrier fluid, iron oxide particles suspended in the fluid, and a stabilizer.

alloys of cobalt with a rare-earth element such as samarium, the copper-nickel alloys, the chromium steels, and other metal alloys.

The atoms of a ferromagnetic material tend to have their magnetic fields closely aligned. Within the crystals of such materials, there are many tiny (usually of microscopic) regions called *domains.* In any such given domain, all the atoms are aligned with their magnetic fields pointing in the same direction. Each domain behaves like a small permanent magnet. However, if the material is not magnetized, it will not have any flux within it, since each tiny domain is oriented randomly. This is illustrated in Figure 3.10(a) where the arrows represent the magnetic-moment direction within each domain. Notice that the domain alignments may be randomly distributed in three dimensions. The size of the domains is such that a single crystal may have many domains, each aligned with an axis of the crystal. If an electric field is applied to such a metal piece, the number of domains aligned with the magnetic field will increase since the individual atoms within each domain will physically switch orientation to align themselves with the magnetic field. This, in turn, increases the flux in the iron as well as the strength of the magnetic field, causing more atoms to switch orientation, as shown in Figure 3.10(b). Figure 3.10(c) shows that all domains are aligned with the magnetic field, and any increase in the strength of the field will not cause any change in orientation. Thus, the material is referred to as *saturated.* If the material becomes saturated as the magnetizing field intensity is increased, the flux density changes very little and eventually not at all.[13] Figure 3.11 shows a typical dc *magnetization curve*[14] of a ferromagnetic material. It shows the behavior of the three regions of domain: the linear region, the knee region, and the saturation region. Figure 3.12 shows the magnetization curves of two typical ferromagnetic materials used in the manufacture of power apparatus.

Consider the magnetic circuit shown in Figure 3.6 and assume that its magnetic core is initially unmagnetized. Assume that, instead of applying a *dc* current to the coil, an ac current is applied. Since the core is initially unmagnetized, the flux in the core is zero. As shown in Figure 3.13(a), if the current in the coil is increased, the flux in the core will increase. As a result, its magnetic field intensity H will also increase and follow the initial magnetization curve (along 0a) until saturation is reached. At the saturation point a, the flux density has reached its maximum value, B_{sat}, and the magnetic material is fully saturated. The corresponding value of the magnetic field intensity is H_{sat}. If the current is now decreased in the coil thereby decreasing the magnetizing force (i.e., magnetic field intensity) H the initial curve will not be retraced. A different path (along the *ab*

---

[13] If the external magnetic field is removed, the orientation of individual domains will not become totally randomized again since shifting back the orientation of atoms will need additional energy that may not be available. Therefore, the metal piece will remain magnetized. Such energy requirement may be fulfilled by applying (1) mmf in the opposite direction, (2) a large enough mechanical shock, or (3) heat. For example, a permanent magnet can easily be demagnetized if it is hit by a hammer, dropped, or heated. Such additional energy can cause domains to lose their alignment. Also note that at a very high temperature known as the *Curie point,* magnetic moments cease to exist, and therefore, the magnetic material involved loses its magnetic properties. For example, the Curie point for iron is about 775°C.

[14] The terms *magnetization curve* and *saturation curve* are used interchangeably in practice.

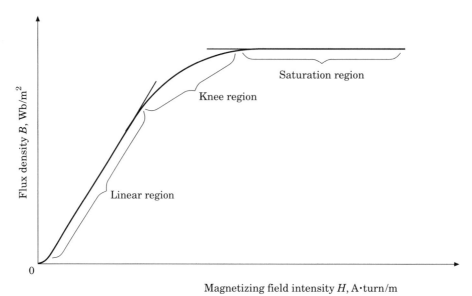

Figure 3.11   Typical magnetized curve showing the behavior of the three regions of domain behavior for a ferromagnetic core.

curve) will be followed indicating that there is a lag or delay in the reversal of domains. (Note that B does not decrease as quickly as it increased.) This irreversibility is called *hysteresis*,[15] which simply means that the flux density B *lags behind* the field intensity H.[16] When the current in the coil is zero, i.e., H is equal to zero, there is still a residual value of magnetic flux density B_{res} in the core, the magnitude of which depends on the material. This is called **residual flux density** or **remanence** and in effect creates a permanent magnet. To decrease the flux density B to zero requires a **coercive field intensity** (also called **coercive force**) $-H_c$. Any further increase in H in the reverse direction causes the magnetic core to be magnetized with the opposite polarity. This is achieved with a reversed current flow in the coil. Increasing the current in the negative direction further will result in a saturation level at which point (i.e., point d) it can be assumed that all domains in the magnetic core are aligned in the opposite direction. As the current or the field intensity H is brought to zero, the magnetic flux density B in the magnetic core will again be equal to its residual magnetism (at point e). Again, the direction of the current in the coil has to be reversed to make the magnetic flux density in the core equal to zero. Therefore, if the process continues in this manner, the *hysteresis loop* shown in Figure 3.13(a) will be traced out.

---

[15] The term originates from the Greek word *hysterein* to be behind or to lag. It represents the failure to retrace the initial magnetization curve.

[16] Since any change of B lags behind the change of magnetizing field intensity H which produces it, there will be an angular displacement between the rotating mmf wave of the stator winding and the alternating field induced in the rotor iron. As a result of this, there will be a hysteresis torque whenever the iron is moving relative to the inducing mmf wave. This is the principle on which small *hysteresis motor* operation is based.

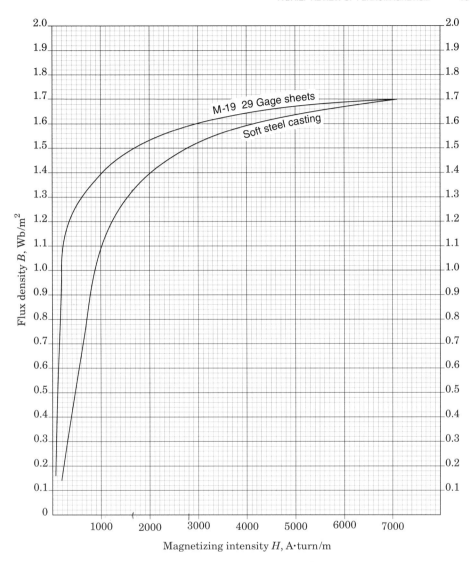

Figure 3.12   Magnetization (or B-H) curves of two typical ferromagnetic materials.

Figure 3.13(b) shows that for each maximum value of the ac magnetic field intensity cycle, there is a separate steady-state loop. Therefore, the complete magnetization characteristic is made up of a set (or family) of loops for different peak values of excitation. In Figure 3.13(b), the dashed curve that connects the tips of the loops is the dc magnetization curve of the magnetic material.

The shape of the hysteresis loop is a function of the type of magnetic material. As stated previously, magnetically soft materials have very low residual flux density and low coercive field intensity. Therefore, these materials (such as silicon iron with 3 to 4 percent S_i content) are used in the manufacture of electric machines and transformers. In such materials, the magnetization can

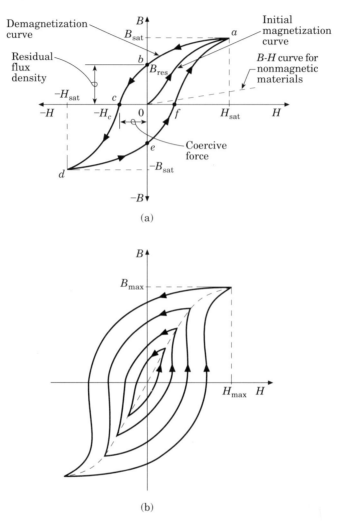

Figure 3.13   (a) Hysteresis loop for a typical ferromagnetic core; (b) family of hysteresis loops corresponding to different peak values of excitation.

be changed quickly without much friction. Thus, their hysteresis loops usually have a tall narrow-shaped small area. At 50 or 60 Hz, such hysteresis loops have a narrower shape than the ones for higher frequencies. In other words, the higher the frequency is, the broader the associated hysteresis loop will be. Therefore, in dc current the hysteresis loop almost turns into a curve known as the **magnetization curve**.

3.7 MAGNETIC CORE LOSSES

When a magnetic material is subjected to a time-varying flux, there is some energy loss in the material in the form of magnetic losses. Such magnetic losses are also called **iron** or **core losses**.[17] Tthe cores of the armatures of dc and ac machines, transformers, and reactors are subject to these core losses. In general, core losses are defined as the sum of hysteresis and eddy-current losses.

3.7.1 HYSTERESIS LOSS

The hysteresis loss is caused by continuous reversals in the alignment of the magnetic domains in the magnetic core. Succinctly put, the energy that is required to cause these reversals is the hysteresis loss. The area of the hysteresis loop represents the energy loss during one cycle in a unit cube of the core material. According to Charles P. Steinmetz, the hysteresis loss can be determined empirically from

$$P_h = v f k_h B_m^n \quad \text{W} \qquad (3.44)$$

where v = volume of ferromagnetic material, m^3

f = frequency, Hz

k_h = proportionality constant depending upon the core material (typically, soft iron, silicon sheet steel, and permalloy are 0.025, 0.001, and 0.0001, respectively)

B_m = maximum flux density

n = the Steinmetz exponent, which varies from 1.5 to 2.5 depending upon the core material, varies from 1.5 to 2.5 (typically, a value of 2.0 is used for estimating purposes)

3.7.2 EDDY-CURRENT LOSS

Because iron is a conductor, time-varying magnetic fluxes induce opposing voltages and currents called **eddy currents** that circulate within the iron core, as shown in Figure 3.14(b).[18] In the solid iron core these undesirable circulating currents flow around the flux and are relatively large because they encounter very little resistance. Therefore, they produce power losses with associated heating effects and cause demagnetization. As a result of this demagnetization, the flux distribution in the core becomes nonuniform, since most of the flux is pushed toward the outer surface of the magnetic material. As shown in Figure 3.14(a), the eddy currents always tend to flow perpendicular to the flux and in a direction that opposes any change in the magnetic field due to Lenz's law. In other words, the induced eddy currents tend to establish a flux that opposes the original change imposed by the source.

---

[17] No core losses take place in iron cores carrying flux that does not vary with time.

[18] At very high frequencies the interior of the magnetic core is practically unused because of the large (and circulating) eddy currents induced and their inhibiting effect. This phenomenon that takes place in magnetic circuits is known as the *magnetic skin effect*.

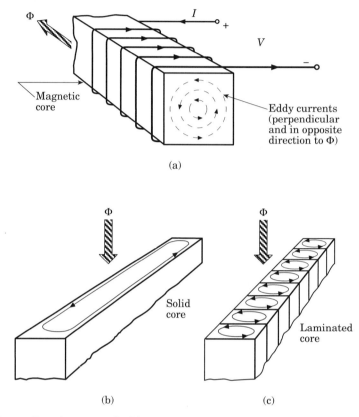

Figure 3.14   Development of eddy currents in magnetic cores.

To significantly increase the resistance encountered by these eddy currents so that the associated power losses can be minimized,[19] the magnetic core is usually built up from stackings of thin steel sheet *laminations,* as shown in Figure 3.14(c). The surfaces of such sheet laminations are coated with an oxide or a very thin layer of electrical insulation (usually an insulating varnish or sometimes paper). As a general rule, the thinner the laminations are, the lower the losses are, since the eddy current losses are proportional to the square of the lamination thickness. In addition, as previously stated, the resistivity of steel laminations is substantially increased by the addition of a small amount of silicon.

Figure 3.15 shows the various shapes of steel laminations that are in use. The use of laminated cores makes the *actual* (or effective) cross-sectional area of the magnetic core being less than the *gross* (or *apparent*) cross-sectional area of the core represented by the stack of laminations. In actual calculations, this is taken into account by using the following stacking factor:

---

[19] However, there are devices that are built based on the use of such eddy currents, such as eddy-current brakes, automobile speedometers (i.e., drag cup tachometers) and others.

$$Stacking\ factor = \frac{Actual\ magnetic\ cross\text{-}sectional\ area}{Gross\ magnetic\ cross\text{-}sectional\ area} \qquad (3.45)$$

As the lamination thickness increases, the stacking factor approaches unity. For example, lamination thickness ranges from 0.0127 mm to 0.36 mm with corresponding stacking factors (at 60 Hz) that range from 0.50 to 0.95. Thus, the stacking factor approaches 1.0 as the lamination and the lamination surface insulation thicknesses increase. According to Charles P. Steinmetz, the eddy-current loss can be determined empirically from

$$P_e = k_e v \left(f t_\ell B_m \right)^2 \quad \text{W} \qquad (3.46)$$

where k_e = proportionality constant depending upon the core material

t_ℓ = lamination thickness

The definitions of the other variables are the same as the ones given for equation (3.44).

Figure 3.15   Various shapes of steel laminations.

Example 3.4

Consider the magnetic core shown in Figure 3.9(a). Assume that it is made up of a square-shaped, uniform cross-sectional area with an air gap and a core of soft steel casting. The square-shaped cross-sectional area of the core is equal to that of the air gap and is 1.4×10^{-3} m^2. The mean length of flux path through the steel core of the magnetic circuit is 0.4 m and the air-gap length ℓ_g is 2 mm (or 0.002 m). If a flux of 2×10^{-3} Wb is needed, determine the coil ampere turns (i.e., coil mmf) that are necessary to produce such flux. Neglect the flux fringing at the air gap.

SOLUTION:

Since the fringing of the flux across the air gap is neglected, the cross-sectional areas of the core and the gap are the same. Thus,

$$A_c = A_g = 1.4 \times 10^{-3} \text{ m}^2$$

The reluctance of the air gap is

$$\mathscr{R}_g = \frac{\ell_g}{\mu_0 A_g}$$

$$= \frac{2 \times 10^{-3} \text{m}}{\left(4\pi \times 10^{-7}\right)\left(1.7153 \times 10^{-3} \text{m}^2\right)}$$

(handwritten annotation: 1.4×10^{-3} under the 1.7153×10^{-3})

$$= 1{,}136{,}821.0221 \text{ A} \cdot \text{turns/Wb}$$

Thus, the needed mmf for the air gap is

$$\mathscr{F}_g = \Phi \mathscr{R}_g$$

$$= (2 \times 10^{-3} \text{Wb})(1{,}136{,}821.0221 \text{ A} \cdot \text{turns/Wb})$$

$$= 2{,}273.642 \text{ A} \cdot \text{turns}$$

The flux density in the steel core can be found from

$$B_c = \frac{\Phi}{A_c} = \frac{2 \times 10^{-3} \text{Wb}}{1.4 \times 10^{-3} \text{m}^2} = 1.4286 \text{ Wb/m}^2$$

From the magnetization curve[20] for soft steel casting shown in Figure 3.12, the corresponding magnetizing intensity is found as

---

[20] In practice, it is also referred to as the *B & H curve*.

$$H_c = 2{,}300 \ \text{A} \cdot \text{turns/m}$$

Therefore, the required mmf to overcome the reluctance of the core can be found as

$$\mathcal{F}_c = H_c \ell_c = (2{,}300 \ \text{A} \cdot \text{turns/m})(0.4 \ \text{m}) = 920 \ \text{A} \cdot \text{turns}$$

Hence, the total required mmf from the coil is

$$\mathcal{F}_{coil} = \mathcal{F}_g + \mathcal{F}_c = 2{,}273.642 + 920 = 3{,}193.642 \ \text{A} \cdot \text{turns/m}$$

Example 3.5

Repeat Example 3.4 assuming that the core is made up of M-19 29-Gage sheets. Use a stacking factor of 0.90 for the laminations to determine the actual (i.e., the effective) cross-sectional area of the core. In the air gap the cross-sectional area of flux is larger than in the iron core. To correct this **fringing** in the air gap, add the gap length to each of the two dimensions which make up its area.

SOLUTION:
From equation (3.45), the actual area is found as

$$A_{c, \, actual} = (A_{c, \, gross}) f_{stacking}$$

$$= (1.4 \times 10^{-3} m^2)(0.90)$$

$$= 1.26 \times 10^{-3} \, \text{m}^2$$

Thus, the flux density in the core is

$$B_c = \frac{\Phi}{A_{c, \, actual}}$$

$$= \frac{2 \times 10^{-3} \, \text{Wb}}{1.26 \times 10^{-3} \, \text{m}^2}$$

$$= 1.5873 \ \text{Wb/m}^2$$

From the magnetism curve for M-19 Gage sheets shown in Figure 3.12, the corresponding magnetizing intensity is found as

$$H_c = 3300 \ \text{A} \cdot \text{turns/m}$$

Thus, the required mmf to overcome the reluctance of the core is

$$\mathscr{F}_c = H_c \ell_c$$

$$= (3300 \text{ A} \cdot \text{turns/m})(0.4\text{m})$$

$$= 1{,}320 \text{ A} \cdot \text{turns}$$

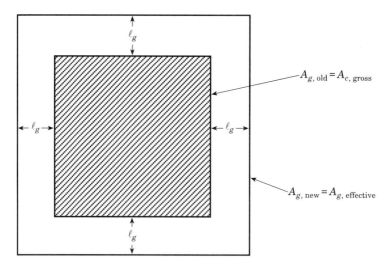

Figure 3.16

At the air gap, the cross-sectional area increases due to the flux fringing. Therefore, the *new* (i.e., *effective*) area in the air gap can be found by adding the air-gap length to each of the two dimensions which make up the square-shaped cross-sectional area, as shown in Figure 3.16. Thus, by using equation (3.43), the new air gap area is found from

$$A_{g,\,new} = (a + \ell_g)(b + \ell_g) \tag{3.43}$$

but $a = b$, since the area is a square. Therefore,

$$A_{g,\,new} = (a + \ell_g)^2 \tag{3.47}$$

where

$$a = \left(A_{g,\,old}\right)^{1/2} = \left(A_{c,\,gross}\right)^{1/2} \tag{3.48}$$

Hence

$$A_{g,\,new} = \left[\left(A_{g,\,old}\right)^{1/2} + 2\,\ell_g\right]^2 \tag{3.49}$$

Thus

$$A_{g,new} = \left[\left(1.4 \times 10^{-3}\right)^{1/2} + 2(0.002)\right]^2 = 1.7153 \times 10^{-3}\,\text{m}^2$$

Therefore, the resulting reluctance of the air gap is

$$\mathscr{R}_g = \frac{\ell_g \; \not{g}}{\mu_0 A_{g,new}} = \frac{2 \times 10^{-3}\,\text{m}}{\left(4\pi \times 10^{-7}\right)\left(1.7153 \times 10^{-3}\,\text{m}^2\right)}$$

$$= 927{,}837.2251\,\text{A}\cdot\text{turns/Wb}$$

Hence, the required mmf to overcome the reluctance of the air gap is

$$\mathscr{F}_g = \Phi \mathscr{R}_g = (2 \times 10^{-3}\,\text{Wb})(927{,}837.2251\ \text{A}\cdot\text{turns/Wb})$$

$$= 1{,}855.6745\ \text{A}\cdot\text{turns}$$

Thus, the total mmf required from the coil is

$$\mathscr{F}_{coil} = \mathscr{F}_g + \mathscr{F}_c = 1{,}855.655.6745 + 1{,}320 = 3{,}175.6745\ \text{A}\cdot\text{turns}$$

Alternatively, the permeability of the core is calculated as

$$\mu_c = \frac{B_c}{H_c}$$

$$= \frac{1.5873\ \text{Wb/m}^2}{3300\ \text{A}\cdot\text{turns/m}}$$

$$= 48 \times 10^{-5}\ \text{Wb/A}\cdot\text{turns}\cdot\text{m}\ (= \text{H/m})$$

Thus, the reluctance of the core is

$$\mathscr{R}_c = \frac{\ell_c}{\mu_c A_{c,\underset{\text{actual}}{new}}} = \frac{0.4\ \text{m}}{\left(48 \times 10^{-5}\right)\left(1.26 \times 10^{-3}\,\text{m}^2\right)}$$

$$= 660{,}000\ \text{A}\cdot\text{turns/Wb}$$

Therefore, the total reluctance is

$$\mathscr{R}_{tot} = \mathscr{R}_c + \mathscr{R}_g = 660{,}000 + 927{,}837.2251$$

$$= 1{,}587{,}837.2251\ \text{A}\cdot\text{turns/Wb}$$

Hence, the total required mmf from the coil is

$$\mathcal{F}_{coil} = \mathcal{R}_{tot}\,\Phi = (1{,}587{,}837.2251)(2 \times 10^{-3}) = 3{,}175.6745 \text{ A·turns}$$

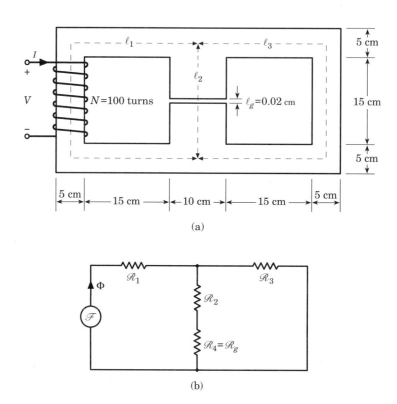

Figure 3.17   (a) Magnetic core for Example 3.6; (b) its equivalent circuit (analog).

Example 3.6

Consider the solid ferromagnetic core shown in Figure 3.17(a). The depth of the solid core is 5 cm. Assume that the relative permeability of the core is 1,000 and remains constant and that the current value in the coil is 4 A. Ignore the fringing effects at the air gap and determine the following:
(a) The equivalent circuit of the given core shown in Figure 3.17(a).
(b) The individual values of all reluctances.
(c) The value of the total reluctance of the core.
(d) The value of the total flux that exists in the core.
(e) The individual value of each flux that exists in each leg of the core.
(f) The value of flux density in each leg.

SOLUTION:
(a) The equivalent circuit is shown in Figure 3.17(b).
(b) The reluctance values are

$$\mathscr{R}_1 = \frac{\ell_1}{\mu A_1} = \frac{\ell_1}{\mu_r \mu_0 A_1}$$

$$= \frac{2\left[(0.05/2) + 0.15 + 0.05\right] + 0.20}{1,000\left(4\pi \times 10^{-7}\right)\left(0.05 \times 0.05\right)}$$

$$= 206{,}901.426 \ \text{A·turns/Wb}$$

$$\mathscr{R}_2 = \frac{\ell_2}{\mu_r \mu_0 A_2} = \frac{0.20 - 0.0002}{\left(1{,}000\right)\left(4\pi \times 10^{-7}\right)\left(0.10 \times 0.05\right)}$$

$$= 31{,}799.1576 \ \text{A·turns/Wb}$$

$$\mathscr{R}_3 = \mathscr{R}_1 = 206{,}901.426 \ \text{A·turns/Wb}$$

$$\mathscr{R}_4 = \frac{\ell_4}{\mu_r \mu_0 A_4}$$

$$= \frac{0.0002}{1\left(4\pi \times 10^{-7}\right)\left(0.10 \times 0.05\right)}$$

$$= 31{,}830.9886 \ \text{A·turns/Wb}$$

(c) The total reluctance of the core is

$$\mathscr{R}_{tot} = \mathscr{R}_1 + \frac{\mathscr{R}_3(\mathscr{R}_2 + \mathscr{R}_4)}{\mathscr{R}_2 + \mathscr{R}_3 + \mathscr{R}_4} = 255{,}565.4983 \ \text{A·turns/Wb}$$

(d) $$\Phi_{tot} = \frac{NI}{\mathscr{R}_{tot}} = \frac{100 \times 4}{255{,}565.4983} = 0.0015652 \ \text{Wb}$$

(e) $$\Phi_1 = \Phi_{tot} = 0.0015652 \ \text{Wb}$$

$$\Phi_2 = \Phi_{tot} \times \left(\frac{\mathscr{R}_3}{\mathscr{R}_2 + \mathscr{R}_3 + \mathscr{R}_4}\right) = 0.001197 \ \text{Wb}$$

$$\Phi_3 = \Phi_{tot} \times \left(\frac{\mathscr{R}_2}{\mathscr{R}_2 + \mathscr{R}_3 + \mathscr{R}_4}\right) = 0.0003681 \ \text{Wb}$$

(f) $$B_1 = \frac{\Phi_1}{A_1} = \frac{0.0015652}{0.0025} = 0.6261 \ \text{Wb/m}^2$$

$$B_2 = \frac{\Phi_2}{A_2} = \frac{0.001197}{0.005} = 0.2394 \text{ Wb/m}^2$$

$$B_3 = \frac{\Phi_3}{A_3} = \frac{0.0003681}{0.0025} = 0.1472 \text{ Wb/m}^2$$

Example 3.7

Figure 3.18(a) shows a cross-section of the magnetic structure of a four-pole dc machine. On each of the four stator poles there is a coil with equal turns. Since the four coils are connected in series, all carry the same current. The stator poles and rotor are made up of laminations of silicon steel sheets, while the stator yoke is made up of cast steel. Based on the given information, do the following:
(a) Draw an equivalent magnetic circuit.
(b) Derive an equation to determine the mmf produced by each winding.

SOLUTION:
(a) The equivalent magnetic circuit of the dc machine is shown in Figure 3.18(b). In the figure, the subscripts r, s, p and g denote rotor, stator yoke, stator pole, and air gap, respectively. Since the magnetic structure is symmetric, an analysis of one quarter is sufficient. Therefore, the mmf produced by each winding is $\mathcal{F} = NI$ and provides the required flux on a per pole basis. If the flux in the air-gap region is known, the flux densities in all sections of the machine can be found.
(b) As can be observed in Figure 3.18(a), the flux supplied by each pole is the same in the pole, the pole face, and the air-gap area. Because the mmf drop in both halves of the yoke or rotor must be the same, the flux is divided equally when it flows through the stator yoke or the rotor. The equivalent magnetic circuit of the dc machine can be represented in terms of the reluctances, as shown in Figure 3.18(b). The required mmf per pole can be determined by using Ampère's law for any one of the flux paths shown in Figure 3.18(b). Therefore,

$$2\mathcal{F} = \Phi(2\mathcal{R}_p + 2\mathcal{R}_g) + \frac{\Phi}{2}(\mathcal{R}_r + \mathcal{R}_s) \tag{3.50}$$

or

$$\mathcal{F} = \Phi[\mathcal{R}_p + \mathcal{R}_g + 0.25(\mathcal{R}_r + \mathcal{R}_s)] \tag{3.51}$$

where $\mathcal{R}_p$ = reluctance of the stator pole
$\mathcal{R}_g$ = reluctance of the air gap
$\mathcal{R}_r$ = reluctance of the rotor
$\mathcal{R}_s$ = reluctance of the stator yoke

Thus, the flux Φ is known and the reluctance of the magnetic circuit can be calculated from its physical dimensions and known permeability.

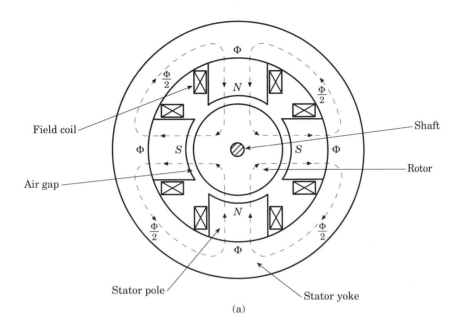

(a)

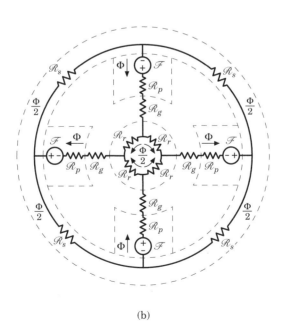

(b)

Figure 3.18 (a) Magnetic structure of a four-pole dc machine; (b) its equivalent circuit.

3.8 HOW TO DETERMINE FLUX FOR A GIVEN MMF

In the previous examples, the problem was: Given a magnetic circuit, find the mmf required to produce a given flux. The nonlinearity of iron presents the following more difficult problem: Given an applied mmf, find the flux in a magnetic circuit. This problem can be solved by the following methods:

- The trial-and-error method.
- The graphical method.
- The magnetization curve method.

3.8.1 THE TRIAL-AND-ERROR METHOD

In the *trial-and-error method,* a value for Φ is selected and the corresponding mmf is computed. It is compared with NI, then a new value of Φ is selected and the corresponding new mmf value is computed. This procedure is repeated until the determined mmf is equal (or almost equal) to NI.

3.8.2 THE GRAPHICAL METHOD

This procedure is also called the *load line method.* Consider the magnetic circuit shown in Figure 3.9(a). For a magnetic circuit with a core length ℓ_c and an air gap length ℓ_g, for a given value of mmf,

$$\mathscr{F} = NI = H_c \ell_c + H_g \ell_g \tag{3.52}$$

since

$$H_g = \frac{B_g}{\mu_0} \tag{3.53}$$

Substituting it into equation (3.52)

$$NI = H_c \ell_c + \frac{B_g}{\mu_0} \ell_g \tag{3.54}$$

and rearranging equation (3.54),

$$B_g = B_c = -\mu_0 \frac{\ell_c}{\ell_g} H_c + \frac{NI\mu_0}{\ell_g} \tag{3.55}$$

Equation (3.52) represents a straight line since it is in the form of

$$y = mx + c \tag{3.56}$$

The resulting straight line is called the *load line* and can be plotted on the magnetization (i.e., the *B & H*) curve of the core. The slope of such a line can be expressed as

$$m = -\mu_0 \frac{\ell_c}{\ell_g} \tag{3.57}$$

The intersection of this line on the *B* ordinate is

$$c = \frac{NI\mu_0}{\ell_g} \tag{3.58}$$

Also, its intersection on the *H* axis is

$$H_c = \frac{NI}{\ell_c} \tag{3.59}$$

The intersection of the load line with the magnetization curve provides the value of B_c. Therefore, the value of the flux is found from

$$\Phi = B_c A_c \tag{3.60}$$

An alternative method of developing the load line is based on two steps:

Step 1. Assume that all the mmf is in the air gap, i.e., $H_c = 0$. Therefore, the air gap flux density can be expressed as

$$B_g = \frac{NI}{\ell_g} \mu_0 \tag{3.61}$$

The resulting value of B_g is the intersection of the load line on the *B* ordinate.

Step 2. Assume that all the mmf is in the core, i.e., $B_g = 0$. Thus, the magnetizing intensity of the core can be expressed as

$$H_c = \frac{NI}{\ell_c} \tag{3.62}$$

The resulting value of H_c is the intersection of the load line on the *H* axis.

3.8.3 THE MAGNETIZATION CURVE METHOD

In this method, various values of flux Φ are chosen and the corresponding values of mmf are determined. The values of Φ versus mmf are plotted. The resulting curve is called the *magnetization curve* of the apparatus. Finally, by using the magnetization curve and the given value of current I, the value of flux Φ corresponding to $\mathscr{F} = NI$ is determined.

Example 3.8

Consider the magnetic core shown in Figure 3.9(a). Assume that it is made up of a square-shaped uniform cross-sectional area with an air gap and that a core of soft steel casting. The square-shaped cross-sectional area of the core is equal to that of the air gap and is 1.5×10^{-3} m^2. The mean length of the flux path through the steel core of the magnetic circuit is 0.5 m and the air-gap length is 3 mm (or 0.003 m). If the coil has 500 turns and the coil current is 7 A, find the flux density in the air gap. Neglect the flux fringing at the air gap and use the *trial-and-error method*.

SOLUTION:
The following steps can be used:

Step 1. Assume a flux density (since $B = \Phi/A$).

Step 2. Find H_c (from the B & H curve) and H_g (from $H_g = B_g/\mu_0$).

Step 3. Find $\mathscr{F}_c$ (from $\mathscr{F}_c = H_c \ell_c$), $\mathscr{F}_g$ (from $\mathscr{F}_g = H_g \ell_g$, and $\mathscr{F}$ (from $\mathscr{F} = \mathscr{F}_c + \mathscr{F}_g$).

Step 4. Find I from $I = \mathscr{F}/N$.

Step 5. If the found I is different from the given current, select a new appropriate value for the flux density. Continue this process until the calculated value of current is close to the given current value of 7 A.

Note that if all the mmf were only in the air gap, the resulting flux density would be

$$B = \frac{NI\mu_0}{\ell_g} = 1.4661 \text{ Wb/m}^2$$

However, since this is not the case, the actual flux density will be less than this value. This calculation process is illustrated in Table 3.1 The value of the flux is found from

$$\Phi = B_g A_g = B_c A_c = (1.2 \text{ Wb/m}^2)(1.5 \times 10^{-3} \text{ m}^2) = 0.0018 \text{ Wb}$$

TABLE 3.1 Table for Example 3.8

B	H_c	H_g	$\mathscr{F}_c$	$\mathscr{F}_g$	$\mathscr{F}$	I
1.3	1630	1.0345×10^{-6}	815	3103.5	3918.5	7.84
1.2	1280	0.9549×10^{-6}	640	2864.8	3504.8	7.0

Example 3.9

Solve Example 3.8 using the *graphical method.*

SOLUTION:
The intersection of the load line on the B ordinate is found using equation (3.58) as

$$c = \frac{NI\mu_0}{\ell_g}$$

$$= \frac{(500 \text{ turns})(7 \text{ A})\left(4\pi \times 10^{-7}\right)}{0.003 \text{ m}} = 1.4661 \text{ Wb/m}^2$$

The intersection of this line on the H axis is

$$H_c = \frac{NI}{\ell_c} \tag{3.62}$$

$$= \frac{(500 \text{ turns})(7 \text{ A})}{0.5 \text{ m}} = 7{,}000 \text{ A} \cdot \text{turns/m}$$

Its slope is

$$m = -\mu_0 \frac{\ell_c}{\ell_g} \tag{3.57}$$

$$(4\pi \times 10^{-7})\frac{0.5 \text{ m}}{0.003 \text{ m}} = -2{,}094 \times 10^{-4}$$

As shown in Figure 3.19, the intersection of the load line with the magnetization curve gives the value of flux density in the air gap as

$$B_g = B_c = 1.2 \text{ Wb/m}^2$$

Thus, the value of flux is found from

$$\Phi = B_c A_c = B_g A_g$$

$$= (1.2 \text{ Wb/m}^2)(1.5 \times 10^{-3} \text{m}^2)$$

$$= 0.0018 \text{ Wb}$$

3.9 PERMANENT MAGNETS

In general, after the removal of the excitation current, all ferromagnetic cores retain some flux density called the *residual flux density B_r*. To return the magnetic core to its original state, it has to be demagnetized by applying the magnetizing field intensity in the opposite direction. The value of the field intensity needed to decrease the residual flux density to zero is called the *coercive force*. Materials suitable for permanent magnets are known as *magnetically hard* materials because they are difficult to magnetize, but have high residual flux density and high coercive force. The various categories of permanent-magnet materials include (1) ductile metallic

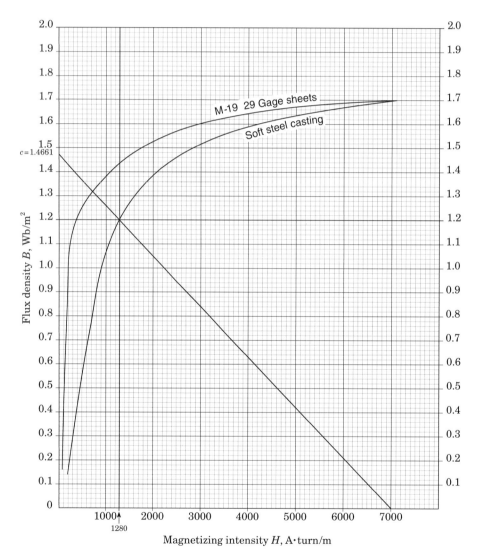

Figure 3.19   B-H curves for two materials.

magnets, such as Cunife; (2) ceramic magnets, such as Indox; (3) brittle metallic magnets, such as Alnico (cast and sintered); and (4) rare-earth cobalt magnets, such as samarium-cobalt. Among the many applications of permanent magnets are loudspeakers, small generators, magnetic clutches and couplings, measuring instruments, magnetrons, television focusing units, video recording and information storage in computers.

In magnet design, the shape of a permanent magnet affects the amount of flux produced. Equal volumes of magnetic core will provide different amounts of flux as a function of their shape. For example, the most common shapes of Alnico magnets are rods, bars, and U shapes. However, since these materials are difficult to machine, their shapes are usually made simple, and soft iron parts are added to the magnetic circuit in the more complex shapes, as shown in Figure 3.20(a). It is interesting to note that such a magnet would be excited by first placing a magnetically soft iron part in the air gap. This part, known as the *keeper*, is removed during use and replaced afterwards.

Figure 3.20(b) shows the demagnetization curve and corresponding energy product curve (or B & H curve) of the permanent magnet, for the magnetic circuit shown in Figure 3.20(a). Ignoring the reluctance of the soft-iron parts and applying Ampère's law

$$H_c \ell_c + H_g \ell_g = 0 \tag{3.63}$$

where H_c is the magnetizing force within the core. If fringing at the air gap is ignored, the flux density inside the core is equal to that in the air gap. Therefore,

$$B_c = B_g = \mu_0 H_g = -\mu_0 H_0 \, \frac{\ell_c}{\ell_g} \tag{3.64}$$

This is the equation of a straight line. Its intersection with the magnetization characteristic provides the optimum operating point[21] P, as shown in Figure 3.20(b). It determines the values of B_c and H_c for the permanent magnet. Here, having an air gap has the same effect as inserting a negative field inserted into the magnetic circuit. From equation (3.61), the magnetizing force of the magnet can be expressed as

$$H_c = - H_g \, \frac{\ell_g}{\ell_c} \tag{3.65}$$

which can also be expressed in terms of the flux density B_c of the magnet by

$$H_c = - \frac{B_c A_c \ell_g}{\mu_0 A_g \ell_c} \tag{3.66}$$

---

[21] To establish the maximum energy in the air gap, the point of operation must correspond to the maximum energy product of the magnet.

Because magnetic leakage is negligibly small, the flux has to be the same in all parts of the magnetic circuit. Therefore,

$$\Phi = B_c A_c = B_g A_g \tag{3.67}$$

Furthermore, based on the assumption that the cross-sectional area of the magnet is uniform, the volume is found from

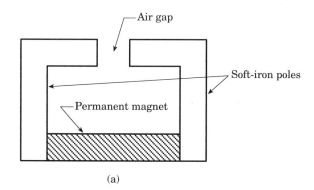

(a)

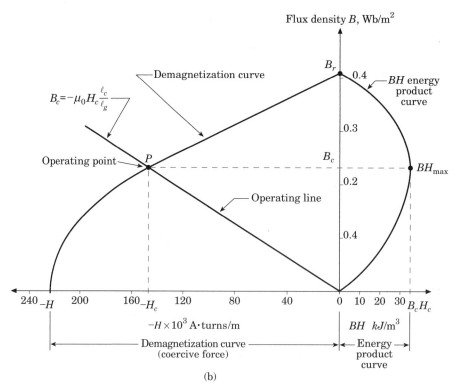

(b)

Figure 3.20   (a) Typical configuration made up of a permanent magnet, soft-iron pole pieces and air gap; (b) typical demagnetization and associated energy-product curves.

$$A_c \ell_c = A_c \left(-\ell_g \frac{H_g}{H_c} \right) = -A_g \ell_g \frac{H_g}{H_c} \qquad (3.68)$$

if fringing and leakage are ignored. Also note that the product of B and H represents the energy density within the core. Thus, at the maximum energy density, the volume of the core is

$$Vol_c = \left(\frac{B_g^2}{\mu_0 B_c H_c} \right)(Vol_g) \qquad (3.69)$$

where Vol_g is the volume of the air gap.

PROBLEMS

3.1   Consider the toroid shown in Figure 3.5(b) with outside and inside radii at 6 and 7 cm, respectively. Assume that 250 turns are wound around the toroid of soft steel casting to produce a total flux of 35.1741×10^{-4} Wb 3.5609×10^{-5} in the core. If the magnetic field intensity in the core is 1000 A·turns/m, determine:
   (a) The length of the average flux path in the toroid and the cross-sectional area perpendicular to the flux.
   (b) The flux density in the core.
   (c) The required magnetomotive force.
   (d) The amount of current that must flow in the turns of the toroid.

3.2   Consider the results of Problem 3.1 and determine:
   (a) The flux linkage in the core.
   (b) The inductance of the coil by using equation (3.30).
   (c) The inductance of the coil by using equation (3.31).
   (d) The permeance of the coil.

3.3   Consider the magnetic core shown in Figure 3.9(a). Assume that it is made up of a square-shaped, uniform cross-sectional area with an air gap and a core of soft-steel casting. The square-shaped cross-sectional area of the core is equal to that of the air gap and is $2 \times 10^{-3}\,\text{m}^2$. The mean length of the flux path through the steel core of the magnetic circuit is 0.6 m and the air gap length ℓ_g is 3 mm (or 0.003 m). If a flux of $2^3 \times 10^{-3}$ Wb is needed, determine the coil ampere turns (i.e., coil mmf) necessary to produce such flux. Neglect the flux fringing at the air gap.

3.4   Solve Problem 3.3 but assume that the core is made up of M-19 29 Gage sheets. Therefore, use a stacking factor of 0.95 for the laminations to determine the actual (i.e., the effective) cross sectional area of the core. The cross-sectional area of the flux in the air gap is larger than the one in the iron core. To correct this *fringing* in the air gap, add the gap length to each of the two dimensions which make up its area.

3.5   Consider Example 3.4 but ignore the iron reluctance. Find the amount of flux that flows in the magnetic circuit if the mmf of the coil is:
(a)   2000 A·turns.
(b)   1000 A·turns.

3.6   Assume that the magnetic core shown in Figure 3.6(a) is made up of soft-steel casting with a cross-sectional area of 200 cm^2 and an average flux length of 100 cm. If the coil has 500 turns, determine:
(a)   The amount of current required to produce 0.02 Wb of flux in the core.
(b)   The relative permeability of the core at the current level found in Part(a).
(c)   The reluctance of the core.

3.7   Solve Problem 3.6 but assume that the amount of flux in the core is 0.03 Wb.

3.8   Solve Problem 3.6 but assume that the cross-sectional area of the core is 50 cm^2 and that the average flux length is 50 cm. Also assume that the amount of flux needed is 55×10^{-4} Wb and that the coil has 200 turns.

3.9   Solve Problem 3.6 but assume that the magnetic core is made up of M-19 29 Gage sheets with an actual cross-sectional core area of 25 cm^2 and an average flux length of 60 cm. The amount of flux that is needed in the core is 39×10^{-4} Wb.

3.10   Solve Problem 3.6 but assume that the cross-sectional area of the core is 100 cm^2 and that the average flux length is 70 cm. Also assume that the amount of flux needed is 125×10^{-4} Wb and that the coil has 800 turns.

3.11   Solve Problem 3.10 but assume that the magnetic core is made up of M-19 29 Gage sheets rather than soft-steel casting. Use an actual cross-sectional core area of 100 cm^2.

3.12   Solve Problem 3.10, but assume that the coil has 200 turns rather than 800 turns.

3.13   Solve Problem 3.6 but assume that the cross-sectional area of the core is 150 cm^2 and that the average flux length is 60 cm. Also assume that the amount of flux needed is 25×10^{-3} Wb and that the coil has 500 turns.

3.14   Solve Problem 3.13 but assume that the magnetic core is made up of M-19 29 Gage sheets rather than soft-steel casting. Use an actual cross-sectional area of 150 cm^2.

3.15   Solve Problem 3.13 but assume that the coil has 250 turns rather than 500 turns.

3.16   Consider the magnetic core shown in Figure 3.9(a). Assume that it is made up of M-19 29 Gage steel laminations with a stacking factor of 0.9. Let the gross cross-sectional area of the core of 0.02 m^2 be equal to the cross-sectional area at the air gap, ignoring the fringing of fluxes around the gap. Also let the lengths of the gap and the average flux path in the iron be 0.001 m and 0.4 m, respectively. Assume that the coil has 2,000 turns and determine:
   (a)   The current needed to produce a flux of 0.03 Wb in the air gap, ignoring iron reluctance.
   (b)   Repeat Part (a) without ignoring iron reluctance.

3.17   Assume that a magnetic circuit made up of soft-steel casting has a uniform cross-sectional area containing an air gap, as shown in Figure 3.9(a). Let the cross-sectional areas of iron core and air gap be 1.5×10^{-3} m^2, ignoring the fringing of fluxes across the air gap. If the lengths of the average flux path in the iron and the gap are 0.4 m and 0.002 m, respectively, determine the coil mmf to produce a flux of 2.5×10^{-3} Wb.

3.18   Assume that a magnetic circuit of uniform cross-sectional area of 1.75×10^{-3} m^2 has an air gap, and that the cross-sectional areas of the core and the air gap are the same. Ignore any effects of fringing around the gap. The magnetic core is made up of M-19 29 Gage sheets, the average length of flux path through the steel part of the magnetic circuit is 0.6 m and the air gap length is 2×10^{-3} m. The permeability of air is $4\pi \times 10^{-7}$ H/m. If a flux of 3×10^{-3} Wb is needed, determine:
   (a)   Air-gap reluctance.
   (b)   Air-gap mmf.
   (c)   Flux density in the air gap.
   (d)   Flux density in the iron.
   (e)   Field intensity in the iron.
   (f)   Magnetomotive force in the iron.
   (g)   Magnetomotive force in the coil.
   (h)   The required current in the coil, if the coil has 3,000 turns.

3.19   Consider Problem 3.18 and assume that the uniform cross-sectional area is 1.5×10^{-3} m^2, the average length of flux path through the core is 0.4 m, and the air gap length is 2.5×10^{-3} m. Assume that the core is made up of soft-steel casting and that the permeability of air is $4\pi \times 10^{-7}$ H/m. If a flux of 2×10^{-3} Wb is needed, answer the questions in Problem 3.18.

3.20 Consider the magnetic core shown in Figure P3.20 and notice that three sides of the core are of uniform width, whereas the fourth side is somewhat thinner. The depth of the core into the page is 10 cm. The coil has 300 turns and the relative permeability of the core is 2,000. Find the amount of flux that will be produced in the core by a 5 A input current.

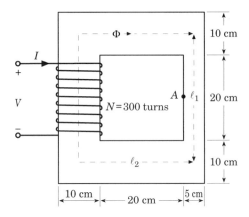

Figure P3.20

3.21 Consider the magnetic core given in Problem 3.20 and assume that there is a small gap of 0.06 cm at point A (i.e., at mid point of ℓ_1 distance) shown in Figure P3.20. Assume that due to fringing the effective cross-sectional area of the air gap has increased by 6 percent. Use the given information here and in Problem 3.20 and determine:
  (a) The total reluctance of the flux path (i.e., including the iron core and the air gap).
  (b) The current necessary to produce a flux density of 1 Wb/m^2 in the air gap.

3.22 Consider the magnetic core shown in Figure P3.22 and notice that three sides of the core are of uniform width, whereas the fourth side is somewhat thinner. The depth of the core into the page is 15 cm. The coil has 250 turns and the relative permeability of the core is 2,500. Find the amount of flux that will be produced in the core by a 10 A input current.

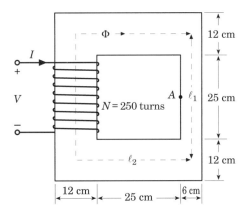

Figure P3.22

3.23 Consider the magnetic core given in Problem 3.22 and assume that there is a small gap of 0.04 cm at point A (i.e., at the midpoint of distance ℓ_1) shown in Figure P3.22. Assume that due to fringing the effective cross-sectional area of the air gap has increased by 4 percent. The core is made up of soft-steel casting. Use the information given here and in Problem 3.22 and determine:
  (a) The total reluctance of the flux path (i.e., including the iron

core and the air gap).

(b) The current necessary to produce a flux density of 2 Wb/m^2 in the air gap.

3.24 Consider the magnetic core and its equivalent shown in Figures P3.24 (a) and (b), respectively, and assume that it represents an elemental stator and rotor setup of a dc motor. The stator has a square-shaped cross-sectional area (A_c) of 25 cm^2 and an average path length (ℓ_c) of 100 cm. The rotor has a cross-sectional area (A_r) of 25 cm^2 and an average path length [i.e., the diameter of the cylindrical rotor (ℓ_r)] of 5 cm. Each air gap (on each side of the rotor) is 0.03 cm wide. The cross-sectional area of the air gap with fringing is 27.5625 cm^2. The relative permeability of the iron core used for both the stator and rotor is 3,000. The coil located on the stator has 400 turns. If the current in the coil is 2 A, determine the resulting flux density in the air gaps.

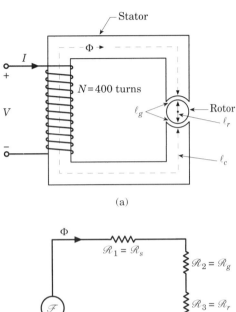

(a)

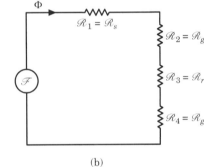

(b)

Figure P3.24

3.25 Solve Problem 3.24 but assume that the length of each air gap (ℓ_g) is 0.01 cm rather than 0.03 cm.

3.26 Solve Problem 3.24 but assume that the length of each air gap (ℓ_g) is 0.06 cm rather than 0.03 cm.

3.27 Solve Problem 3.24 but assume that the number of turns of the coil is 200 turns rather than 400.

3.28 Consider the elemental stator and rotor setup of the dc motor given in Problem 3.24. Assume that the stator has a square-shaped cross-sectional area (A_c) of 150 cm^2 and an average path length (ℓ_c) of 150 cm. The rotor has a cross-sectional area (A_r) of 150 cm^2 and an average path length (i.e., the diameter of cylindrical rotor) (ℓ_r) of 15 cm. Each air gap (on each side of the rotor) is 0.02 cm wide. The cross-sectional area of the air gap with fringing is 158 cm^2. The relative permeability of the iron core, used for both the stator and rotor, is 4,000. The coil located on the stator has 600 turns. If the current in the coil is 4 A, determine the resulting flux density in the air gaps.

3.29 Solve Problem 3.28 but assume that the length of each air gap (ℓ_g) is 0.01 cm rather than 0.02 cm.

3.30 Solve Problem 3.23 but assume that the length of each air gap (ℓ_g) is 0.04 cm rather than 0.02 cm.

3.31 Solve Problem 3.28 but assume that the number of turns of the coil is 300 turns rather than 600.

3.32 Consider the magnetic core shown in Figure P 3.32. Assume that the depth of the core is 10 cm and that the relative permeability of the solid core is 2,000. Determine the following:

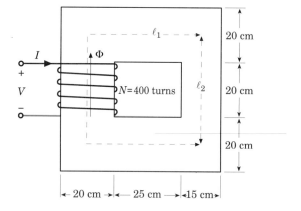

Figure P3.32

(a) The value of the current that will produce a flux of 0.01 Wb in the core.

(b) The flux density at the bottom of the core, if the current amount found in Part (a) is used.

(c) The flux density at the right side of the core, if the current amount found in Part (a) is used.

3.33 Consider the magnetic core shown in Figure P3.33. Assume that the depth of the solid core is 5 cm and that the relative permeability of the core is 1,000. Also assume that the air gaps on the left and right legs are 0.03 cm and 0.06 cm, respectively. Take the fringing effects at each gap into account by calculating the effective area of each air gap as 5 percent greater than its actual physical size. If the coil has 100 turns and 2 A current in it, determine:

(a) The total reluctance of the core.
(b) The total flux in the core.
(c) The flux in the left leg of the core.
(d) The flux in the center leg of the core.
(e) The flux in the right leg of the core.

(f)   The flux density in the left air gap.
(g)   The flux density in the right air gap.

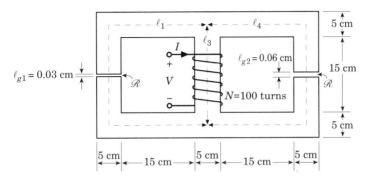

Figure P3.33

3.34   Consider the magnetic core shown in Figure P3.34. Notice the way each coil is wound on each leg. Use a relative permeability of 2,000 for the solid core and assume that it remains constant. If the value of current $I_1 = 1.0$ A and current $I_2 = 1.5$ A, find the total flux produced by them in the core. Use a core depth of 5 cm.

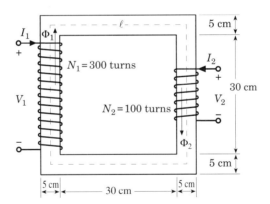

Figure P3.34

CHAPTER 4

TRANSFORMERS

Education is the best provision for old age.
Aristotle, 365 B.C.

A teacher affects eternity.
Author Unknown

Education is...hanging around until you've caught on.
Will Rogers

4.1 INTRODUCTION

In general, a transformer is a *static*[1] electromagnetic machine (i.e., it has no moving parts). Transformers are commonly used for changing the voltage and current levels in a given electrical system, establishing electrical isolation, impedance matching, and measuring instruments. Power and distribution transformers are used extensively in electrical power systems to generate the electrical power at the most economical generator voltage level; to transmit and distribute electrical power at the most economical voltage level; and to utilize power at the most economical, suitable, and safe voltage level.

Isolating transformers are used to electrically isolate electric circuits from each other or to block dc signals while maintaining ac continuity between the circuits, and to eliminate electromagnetic noise in many types of circuits. Transformers are widely used in communication systems that vary in frequency from audio to radio to video levels. They perform various tasks, such as impedance-matching for improved power transfer, and are used as input transformers, output transformers, and insulation apparatus between electric circuits, and interstage transformers. Transformers are used in the whole frequency spec-

---

[1] However, there are some special transformers in which some motion takes place in components of the electromagnetic structure. Examples include the variable autotransformer, which has a tap that moves between primary and secondary, as well as some types of voltage regulators that are employed in distribution systems.

trum in electrical circuits, from near dc to hundreds of megahertz, including both continuous sinusoidal and phase waveforms. For example, they can be found in use at power-line frequencies (between 60 and 400 Hz), audio frequencies (20 to 20,000 Hz), ultrasonic frequencies (20,000 to 100,000 Hz), and radio frequencies[2] (over 300 kHz).

Transformers are also used in measuring instruments. Instrument transformers are used to measure high voltages and large currents with standard small-range voltmeters (120 V), ammeters (5 A), and wattmeters, and to transform voltages and currents to activate relays for control and protection. *Voltage transformers*[3] (VTs) (also known as PTs, i.e., *potential transformers*) are single-phase transformers that are used to step down the voltage to be measured to a safe value. *Current transformers* (CTs) are used to step down currents to measurable levels. The secondaries of both voltage and current transformers are normally grounded.

A transformer consists of a primary winding and a secondary winding linked by a mutual magnetic field. Transformers may have an air core, an iron core, or a variable core, depending upon their operating frequency and application. Transformers are also quite different in size and shape depending on the application. In power system applications, the single- or three-phase transformers with ratings up to 500 kVA are defined as *distribution transformers*, whereas those transformers with ratings over 500 kVA at voltage levels of 69 kV and above, are defined as *power transformers*.[4] Figure 4.1 shows a cutaway view of a single-phase, overhead pole-mounted distribution transformer. Notice that it has two high-voltage bushings since it is built to operate under line-to-line voltage rather than line-to-neutral voltage. Figure 4.2 shows a three-phase 345/161 kV autotransformer used as a power transformer. Its power ratings are 214/285/357 MVA for its OA/FA/FOA operations. Note that the **OA/FA/FOA** means *oil-immersed, self-cooled/forced-air-cooled/forced-oil-cooled*.[5]

A transformer is basically made up of two or more windings coupled by a mutual magnetic field. Ferromagnetic cores are employed to develop tight magnetic coupling and high flux densities. When such a coupling exists, the

---

[2] In general, the size of a transformer can be significantly reduced by using it with higher frequencies. (Because of this fact, aircraft generators are designed to produce power at 400 Hz rather than at 50 or 60 Hz.) Also, a transformer designed for use at 50 or 60 Hz can always be used at higher frequencies, whereas a transformer designed for use at 400 Hz would not operate properly at lower frequencies, because its core would saturate and the secondary voltage would not be similar to nor proportional to the primary voltage.

[3] They are also called *pot transformers*.

[4] For further information, see Gönen [33].

[5] Today, various methods are in use to get the heat out of the tank more effectively. Historically, as the transformer sizes increased, the losses outgrew any means of self-cooling that was available at the time, thus a water-cooling method was put into practice. This was done by placing a coil of metal tubing in the top oil, around the inside of the tank. Water was pumped through this cooling coil to get rid of the heat from the losses. Another method was circulating the hot oil through an external oil-to-water heat exchanger. This method is called forced-oil-to-water cooling (*FOW*). Today, the most common of these forced-oil-cooled transformers employs an external bank of oil-to-air heat exchangers through which the oil is continuously pumped. It is known as

transformer is called an *iron-core transformer*. Most distribution and power transformers are immersed in a tank of oil for better insulation[6] and cooling purposes. The leads of the windings are brought to the outside of the tank

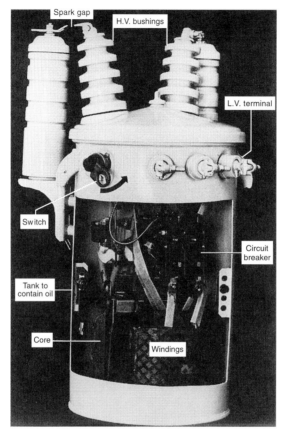

Figure 4.1   Cutaway view of a single-phase distribution transformer. (Courtesy of ABB Corporation.)

---

type *FOA*. In present practice fans are automatically used for the first stage and pumps for the second, in triple-rated transformers which are designated as type *OA/FA/FOA*. These transformers carry up to about 60% of maximum nameplate rating (i.e., *FOA* rating) by natural circulation of the oil (*OA*) and 80% of maximum nameplate rating by forced cooling which consists of fans on the radiators (*FA*). Finally, at maximum nameplate rating (*FOA*), not only is oil forced to circulate through external radiators, but fans are also kept on to blow air onto the radiators as well as into the tank itself.

[6] The National Electric Manufacturers Association (NEMA) has grouped various types of insulation into classes and assigned a maximum permissible hottest spot temperature to each class. Since the hottest spot temperature is usually at some inaccessible spot within a coil, the maximum permissible average temperature (determined by measuring the resistance of the coil) is somewhat lower. The difference between the ambient temperature and the average temperature of the coil is its temperature rise. The sum of the ambient temperature plus the maximum temperature rise plus the hot-spot allowance equals the maximum temperature rating of the insulation.

through insulating bushings which are attached to the tank, as shown in Figure 4.1. Such transformers are used in high-power applications.[7] When there is no ferromagnetic material but only air present, such a transformer is called an *air-core transformer*.[8] These transformers have poor magnetic coupling and are usually used in lower-power applications such as in electronic circuits. In this chapter the focus is set exclusively on *iron-core transformers*.

Figure 4.2   Installed view of a three-phase 345/161-kV autotransformer with power ratings of 214/285/357 MVA for OA/FA/FOA. (Courtesy of North American Transformer.)

4.2 TRANSFORMER CONSTRUCTION

The magnetic cores of transformers used in power systems are built either in core type or shell type, as shown in Figure 4.3. In either case, the magnetic cores are made up of stacks of laminations cut from silicon-steel sheets. Silicon-steel sheets usually contain about 3 percent silicon and 97 percent steel. The silicon content decreases the magnetizing losses, especially the ones due to hysteresis loss. The laminations are coated with a nonconducting and insulat-

---

[7] Today, it is technically possible to manufacture large power transformers having sheet-wound coils that are insulated by compressed gas (for example, SF_6, i.e., *sulfur hexafuoride*) and cooled by forced circulation of liquid.

[8] They are also known as *dry-type transformers*.

ing varnish on one side. Such a laminated core substantially reduces the core loss due to eddy currents. Most laminated materials are cold rolled and often specially annealed to orient the grain or iron crystals. This causes a very high permeability and low hysteresis to flux in the direction of rolling. Thus, in turn, it requires a lower exciting current. The laminations for the core-type transformer, shown in Figure 4.3(a), may be made up of L-shape, or U- and I-shaped laminations. The core for the shell-type transformer, shown in Figure 4.3(b), is usually made up of E- and I-shaped laminations. It is necessary to clamp laminations and impregnate the coils because of the cyclic magnetic forces and other forces that exist between parallel conductors carrying current. The lack of clamping, or improper clamping may cause an objectionable audible noise[9] that can be characterized as a humming sound. To mini-

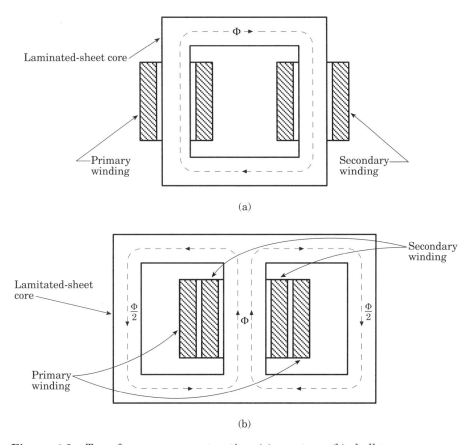

Figure 4.3   Transformer core construction: (a) core type; (b) shell type.

---

[9] The source of the audible sound is mechanical vibration of the core, produced by a steel characteristic known as magnetostriction. Because the magnetostrictive motion grows with increased flux density, the audible sound can be minimized by decreasing flux density.

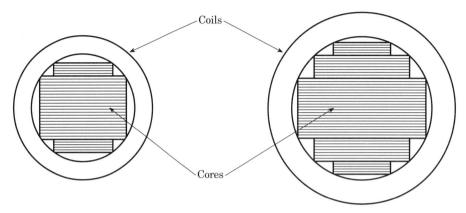

Figure 4.4   Stepped transformer cores.

mize the use of copper and decrease copper loss, the magnetic cores of large transformers are built in stepped cores, as shown in Figure 4.4.

Figure 4.3(a) shows a *core-type construction* that has the total number of primary winding turns located on one leg of the core and the total number of secondary winding turns placed on the other leg. This design causes large leakage flux and therefore results in a smaller mutual flux for a given primary voltage. To keep the leakage flux within a few percent of the mutual flux, each winding may be divided into two coils; the two half coils are then mounted on two sides of the rectangle.[10] A larger reduction in leakage flux can be obtained by further subdividing and sandwiching the primary and secondary turns, however, at considerable cost. Leakage flux can be greatly decreased by using the *shell-type construction* shown in Figure 4.3(b). However, the steel-to-copper weight ratio is greater in the shell-type transformer. It is more efficient but more costly in material. The coils employed in shell-type transformers are usually of a "*pancake*" form unlike the cylindrical forms used in the core-type transformer, where the coils are placed one on top of the other the low-voltage winding is placed closer to the core with the high-voltage winding on top. This design simplifies the problem of insulating the high-voltage winding from the core and reduces the leakage flux considerably.

4.3 A BRIEF REVIEW OF FARADAY'S AND LENZ'S LAWS OF INDUCTION

According to **Faraday's law of induction**, *whenever a flux passes through a turn of a coil, a voltage* [i.e., *an electromotive force (emf)*] *is induced, in each turn of that coil, that is directly proportional to the rate of change in the flux with respect to time.* Therefore, induced voltage can be found from

---

[10] Such design is especially beneficial for laboratory use because each pair of coils can be connected in series or in parallel, and therefore four different primary and secondary potential differences can be provided.

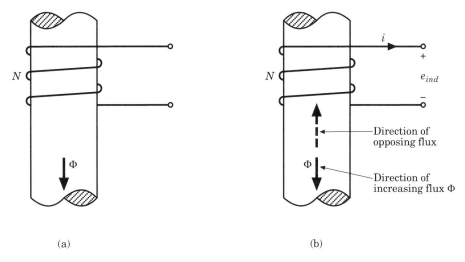

Direction of opposing flux

Direction of increasing flux Φ

(a)   (b)

Figure 4.5   Illustration of Lenz's law.

$$e_{ind} = \frac{d\Phi}{dt} \qquad (4.1)$$

where Φ is the flux that passes through the turn. If such a coil has N turns, as shown in Figure 4.5(a), and the same flux passes through all of them, the resulting induced voltage between the two terminals of the coil becomes

$$e_{ind} = N\frac{d\Phi}{dt} \qquad (4.2)$$

However, according to **Lenz's law of induction**, *if the coil ends were connected together, the voltage built-up would produce a current that would create a new flux opposing the original flux change.* Therefore, such a voltage buildup in the coil has to be in the proper direction to facilitate this, as shown in 4.5(b). Thus, equations (4.1) and (4.2) can be reexpressed as

$$e_{ind} = -\frac{d\Phi}{dt} \qquad (4.3)$$

and

$$e_{ind} = -N\frac{d\Phi}{dt} \qquad (4.4)$$

where the negative sign in the equations signifies that the polarity of the induced voltage opposes the change that produced it.[11]

---

[11] Therefore, Lenz's Law is very useful in determining the polarity of the voltage induced in the secondary winding of a transformer.

Alternatively, the magnitude of the induced voltage can be determined using the *flux linkage* λ of a given coil. Thus,

$$e_{ind} = \frac{d\lambda}{dt} \tag{4.5}$$

where

$$\lambda = \sum_{i=1}^{N} \Phi_i = Li \tag{4.6}$$

Furthermore, because the induced voltage equals the rate of change of flux linkages, an applied sinusoidal voltage has to produce a sinusoidally changing flux, provided that the resistive voltage drop is negligible. Thus, if the flux as a function of time is given as

$$\Phi = \Phi_m \sin \omega t \tag{4.7}$$

where Φ_m is the maximum value of the flux, and ω is $2\pi f$, f is the frequency in hertz, then the induced voltage is given as

$$e(t) = N\frac{d\Phi}{dt} = \frac{d(Li)}{dt} = \omega N \Phi_m \cos \omega t \tag{4.8}$$

Notice that the induced emf leads the flux by 90°. The rms value of the induced emf is given as

$$E = \frac{2\pi}{\sqrt{2}} f N \Phi_m = 4.44 f N \Phi_m \tag{4.9}$$

If the resistance drop in the winding is neglected, the counter emf equals the applied voltage. Therefore,

$$V = E = 4.44 f N \phi_m \tag{4.10}$$

and

$$\phi_m = \frac{V}{4.44 f N} \tag{4.11}$$

where V is the rms value of the applied voltage. [Note that equation (4.10) is known as the **emf equation**[12] of a transformer.] The flux is determined solely by the applied voltage, the frequency of the applied voltage, and the number of

---

[12] It is also known as the *general transformer equation.*

turns in the winding. The excitation (or exciting) current adjusts itself to produce the maximum flux required. Therefore, if the maximum flux density takes place in a saturated core, the current has to increase disproportionately during each half period to provide this flux density. For this reason, inductors with ferromagnetic cores end up having nonsinusoidal excitation currents.

If the core is unsaturated and the resistance of the coil is negligible, the maximum value of the magnetizing current[13] can be found from

$$I_m = \frac{N\phi_m}{L} = \frac{N}{L}\frac{V}{4.44fN} = \frac{\sqrt{2}V}{\omega L} \tag{4.12}$$

In the phasor form, the magnetizing current that produces the mutual flux is

$$I_m = \frac{V}{j\omega L} = \frac{V}{jX_m} \tag{4.13}$$

where X_m is the magnetizing reactance of the coil.

4.4 THE IDEAL TRANSFORMER

Consider a transformer with two windings, a primary winding of N_1 turns and a secondary winding of N_2 turns, as shown in Figure 4.6(a). The core is made up of a ferromagnetic material. Assume that the transformer is an **ideal transformer**[14] with the following properties:

- The winding resistances are negligible.
- All magnetic flux is confined to the ferromagnetic core and links both windings, i.e., leakage fluxes do not exist.
- The core losses are negligible.
- The permeability of the core material is almost infinite so that negligible net mmf is required to establish the flux in the core. In other words, the excitation current required to establish flux in the core is negligible.
- The magnetic core material does not saturate.

If the primary winding is connected to an energy source with a time-varying voltage v_1, a time-varying flux Φ and a flux linkage λ_1 of winding N_1 is estab-

---

[13] It may be important to keep in mind that for a voltage to be induced across the secondary winding, there must be a changing current in the primary winding. In the event that a dc source were connected to the primary winding, the current would become so large that the transformer would burn out. This is because on dc (i.e., when $f = 0$ Hz), the primary winding acts like a low resistance due to the fact that once the current reaches a steady-state value, the inductive reactance is equal to zero ohms.

[14] The ideal transformer, although it is fictitious, is a very useful device in power and communication systems analysis.

lished in the core. If v_1 varies over time, then i_1, Φ, and λ_1 will vary over time, and an emf e_2 will be induced in winding N_1. Therefore,

$$v_1 = e_1 = \frac{d\lambda_1}{dt} = N_1 \frac{d\Phi}{dt} \qquad (4.14)$$

Because there is no leakage flux, the flux Φ must link all N_2 turns of the secondary winding. Since the resistance of the secondary winding is assumed to be zero in an ideal transformer, it induces a voltage e_2 which is the same as the terminal voltage v_2. Thus,

$$v_2 = e_2 = \frac{d\lambda_2}{dt} = N_2 \frac{d\Phi}{dt} \qquad (4.15)$$

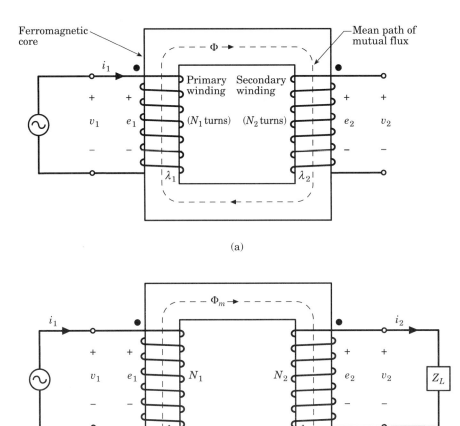

(a)

(b)

Figure 4.6   Ideal transformer: (a) with no load; (b) with load.

From equations (4.14) and (4.15),

$$\frac{v_1}{v_2} = \frac{e_1}{e_2} = \frac{N_1}{N_2} = a \tag{4.16}$$

which may also be written in terms of rms values as

$$\frac{V_1}{V_2} = \frac{E_1}{E_2} = \frac{N_1}{N_2} = a \tag{4.17}$$

where a is known as the **turns ratio**.[15] Note that the potential ratio is equal to the turns ratio. [Here, lowercase letters are used for instantaneous values and uppercase letters are used for rms values.] From equation (4.17),

$$V_1 = \frac{N_1}{N_2} V_2 = a V_2 \tag{4.18}$$

Assume that a load (energy sink) with an impedance $\mathbf{Z}_L$ is connected at the terminals of the secondary winding, as shown in Figure 4.6(b). Therefore, a load current (i.e., secondary current) will flow in the secondary winding. Since the core of an ideal transformer is infinitely permeable, the net mmf will always be zero. Thus,

$$\mathcal{F}_{net} = N_1 i_1 - N_2 i_2 = \Phi \mathcal{R} = 0 \tag{4.19}$$

where $\mathcal{R}$ is the reluctance of the magnetic core. Since the reluctance of a magnetic core of a well-designed modern transformer is very small (almost zero) before the core is saturated, then

$$N_1 i_1 - N_2 i_2 = 0 \tag{4.20}$$

or

$$N_1 i_1 = N_2 i_2 \tag{4.21}$$

That is, the primary and secondary mmfs are equal and opposite in direction.[16] From equation (4.21),

$$\frac{i_1}{i_2} = \frac{N_2}{N_1} = \frac{1}{a} \tag{4.22}$$

which may be written in terms of rms values as

---

[15] It is also known as the *ratio of transformation*.
[16] The net mmf acting on the core is thus zero.

$$\frac{I_1}{I_2} = \frac{N_2}{N_1} = \frac{1}{a} \tag{4.23}$$

Hence, the currents in the windings are inversely proportional to the turns of the windings. From equation (4.23),

$$I_1 = \frac{N_2}{N_1} I_2 = \frac{I_2}{a} \tag{4.24}$$

From equations (4.16) and (4.22),

$$v_1 i_1 = v_2 i_2 \tag{4.25}$$

or in terms of rms values

$$V_1 I_1 = \left(\frac{N_1}{N_2} V_2\right)\left(\frac{N_2}{N_1} I_2\right) = V_2 I_2 \tag{4.26}$$

That is, in an ideal transformer, the input power (VA) is equal to the output power (VA). In other words, the value of the apparent power remains the same.[17] This is the *power invariance principle* which means that the volt-amperes are conserved. Furthermore, the complex power supplied to the primary is equal to the complex power delivered by the secondary to the load. Thus,

$$V_1 I_1^* = V_2 I_2^* \tag{4.27}$$

In the event that the primary and secondary turns are equal, these transformers are usually known as the **isolating transformers**, as previously stated. In power systems, if the number of turns of the secondary winding are greater than the number of turns of the primary winding, the transformer is known as a **step-up transformer**. On the other hand, if the number of turns of the primary winding are greater than those of the secondary winding, the transformer is known as a **step-down transformer**.

Example 4.1

Determine the number of turns of the primary and the secondary windings of a 60-Hz 240/120-V ideal transformer, if the flux in its magnetic core is no more than 6 mWb.

---

[17] This transfer of power happens without any direct electrical connection between the primary and secondary windings. Such *electrical isolation* is, in certain applications, mandated for safety reasons. Examples include some medical apparatus and instrumentation designs for systems operating at a very high voltage.

SOLUTION:

From equation (4.10), the number of turns that the primary winding must have is

$$N_1 = \frac{V_1}{4.44\,f\Phi_m} = \frac{240\,\text{V}}{4.44\,(60\,\text{Hz})(5\times 10^{-3}\,\text{Wb})} \cong 180 \text{ turns}$$

and the number of turns that the secondary winding must have is

$$N_2 = \frac{V_2}{4.44\,f\Phi_m} = \frac{120\,\text{V}}{4.44\,(60\,\text{Hz})(5\times 10^{-3}\,\text{Wb})} \cong 90 \text{ turns}$$

or simply,

$$N_2 = \frac{N_1}{a} = \frac{180}{2} = 90 \text{ turns}$$

4.4.1 THE DOT CONVENTION IN TRANSFORMERS

Notice that the primary and secondary voltages that are shown in Figure 4.6(a) have the same polarities. The dots near the upper end of each winding are known as the **polarity marks**. Such dots point out that the upper or marked terminals have the same polarities, at a given instant of time when current enters the primary terminal and leaves the secondary terminal. In other words, the **dot convention** implies that (1) *currents entering at the dotted terminals will result in mmfs that will produce fluxes in the same direction*, and (2) *voltages from the dotted to undotted terminals have the same sign.*[18] Therefore, in Figure 4.6(a), since the current i_1 flows into the dotted end of the primary winding and the current i_2 flows out of the dotted end of the secondary winding, the mmfs will be subtracted from each other. Thus it can be said that the transformer has a **subtractive polarity**. Here, current i_2 is flowing in the direction of the induced current, according to Lenz's law.

As shown in Figure 4.7(a), for single-phase transformer windings, the terminals on the high-voltage side are labeled H_1 and H_2, while those on the low-voltage side are identified as X_1 and X_2. The terminal with subscript 1 in this convention (known as the *standard method of marking transformer terminals*) is equivalent to the dotted terminal in the dot-polarity notation. In a transformer where H_1 and X_1 terminals are adjacent, as shown in Figure 4.7(a), the transformer is said to have **subtractive polarity**. If terminals H_1 and X_1 are diagonally opposite, the transformer is said to have **additive polarity**.[19] Note

---

[18] Once a dot is assigned arbitrarily to a terminal of a given coil, the dotted terminals of all other coils coupled to it are found by Lenz's law, and therefore cannot be chosen at random. Sometimes, the identical terminals are marked by the ± sign instead of the dot sign.

[19] Polarities result from the relative directions in which the two windings are wound on the core.

that having the polarity markings in both dot convention as well as standard marking is really an unnecessary duplication.

Transformer polarities can be found by performing a simple test in which two adjacent terminals of high- and low-voltage windings are connected together and a small voltage is applied to the high-voltage winding, as shown in Figure 4.7(b). Then the voltage between the high- and low-voltage winding terminals that are not connected together are measured. *The polarity is **subtractive** if the voltage V reading is less than the voltage V_1 which is applied to the high voltage winding. The polarity is **additive**[20] if the voltage V reading is greater than the applied voltage V_1.*

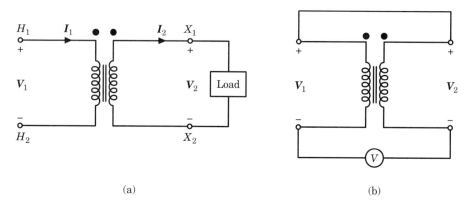

(a)   (b)

Figure 4.7   Polarity determination: (a) polarity markings of a single-phase two-winding transformer; (b) polarity test.

4.4.2 IMPEDANCE TRANSFER THROUGH A TRANSFORMER

Consider Figure 4.8(a) which shows an ideal transformer with a load impedance Z_L (of an apparatus or a circuit element) connected to its secondary terminals. Assume that all variables involved are given in phasors. Therefore, the impedance Z_L is defined as the ratio of the phasor voltage across it to the phasor current flowing through it. Hence,

$$Z_L = \frac{V_2}{I_2} \tag{4.28}$$

Here, not only the voltages V_1 and V_2 are in phase, but also the currents I_1 and I_2. As shown in Figure 4.8(b), the apparent impedance of the primary circuit of the transformer is

---

[20] According to the American National Standards Institute (ANSI) additive polarities are required in large (greater than 200-kVA) high-voltage (higher than 8,660 V) power transformers, To reduce voltage stress between adjacent leads, small transformers have subtractive polarities. For further information, see Gönen [33].

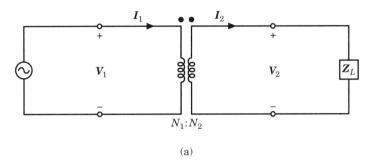

(a)

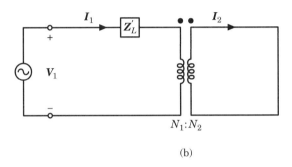

(b)

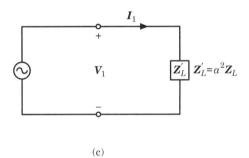

(c)

Figure 4.8   Illustration of impedance transfer across an ideal transformer.

$$Z'_L = \frac{V_1}{I_1} \tag{4.29}$$

where the primary voltage and current, respectively, are

$$V_1 = aV_2 \tag{4.18}$$

$$I_1 = \frac{I_2}{a} \tag{4.24}$$

Substituting equations (4.18) and (4.24) into equation (4.29), the apparent impedance of the primary becomes

$$Z'_L = \frac{V_1}{I_1} = \frac{aV_2}{I_2/a} = a^2 \frac{V_2}{I_2} \tag{4.30}$$

or

$$Z'_L = a^2 Z_L = \left(\frac{N_1}{N_2}\right)^2 Z_L \tag{4.31}$$

The resulting equivalent circuit is shown in Figure 4.8(c). Note that the a^2 is known as the **impedance ratio** of the transformer. Therefore, as far as the source is concerned, the three circuits shown in Figure 4.8 are the same. The impedance Z'_L is simply the result of impedance transformation of the load impedance Z_L through the transformer. Transferring an impedance from one side of the transformer to the other in this manner is known as **referring the impedance**[21] to the other side. Thus, Z'_L is known as the *load impedance referred to the primary side*. Using equations (4.18) and (4.24), voltages and currents can also be referred to one side or the other.

Similarly, an impedance located at the primary side of a transformer can also be referred to the secondary side as

$$Z'_1 = \frac{Z_1}{a^2} = \left(\frac{N_2}{N_1}\right)^2 Z_1 \tag{4.32}$$

Impedance transfer is very beneficial in calculations since it helps to get rid of a coupled circuit in an electrical circuit and thus simplifies the circuit. Furthermore, it can be used in *impedance matching* to determine the *maximum power transfer* from a source with an internal impedance Z_s to a load impedance Z_L. Here, it is necessary to select the turns ratio so that

$$Z'_L = \left(\frac{N_1}{N_2}\right)^2 Z_L = a^2 Z_L = Z_s \tag{4.33}$$

For maximum power transfer when Z_L may be complex,

$$Z'_L = \left(\frac{N_1}{N_2}\right)^2 Z_L = a^2 Z_L = Z_s^* \tag{4.34}$$

where Z_s^* is the conjugate of Z_s.

---

[21] It s also known as *reflecting, transferring,* or *scaling* the impedance.

4.4.3 RELATIONSHIP BETWEEN INPUT AND OUTPUT POWERS OF AN IDEAL TRANSFORMER

The input power provided to a transformer by its primary circuit is

$$P_{in} = V_1 I_1 \cos \theta_1 \tag{4.35}$$

where θ_1 is the angle between the primary voltage and the primary current. The power output of a transformer through its secondary circuit to its load is

$$P_{out} = V_2 I_2 \cos \theta_2 \tag{4.36}$$

where θ_1 is the angle between the secondary voltage and the secondary current. In an ideal transformer,

$$\theta_1 = \theta_2 = \theta$$

Therefore, the same power factor is seen by both the primary and secondary windings. Also, since $V_2 = V_1/a$ and $I_2 = aI_1$, substituting them into equation (4.36),

$$P_{out} = V_2 I_2 \cos \theta = \left(\frac{V_1}{a}\right)(aI_1)\cos \theta$$

or

$$P_{out} = V_1 I_1 \cos \theta = P_{in} \tag{4.37}$$

Therefore, in an ideal transformer, the output power is equal to its input power. This makes sense because, by definition, an ideal transformer has no internal power losses. One can extend the same argument to reactive and apparent powers. Therefore,

$$Q_{in} = V_1 I_1 \sin \theta = V_2 I_2 \sin \theta = Q_{out} \tag{4.38}$$

and

$$S_{in} = V_1 I_1 = V_2 I_2 = S_{out} \tag{4.39}$$

Example 4.2

Assume that a 60-Hz 250-kVA 2,400/240-V distribution transformer is an ideal transformer and determine the following:
(a) Its turns ratio. _current if a load_
(b) The value of load impedance connected to its secondary (i.e., low-voltage side) terminals to have the transformer fully loaded. _find I_2_
(c) The value of the primary-side (i.e., high-voltage side) current.
(d) The value of the load impedance referred to the primary-side of the transformer.

SOLUTION:

(a) The turns ratio of the transformer is

$$a = \frac{N_1}{N_2} = \frac{V_1}{V_2} = \frac{2,400 \text{ V}}{240 \text{ V}} = 10$$

(b) Since the transformer is an ideal transformer, it has no losses. Therefore,

$$S = V_1 I_1 = V_2 I_2$$

from which

$$I_2 = \frac{S}{V_2} = \frac{250,000 \text{ V}}{240 \text{ V}} = 1,041.67 \text{ A}$$

(c) The corresponding primary current is

$$I_1 = \frac{I_2}{a} = \frac{1,041.67 \text{ A}}{10} = 104.167 \text{ A}$$

or

$$I_1 = \frac{S}{V_1} = \frac{250,000 \text{ VA}}{2,400 \text{ V}} = 104.167 \text{ A}$$

(d) The value of the load impedance at the secondary side is

$$Z_L = \frac{V_2}{I_2} = \frac{240 \text{ V}}{1,041.67 \text{ A}} = 0.2304 \ \Omega$$

Thus, its value referred to the primary side is

$$Z'_L = \left(\frac{N_1}{N_2}\right)^2 Z_L = a^2 Z_L = 10^2 \times 0.2304 = 23.04 \ \Omega$$

Example 4.3

A single-phase 60-Hz transformer is supplying power to a load of $3 + j5 \ \Omega$ through a short power line with an impedance of $0.2 + j0.6 \ \Omega$, as shown in Figure 4.9(a). The voltage at the generator bus (i.e., bus 1) is $277\angle 0°$ V. Determine the following:

(a) If the current at bus 1 is equal to the current at bus 2 (i.e., $I_{line} = I_{load}$), find the voltage at the load bus and the power losses that take place in the power line.

(b) If two ideal transformers T_1 and T_2 are inserted at the beginning and end of the line, as shown in Figure 4.9(b), find the voltage at the load bus and the power losses in the line.

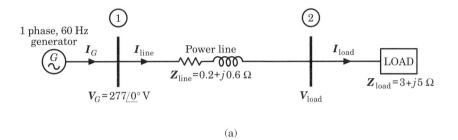

(a)

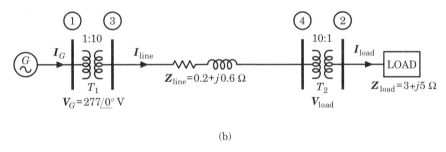

(b)

Figure 4.9   One-line diagram of the power system given in Example 4.3: (a) without the step-up and step-down transformers; (b) with the transformers.

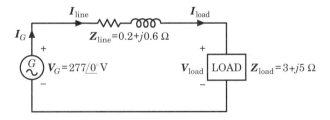

Figure 4.10   The line-to neutral diagram of the system given in Figure 4.9(a).

SOLUTION:
(a) Figure 4.10 shows the line-to-neutral diagram of the one-line diagram of the given system. Using the generator terminal voltage as the reference phasor, the line current can be found as

$$I_{line} = \frac{V_G}{Z_{line} + Z_{load}} = \frac{277\angle 0° \text{ V}}{3.2 + j5.6 \, \Omega} = 42.9471\angle -60.26° \text{ A}$$

Since $I_{load} = I_{line}$, the voltage at the load bus (i.e., bus 2) is

$$V_{load} = I_{load} \, Z_{load}$$

$$= (42.9471 \angle -60.26° \text{ A})(3 + j5 \text{ } \Omega) = 250.4245 \angle -1.22° \text{ A}$$

and the line losses (i.e., the copper losses) are

$$P_{line \, loss} = I_{line}^2 R_{line}$$

$$= (42.9471 \text{ A})^2 (0.2 \text{ } \Omega) = 368.8901 \text{ W}$$

(b) Figure 4.9(b) shows the one-line diagram of the system with the step-up transformer T_1 (with a turns ratio[22] of $a_1 = 1/10$) and the step-down transformer T_2 (with a turns ratio of $a_2 = 10/1$). The load impedance referred to the power-line side of transformer T_2 is

$$Z'_{load} = a_2^2 Z_{load} = \left(\frac{N_1}{N_2} \right)^2 Z_{load}$$

$$= \left(\frac{10}{1} \right)^2 (3 + j5 \text{ } \Omega) = 300 + j500 \text{ } \Omega$$

The resulting equivalent impedance is

$$Z_{eq} = Z_{line} + Z'_{load}$$

$$= (0.2 + j0.6 \text{ } \Omega) + (300 + j500 \text{ } \Omega) = 300.2 + j500.6 \text{ } \Omega$$

Referring this Z_{eq} to the generator-side of transformer T_1, the new equivalent impedance is found as

$$Z'_{eq} = a_1^2 Z_{eq} = \left(\frac{N_1}{N_2} \right)^2 Z_{eq}$$

$$= \left(\frac{1}{10} \right)^2 (300.2 + j500.6 \text{ } \Omega) = 3.002 + j5.006 \text{ } \Omega$$

Thus, the generator current can be calculated from

---

[22] The given numbers in 10:1 and 1:10 simply represent the turns ratios, respectively, rather than representing the actual number of turns in each winding.

$$I_G = \frac{V_G}{Z'_{eq}} = \frac{277\angle 0° \text{ V}}{3.002 + j5.006 \text{ }\Omega} = 47.4549\angle - 59.05 \text{ A}$$

Therefore, working back through transformer T_1,

$$N_1 I_G = N_2 I_{line}$$

and the line current can be found as

$$I_{line} = \frac{N_1}{N_2} I_G$$

$$= \frac{1}{10}(47.4549\angle - 59.05° \text{ A}) = 4.7455\angle - 59.05° \text{ A}$$

Similarly, working back through transformer T_2,

$$N_{line} I_{line} = N_2 I_{load}$$

and the load current can be found as

$$I_{load} = \frac{N_1}{N_2} I_{line}$$

$$= \frac{1}{10}(4.7455\angle - 59.05° \text{ A}) = 47.455\angle - 59.05° \text{ A}$$

Hence, the voltage at the load bus is

$$V_{load} = I_{load} Z_{load}$$

$$= (47.455\angle - 59.05° \text{ A})(3 + j5 \text{ W}) = 276.7093\angle - 0.01° \text{ V}$$

The line losses are

$$P_{line\ loss} = I_{line}^2 R_{line}$$

$$= (4.7455 \text{ A})^2 (0.2 \text{ }\Omega) = 4.5039 \text{ W}$$

Notice that the percent reduction in the line losses, after adding the step-up and the step-down transformers is

$$\text{Reduction in } P_{line\ loss} = \frac{368.8901 \text{ W} - 4.5039 \text{ W}}{368.8901 \text{ W}} \times 100 = 98.78\%$$

Example 4.4

Assume that the impedances of the transformers T_1 and T_2, given in Part(a) of Example 4.3, are not small enough to ignore, and that they are $\boldsymbol{Z}_{T_1} = 0 + j0.15\ \Omega$ and $\boldsymbol{Z}_{T_2} = 0 + j0.15\ \Omega$, respectively. Also assume that they are referred to the high-voltage sides of each transformer, respectively. Solve Part (b) of Example 4.3, accordingly.

SOLUTION:

Figure 4.11(a) shows the one-line diagram of the given system that includes the T_1 step-up and the T_2 step-down transformers. Figure 4.11(b) shows the line-to-neutral diagram of the same system. Figure 4.11(c) shows the load impedance referred to power-line side of transformer T_2 as well as the impedance of the transformer (which is given as already referred to its high-voltage side). Therefore, the resulting equivalent impedance can be found as

$$\boldsymbol{Z}_{eq} = \boldsymbol{Z}_{T_1} + \boldsymbol{Z}_{line} + \boldsymbol{Z}_{T_2} + \boldsymbol{Z}'_{load}$$

$$= \boldsymbol{Z}_{T_1} + \boldsymbol{Z}_{line} + \boldsymbol{Z}_{T_2} + a_2^2 \boldsymbol{Z}_{load}$$

$$= j0.15 + (0.2 + j0.6) + j0.15 + 10^2(3 + j5)$$

$$- 300.2 + j500.9\ \Omega$$

Referring this $\boldsymbol{Z}_{eq}$ to the generator-side of transformer T_1, as shown in Figure 4.11(d), the new equivalent impedance is found as

$$\boldsymbol{Z}'_{eq} = a_1^2 \boldsymbol{Z}_{eq}$$

$$= \left(\frac{1}{10}\right)^2 (300.2 + j500.9) = 3.002 + j5.009\ \Omega$$

Therefore, the generator current can be found from

$$I_G = \frac{V_G}{\boldsymbol{Z}'_{eq}}$$

$$= \frac{277\angle 0°\ \text{V}}{3.002 + j5.009\ \Omega} = 47.4339\angle -59.06\ \text{A}$$

Thus, working back through transformer T_1,

$$I_{line} = \frac{N_1}{N_2} I_G$$

$$= \frac{1}{10}(47.4339\angle - 59.06°) = 4.74339\angle - 59.06° \text{ A}$$

Similarly, working back through transformer T_2,

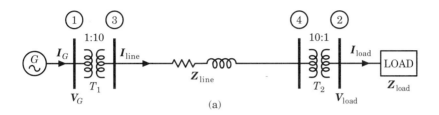

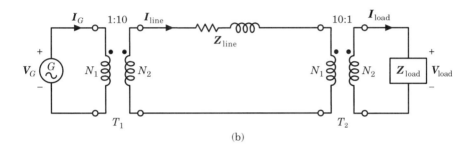

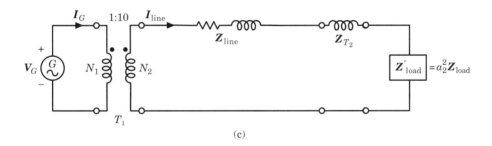

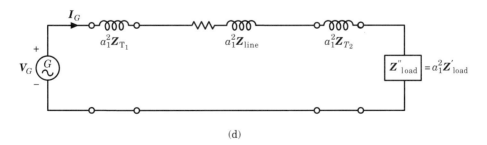

Figure 4.11

$$I_{load} = \frac{N_1}{N_2} I_{line}$$

$$= \left(\frac{10}{1}\right)(4.74339\angle - 59.06°) = 47.4339\angle - 59.06° \text{ A}$$

Hence, the voltage at the load bus is

$$V_{load} = I_{load} Z_{load}$$

$$= (47.4339\angle - 59.06°)(3 + j5) = 276.5873\angle - 0.02° \text{ V}$$

The line losses are

$$P_{line\ loss} = I_{line}^2 R_{line}$$

$$= (4.74339)^2(0.2) = 4.4999 \text{ W}$$

The percent reduction in line losses, after adding the step-up and the step-down transformer, is

$$Reduction\ in\ P_{line\ loss} = \frac{368.8901 \text{ W} - 4.4999 \text{ W}}{368.8901 \text{ W}} \times 100 = 98.78\%$$

4.5 THE REAL TRANSFORMER

A real transformer differs from an ideal transformer in many respects. For example, as illustrated in Figure 4.12(a), (1) the primary and secondary winding resistances R_1 and R_2 are not negligible, (2) the leakage fluxes Φ_{ℓ_1} and Φ_{ℓ_2} exist, (3) the core losses are not negligible, (4) the permeability of the core material is not infinite and therefore a considerable mmf is required to establish mutual flux Φ_m in the core, and (5) the core material saturates. The resulting representation of this transformer is shown in Figure 4.12(b). Here, X_{ℓ_1} and X_{ℓ_2} are the leakage fluxes, respectively. Therefore,

$$X_{\ell_1} = \omega L_{\ell_1} = \omega N_1^2 P_{\ell_1} \qquad (4.40)$$

and

$$X_{\ell_2} = \omega L_{\ell_2} = \omega N_2^2 P_{\ell_2} \qquad (4.41)$$

where $\omega = 2\pi f$

L_{ℓ_1} = leakage inductance of the primary winding

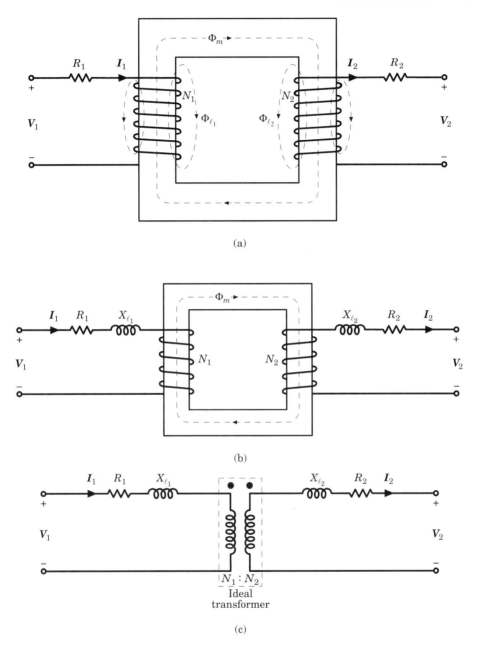

(a)

(b)

(c)

Figure 4.12   Development of transformer-equivalent circuits.

$$= N_1^2 P_{\ell_1} = \frac{N_1 \Phi_{\ell_1}}{I_1} \tag{4.42}$$

L_{ℓ_2} = leakage inductance of the secondary winding

$$= N_2^2 P_{\ell_2} = \frac{N_2 \Phi_{\ell_2}}{I_2} \tag{4.43}$$

P_{ℓ_1} = permeance of the leakage flux path of the primary winding
P_{ℓ_2} = permeance of the leakage flux path of the secondary winding

In such a representation, the transformer windings are tightly coupled by a mutual flux, and represented as shown in Figure 4.12(c).

As illustrated in Figure 4.13(a), the primary current I_1 must be large enough to compensate the demagnetizing effect of the load current (i.e., the secondary current), but also provide for adequate mmf to develop the resultant mutual flux. Note that I'_2 is the load component in the primary and can be expressed as

$$I'_2 = \frac{N_2}{N_1} I_2 = \frac{I_2}{a} \tag{4.44}$$

In other words, I'_2 is the secondary current referred to the primary, as it is in the ideal transformer. Therefore, the primary current can be expressed in terms of phasor summation as

$$I_1 = I'_2 + I_e = \frac{I_2}{a} + I_e \tag{4.45}$$

where I_e is the excitation current (i.e., the additional primary current) needed to develop the resultant mutual flux. Such excitation current I_e is nonsinusoidal and can be expressed as

$$I_e = I_c + I_m \tag{4.46}$$

where I_c is the core-loss component of the excitation current supplying the hysteresis and eddy-current losses in the core, and I_m is the magnetizing component of the excitation current needed to magnetize the core. Here, I_c is in phase with the counter emf E_1 and I_m lags E_1 by 90°. Therefore, the core-loss component and the magnetizing component are modeled by a resistance R_c and an inductance X_m, respectively, that are connected across the primary voltage source. Even though *Faraday's law* dictates the core flux to be very nearly sinusoidal for a sinusoidal terminal voltage, saturation causes the excitation current to be very nonsinusoidal. Thus, in reality, both I_c and I_m are actually nonlinear, and hence the resistance R_c and the reactance X_m are at best approximations.

The ideal transformer[23] shown in Figure 4.13(a) can be eliminated by referring all secondary quantities to the primary. Figures 4.13(b) and 4.13(c)

---

[23] Such a representation is known as the *Steinmetz circuit model* of a transformer. Steinmetz had the brilliant idea of separating the linear phenomenon by which leakage-flux voltage is induced, from the nonlinear phenomenon by which mutual-flux voltage is induced in an iron-core transformer. His approach, based on linear circuit theory, provided an easy solution for developing a circuit model for an iron-core transformer.

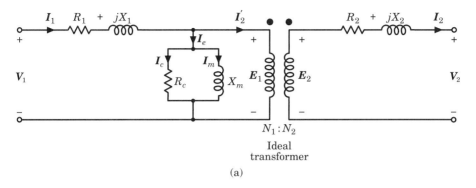

(a)

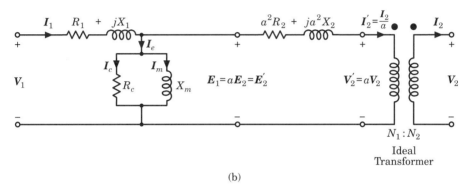

(b)

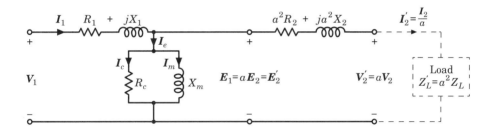

(c)

Figure 4.13 Transformer-equivalent circuits: (a) the real transformer; (b) referred to the primary; (c) referred to the primary (without showing the ideal transformer).

show the first and second steps of this process. Figure 4.13(c) shows the equivalent circuit of a transformer (model) referred to its primary.[24] Figure 4.14 (a)

---

[24] The physical meaning of referring secondary quantities to the primary implies that the real and reactive powers in an impedance Z_L through which the secondary current I_2 flows is the same when the primary current I_1 flows through an equivalent impedance Z_L. Therefore, there

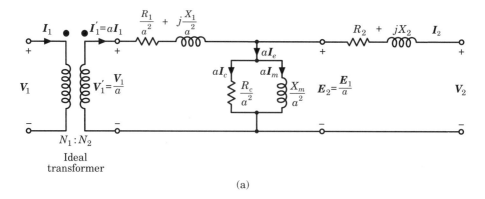

(a)

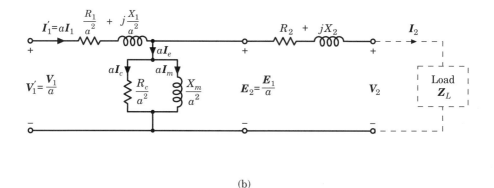

(b)

Figure 4.14   Transformer-equivalent circuits: (a) referred to the secondary; (b) referred to the secondary (without showing the ideal transformer).

shows the equivalent circuit of a transformer referred to its secondary. Figure 4.14(b) shows the transformer equivalent circuit referred to its secondary without showing the ideal transformer.

4.6 APPROXIMATE EQUIVALENT CIRCUIT OF A REAL TRANSFORMER

The transformer equivalent circuits developed in the previous section [and shown in Figures 4.13(c) and 4.14(b)] are often more accurate than is necessary in practice. This is especially true in power system applications. The excitation branch has a very small current in comparison to the load current of the transformer. Of course, such a small excitation current, I_e, causes a negligibly small voltage drop in the primary winding impedance ($R_1 + jX_1$). Therefore, by moving the excitation admittance (i.e., the shunt branch) from the middle of

cannot be any difference in the performance of a transformer determined from an equivalent circuit referred to the primary or the secondary.

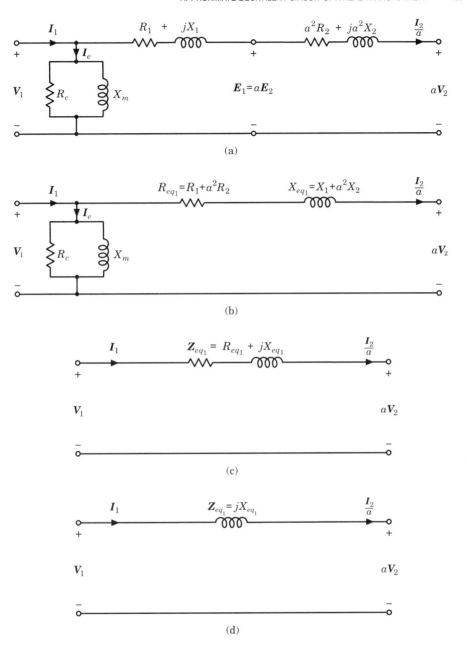

Figure 4.15   Approximate equivalent circuits referred to the primary of an iron-core transformer.

the *T*-circuit to either the left [as shown in Figure 4.15(a)] or the right, the primary and secondary impedances are left in series with each other so that they can be added together as shown in Figure 4.15(b). The equivalent impedance of such an approximate equivalent circuit of the transformer depends on whether its equivalent circuit is referred to the primary or secondary. As shown in Figure 4.15(b), if the equivalent impedance is referred to the primary,

$$\boldsymbol{Z}_{eq_1} = R_{eq_1} + jX_{eq_1} \tag{4.47}$$

where

$$R_{eq_1} = R_1 + a^2 R_2 \tag{4.48}$$

$$X_{eq_1} = X_1 + a^2 X_2 \tag{4.49}$$

Here, the R_{eq_1} and X_{eq_1} are the equivalent resistance and reactance referred to the primary, respectively. As shown in Figure 4.16(b), if the equivalent impedance is referred to the secondary,

$$\boldsymbol{Z}_{eq_2} = R_{eq_2} + jX_{eq_2} \tag{4.50}$$

where

$$R_{eq_2} = \frac{R_1}{a^2} + R_2 \tag{4.51}$$

$$X_{eq_2} = \frac{X_1}{a^2} + X_2 \tag{4.52}$$

Here, the R_{eq_2} and X_{eq_2} are the equivalent resistance and reactance referred to the secondary, respectively. It is interesting to notice that

$$\frac{\boldsymbol{Z}_{eq_1}}{\boldsymbol{Z}_{eq_2}} = \frac{R_{eq_1}}{R_{eq_2}} = \frac{X_{eq_1}}{X_{eq_2}} = a^2 \tag{4.53}$$

A further approximation of the equivalent circuit can be made by removing the excitation branch, as shown in Figure 4.15(c) and 4.16(c). The resultant error is very small since the excitation current $\boldsymbol{I}_e$ is very small in comparison to the rated current of the transformer.

Furthermore, in power transformers, the equivalent resistance R_{eq} is small in comparison to the equivalent reactance X_{eq}. Therefore, the transformer can only be represented by its equivalent reactance X_{eq}, as shown in Figures 4.15(d) and 4.16(d). Thus, the corresponding equivalent impedances can be expressed as

$$\boldsymbol{Z}_{eq_1} = jX_{eq_1} \tag{4.54}$$

and

$$\boldsymbol{Z}_{eq_2} = jX_{eq_2} \tag{4.55}$$

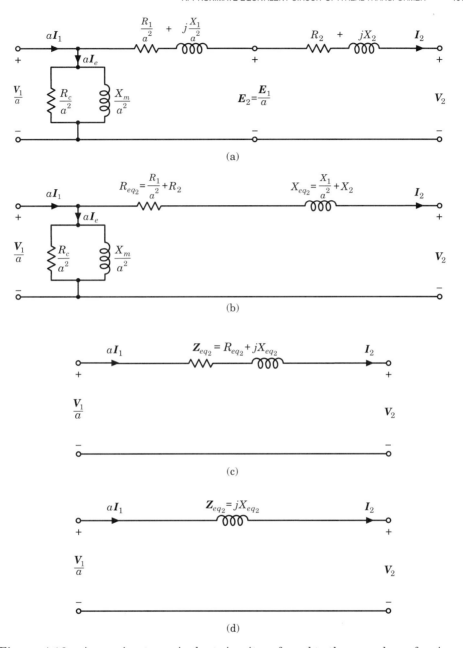

Figure 4.16   Approximate equivalent circuits referred to the secondary of an iron-core transformer.

Example 4.5

Consider a 75-kVA 2,400/240-V 60-Hz distribution transformer with $Z_1 = 0.612 + j1.2\ \Omega$ and $Z_2 = 0.0061 + j0.0115\ \Omega$ for its high-voltage and low-voltage windings, respectively. Its excitation admittance referred to the 240 V side is $Y_{e_2} = 0.0191 - j0.0852$ S. The transformer delivers its rated I_L at 0.9 lagging PF.

(a) Draw the equivalent circuit with the excitation admittance referred to the primary side.

(b) Find the emfs of E_1 and E_2 induced by the equivalent mutual flux, I_e, I_1 at 0.9 lagging PF; and the applied V_1 when the transformer delivers a rated load at rated V_2 voltage.

SOLUTION:

(a) Figure 4.17 shows the equivalent circuit of the transformer, with the excitation admittance referred to the primary side. Therefore,

$$Y_{e_1} = \frac{Y_{e_2}}{a^2}$$

$$= \frac{0.0191 - j0.0852 \, \text{S}}{10^2} = 1.91 \times 10^{-4} - j8.52 \times 10^{-4} \, \text{S}$$

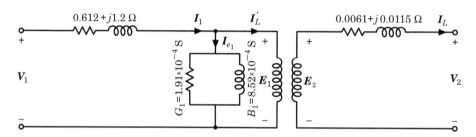

Figure 4.17

(b) Here,

$$I_L = I_2 = \frac{S}{V_2}$$

$$= \frac{75,000 \, \text{VA}}{240 \, \text{V}} = 312.5 \, \text{A}$$

Let V_2 be the reference phasor so that

$$V_2 = 240\angle 0° \, \text{V}$$

Thus,

$$I_L = I_2$$
$$= 312.5(0.9 - j0.4359)$$

$$= 281.25 - j136.2156 \text{ A}$$

Hence,

$$\boldsymbol{E}_2 \quad = \boldsymbol{V}_2 + \boldsymbol{I}_L \left(R_2 + jX_2 \right)$$

$$= 240\angle 0° + (281.25 - j136.2156)(0.0061 + j0.00115)$$

$$= 243.294\angle 0.57° \text{ V}$$

so that,

$$\boldsymbol{E}_1 = a \, \boldsymbol{E}_2$$

$$= 240\angle 0° + (281.25 - j136.2156)(0.0061 + j0.00115)$$

$$= 243.294\angle 0.57° \text{ V}$$

Therefore, the load current referred to the primary side is

$$\boldsymbol{I}'_L = \frac{\boldsymbol{I}_L}{a}$$

$$= \frac{281.25 - j136.2156 \text{ A}}{10} = 28.125 - j13.6216 \text{ A}$$

Notice that

$$\boldsymbol{I}_1 = \boldsymbol{I}'_L + \boldsymbol{I}_{e_1}$$

where $\boldsymbol{I}_{e_1}$ produces the mutual flux in the core. Since

$$\boldsymbol{I}_{e_1} = \boldsymbol{E}_1 \boldsymbol{Y}_{e_1}$$

$$= (2432.94\angle 0.57°)(1.91 \times 10^{-4} - j8.52 \times 10^{-4})$$

$$= 0.4851 - j2.0682 \text{ A}$$

also

$$\boldsymbol{I}_1 = \boldsymbol{I}'_L + \boldsymbol{I}_{e_1}$$

$$= (28.125 - j13.6216) + (0.4851 - j2.0682)$$

$$= 32.6299\angle - 28.74° \text{ A}$$

Therefore,

$$\boldsymbol{V}_1 = \boldsymbol{E}_1 + \boldsymbol{I}_1(R_1 + jX_1)$$

$$= 2432.94 \angle 0.57° + (32.6299\angle - 28.74°$$

$$= 2,469.6396\angle 1.13° \text{ V}$$

4.7 DETERMINATION OF EQUIVALENT-CIRCUIT PARAMETERS

The equivalent circuits of a given transformer can be used to predict and evaluate its performance. If the complete design data of a transformer are available (such data are usually available only to its designer), the necessary parameters can be computed from the dimensions and properties of the materials used. However, once the transformer is manufactured, it may be desirable to verify the accuracy of the performance predictions. This can be achieved by means of two tests designed to determine the parameters of the equivalent circuit. These two tests[25] are known as the *open-circuit test* and the *short-circuit test*.

4.7.1 THE OPEN-CIRCUIT TEST

The purpose of the open-circuit test[26] is to determine the excitation admittance of the transformer-equivalent circuit, the no-load loss, the no-load excitation current, and the no-load power factor. Such an open-circuit test is performed by applying rated voltage to one of the windings, with the other winding (or windings) open circuited. The input power, current, and voltage are measured, as shown in Figure 4.18(a). (*However, for reasons of safety and convenience, usually the high-voltage winding is open-circuited and the test is conducted by placing the instruments on the low-voltage side of the transformer.*) Once such information is collected, one can determine the magnitude and the angle of excitation impedance after finding the open-circuit (i.e., no-load) power factor. Here, the voltage drop in the leakage impedance of the winding (which is excited in the open-circuit test) caused by the normally small excitation current is usually ignored. This results in an approximate equivalent circuit as shown in Figure 4.18(b). Also ignored is the (primary) power loss due to the excitation current. Therefore, the excitation admittance can be expressed as

$$Y_e = Y_{oc} = \frac{I_{oc}}{V_{oc}} \angle -\theta_{oc} \tag{4.56}$$

where θ_{oc} is the angle of the admittance, found from the open-circuit power factor PF_{oc} as

$$PF_{oc} = \cos \theta_{oc} = \frac{P_{oc}}{V_{oc} I_{oc}} \tag{4.57}$$

---

[25] Each test can be done by exciting either winding. However, in large transformers with high levels of both voltage and current, it may be a good idea to excite the low-voltage winding for the open-circuit test and to excite the high-voltage winding for the short-circuit test.

[26] It is also known as the *core-loss test*, the *iron-loss test*, the *no-load test*, the *excitation* test, or the *magnetization test*.

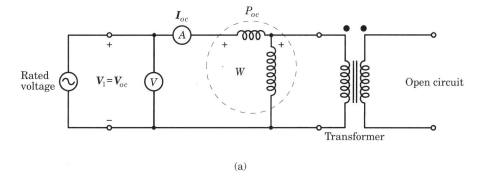

(a)

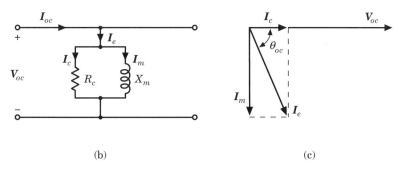

(b)   (c)

Figure 4.18   Open-circuit test: (a) wiring diagram for the open-circuit test; (b) equivalent circuit; (c) no-load phasor diagram.

so that

$$\theta_{oc} = \cos^{-1} \frac{P_{oc}}{V_{oc} I_{oc}} \tag{4.58}$$

or

$$\theta_{oc} = \cos^{-1} PF_{oc} \tag{4.59}$$

For a given transformer, the P_{oc} is always lagging. For this reason, there is a negative sign in front of θ_{oc} in equation (4.56). If the excitation admittance is expressed in rectangular coordinates,

$$\boldsymbol{Y}_e = \boldsymbol{Y}_{oc} = G_c - jB_m \tag{4.60}$$

or

$$\boldsymbol{Y}_e = \boldsymbol{Y}_{oc} = \frac{1}{R_c} - j\frac{1}{X_m} \tag{4.61}$$

from which the R_c and X_m can be determined as

$$R_c = \frac{1}{G_c} \tag{4.62}$$

and

$$X_m = \frac{1}{B_m} \tag{4.63}$$

Alternatively, the core-loss conductance and the susceptance can be found, respectively, from

$$G_c \cong G_{oc} = \frac{P_{oc}}{V_{oc}^2} \tag{4.64}$$

and

$$B_m \cong B_{oc} = (Y_{oc}^2 - G_{oc}^2)^{1/2} \tag{4.65}$$

The values of R_c and X_m can be determined from equations (4.62) and (4.63), respectively, as before. The no-load phasor diagram can be drawn as shown in Figure 4.18(c).

4.7.2 THE SHORT-CIRCUIT TEST

The purpose of the short-circuit test[27] is to determine the equivalent resistance and reactance of the transformer under rated conditions. This test is performed by short-circuiting one winding (usually the low-voltage winding) and applying a *reduced* voltage to the other winding, as shown in Figure 4.19(a). The reduced input voltage is adjusted until the current in the shorted winding is equal to its rated value. The input voltage, current, and power are measured as before. The applied voltage V_{sc} is only a small percentage of the rated voltage and is sufficient to circulate rated current in the windings of the transformer. Usually, this voltage is about 2 to 12 percent of the rated voltage. Therefore, the excitation current is small enough to be ignored. If it is neglected, then one can assume that all the voltage drop will take place in the transformer and is due to the series elements in the circuit, as shown in Figure 4.19(b). The shunt branch representing excitation admittance does not appear in this equivalent circuit. The series impedance Z_{sc} can be found from

---

[27] It is also known as the *impedance test* or the *copper-loss test*. Since the voltage applied under short-circuit conditions is small, the core losses are ignored and the wattmeter reading represents the copper losses in the windings. Therefore, for all practical purposes $P_{cu} = P_{sc}$.

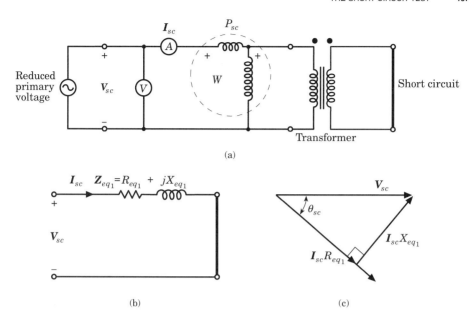

Figure 4.19   Short-circuit test: (a) wiring diagram for the short-circuit test; (b) equivalent circuit; (c) phasor diagram.

$$\boldsymbol{Z}_{sc} = \boldsymbol{Z}_{eq_1} = \frac{\boldsymbol{V}_{sc}}{\boldsymbol{I}_{sc}} = \frac{V_{sc} \angle 0°}{I_{sc} \angle -\theta_{sc}} \qquad (4.66)$$

The short-circuit power factor is lagging and determined from

$$PF_{sc} = \cos \theta_{sc} = \frac{P_{sc}}{V_{sc} I_{sc}} \qquad (4.67)$$

so that

$$\theta_{sc} = \cos^{-1} \frac{P_{sc}}{V_{sc} I_{sc}} \qquad (4.68)$$

or

$$\theta_{sc} = \cos^{-1} PF_{sc} \qquad (4.69)$$

For a given transformer, the PF_{sc} is *always* lagging. For this reason, there is a negative sign in front of θ_{sc} in equation (4.66). Equation (4.66) can be expressed as

$$\boldsymbol{Z}_{eq_1} = \frac{V_{sc}}{I_{sc}} \angle \theta_{sc} = R_{eq_1} + jX_{eq_1} \qquad (4.70)$$

where

$$R_{eq_1} = R_1 + a^2 R_2 \tag{4.71}$$

$$X_{eq_1} = X_1 + a^2 X_2 \tag{4.72}$$

Alternatively, the equivalent circuit resistance and reactance can be found, respectively, from

$$R_{eq_1} \cong R_{sc} = \frac{P_{sc}}{I_{sc}^2} \tag{4.73}$$

and

$$X_{eq_1} \cong X_{sc} = (Z_{eq}^2 - R_{eq}^2)^{1/2} \tag{4.74}$$

where

$$Z_{eq_1} \cong Z_{sc} = \frac{V_{sc}}{I_{sc}} \tag{4.75}$$

Figure 4.19(c) shows the phasor diagram under short-circuit conditions. In a well-designed transformer, when all impedances are referred to the same side (in this case, to the primary side),

$$R_1 = a^2 R_2 = R'_2 \cong \frac{R_{eq_1}}{2} \tag{4.76}$$

and

$$X_1 = a^2 X_2 = X'_2 \cong \frac{X_{eq_1}}{2} \tag{4.77}$$

As mentioned before, it is possible to perform the open-circuit and short-circuit tests on the secondary side (i.e., the low-voltage side) of the transformer. However, the resultant equivalent circuit impedances would be referred to the *secondary side* rather than to the *primary side*. Of course, with large units it may be preferable to excite the low-voltage winding on the open circuit and the high-voltage winding on the short circuit.

4.8 TRANSFORMER NAMEPLATE RATING

Among the information provided by the nameplate of a transformer are its apparent power (in terms of the kVA rating or the MVA rating), voltage ratings, and impedance. For example, a typical transformer may have 25 kVA,

2,400/~~240~~ V. Here, the voltage ratings point out that the transformer has two windings, one rated for 2,400 V, and the other for 120 V. Since the voltage ratio also represents the turns ratio, the turns ratio of the transformer is

$$a = \frac{N_1}{N_2} = \frac{V_1}{V_2} = \frac{2,400 \text{ V}}{120 \text{ V}} = 20$$

Also, the given 25 kVA rating indicates that each winding is designed to carry 25 kVA. Thus, the current rating for the high-voltage winding is 25,000 VA / 2,400 V = 10.42 A, but for the low-voltage winding is 25,000 VA/120 V = 208.33 A. When a current of 208.33 A flows through the secondary winding, there will be a current of 10.42 A in the primary winding, ignoring the small excitation current that flows through the primary winding. The kVA rating always refers to the **output kVA** measured at the secondary (load) terminals. The input kVA will be slightly more due to the losses involved.

Transformer impedance is always provided on the nameplate in percentage. For example, 5%, means 0.05 per unit based on its nameplate ratings. In terms of percentage or per unit, the given figure could be referred to the primary winding or secondary winding. Nevertheless, in either case, it would still be 5%.

Example 4.6

Consider a 15-kVA 7,500/480-V 60-Hz distribution transformer. Assume that the open-circuit and short-circuit tests were performed on the primary side of the transformer and that the following data was obtained:

	Open-Circuit Test (on Primary)	Short-Circuit Test (on Primary)
Voltmeter	V_{oc} = 7,500 V	V_{sc} = 366 V
Ammeter	I_{oc} = 0.2006 A	I_{sc} = 2 A
Wattmeter	P_{oc} = 180 W	P_{sc} = 300 W

Determine the impedance of the approximate equivalent circuit referred to the primary side, and draw the corresponding simplified equivalent circuit.

SOLUTION:

The power factor during the open-circuit test is

$$PF_{oc} = \cos\theta_{oc} = \frac{P_{oc}}{V_{oc}I_{oc}}$$

$$= \frac{180 \text{ W}}{(7{,}500 \text{ V})(0.2006 \text{ A})} = 0.1196 \; lagging$$

The excitation admittance is

$$\mathbf{Y}_e = \mathbf{Y}_{oc} = \frac{\mathbf{I}_{oc}}{\mathbf{V}_{oc}} \; \angle - \cos^{-1} PF_{oc}$$

$$= \frac{0.2006 \text{ A}}{7{,}500 \text{ V}} \angle - \cos^{-1}(0.1196)$$

$$= 0.0000267 \angle - 83.129° \text{ S}$$

$$= 0.0000032 - j0.0000265 \text{ S}$$

$$= \frac{1}{R_c} - j \frac{1}{X_m}$$

Therefore,

$$R_c = \frac{1}{0.0000032} = 312{,}500 \; \Omega \cong 312.5 \text{ k}\Omega$$

and

$$X_m = \frac{1}{0.0000\overset{265}{\cancel{32}}} = 37{,}658.32 \; \Omega \cong 37.7 \text{ k}\Omega$$

The power factor during the short-circuit test is

$$PF_{sc} = \cos\theta_{sc} = \frac{P_{sc}}{V_{sc}I_{sc}}$$

$$= \frac{300 \text{ W}}{(366 \text{ V})(2 \text{ A})} = 0.41 \; lagging$$

The series (i.e., the *equivalent*) impedance is

$$\mathbf{Z}_{eq_1} = \mathbf{Z}_{sc} = \frac{\mathbf{V}_{sc}}{\mathbf{I}_{sc}} \angle \cos^{-1} PF_{sc}$$

$$= \frac{366 \text{ V}}{2 \text{ A}} \angle \cos^{-1} 0.41 = 183 \angle 65.81° \; \Omega$$

$$= 75 + j166.93 \; \Omega$$

Therefore, the equivalent resistance and reactance are

$$R_{eq_1} = 75\ \Omega \text{ and } X_{eq_1} = 166.93\ \Omega$$

The corresponding simplified equivalent circuit is shown in Figure 4.20(a).

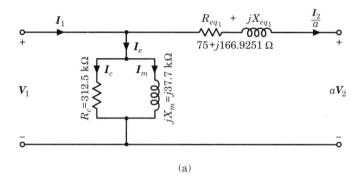

(a)

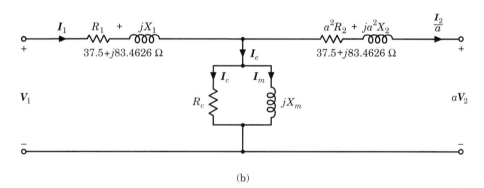

(b)

Figure 4.20

However, if an equivalent T-circuit is needed, the values of individual primary and secondary resistances and leakage reactances, referred to the same side, are usually assumed to be equal. Therefore,

$$R_1 = a^2 R_2 = R'_2 \cong \frac{R_{eq_1}}{2}$$

$$= \frac{75\ \Omega}{2} = 37.5\ \Omega$$

and

$$X_1 = a^2 X_2 = X'_2 \cong \frac{X_{eq_1}}{2}$$

$$= \frac{166.93 \, \Omega}{2} \cong 83.46 \, \Omega$$

where

$$a = \frac{V_1}{V_2} = \frac{7{,}500 \, \text{V}}{480 \, \text{V}} = 15.625$$

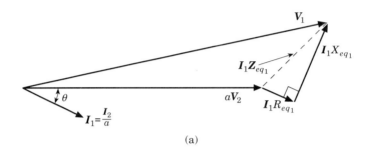

(a)

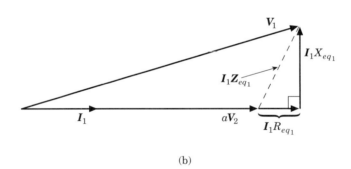

(b)

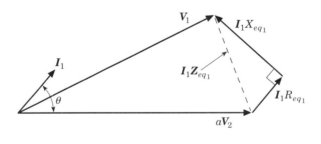

(c)

Figure 4.21   Phasor diagram of a transformer operating at: (a) lagging power factor; (b) unity power factor; (c) leading power. All quantities are referred to the primary side of the transformer.

Therefore, as shown in Figure 4.21(b),

$$R_2 = \frac{R'_2}{a^2}$$

$$= \frac{37.5\,\Omega}{15.625^2} \cong 0.154\,\Omega$$

and

$$X_2 = \frac{X'_2}{a^2}$$

$$= \frac{83.46\,\Omega}{15.63^2} = 0.342\,\Omega$$

4.9 PERFORMANCE CHARACTERISTICS OF A TRANSFORMER

The main use of the equivalent circuit of a given transformer is to determine its performance characteristics, which are basically its voltage regulation and its efficiency.

4.9.1 VOLTAGE REGULATION OF A TRANSFORMER

The **voltage regulation** of a transformer is the change in the magnitude of secondary terminal voltage from no load to full load when the primary voltage is constant. It is usually expressed as a percentage of the full-load value,[28] as

$$\% \ Voltage \ Regulation = \frac{V_{2(no\,load)} - V_{2(full\,load)}}{V_{2(full\,load)}} \times 100 \qquad (4.78)$$

Here, the full load is the rated load of the secondary. At no load, the secondary terminal voltage may change from the rated voltage[29] value due to the effect of the impedance of the transformer. Also, since at no load,

$$V_{2(no\,load)} = \frac{V_1}{a}$$

then

---

[28] For further information, see Chapter 9 of Gönen [33].
[29] It is also called the *nameplate voltage, nominal transformer voltage*, or *full-load voltage*.

$$\% \ Voltage \ regulation = \frac{\dfrac{V_1}{a} - V_{2(full \ load)}}{V_{2(full \ load)}} \times 100 \qquad (4.79)$$

The voltage regulation is affected by the magnitude and power factor of the load as well as by the internal impedance (i.e., the *leakage impedance*) of the transformer. Even though in electric power engineering applications it is usually considered good practice to have a small voltage regulation, under certain circumstances transformers with high impedance and high-voltage regulation are used to decrease the fault currents in a circuit.

As shown in the phasor diagrams of Figure 4.21, depending on the power factor of the load, the voltage regulation can be positive, zero, or negative. Here, all circuit parameters are *referred to the primary side* of the transformer. At a lagging power factor, the voltage regulation is positive, as shown in Figure 4.21(a). With certain exceptions, as mentioned before, it is usually good practice to minimize the voltage regulation. As shown in Figure 4.21(b), the voltage regulation is *zero* at *unity* power factor, whereas it is *negative* at *leading* power factors, as shown in Figure 4.21(c). This means that the secondary terminal voltage is greater under full load than under no load. Such a situation takes place when the power-factor-correction capacitor banks remain in the circuit while the load is low. (This causes a partial resonance between the capacitance of the load and the leakage inductance of the transformer.) The solution is to adjust the capacitor sizes and/or to use some of them as *switchable* capacitors.[30] As can be seen from Figure 4.21(a), the primary voltage can be expressed as

$$\boldsymbol{V}_1 = a\boldsymbol{V}_2 + \boldsymbol{I}_1 \boldsymbol{Z}_{eq_1} \qquad (4.80)$$

or

$$\boldsymbol{V}_1 = a\boldsymbol{V}_2 + \boldsymbol{I}_1 R_{eq_1} + j\boldsymbol{I}_1 X_{eq_1} \qquad (4.81)$$

If all circuit parameters are *referred to the secondary side* of the transformer, then the phasor diagrams corresponding to lagging, unity, and leading power factors are as shown in Figure 4.22. For example, as can be seen from Figure 4.22(a), the primary voltage can be expressed as

$$\frac{\boldsymbol{V}_1}{a} = \boldsymbol{V}_2 + \boldsymbol{I}_2 \boldsymbol{Z}_{eq_2} \qquad (4.82)$$

or

$$\frac{\boldsymbol{V}_1}{a} = \boldsymbol{V}_2 + \boldsymbol{I}_2 R_{eq_2} + j\boldsymbol{I}_2 X_{eq_2} \qquad (4.83)$$

---

[30] For further information, see Chapter 8 of Gönen [33].

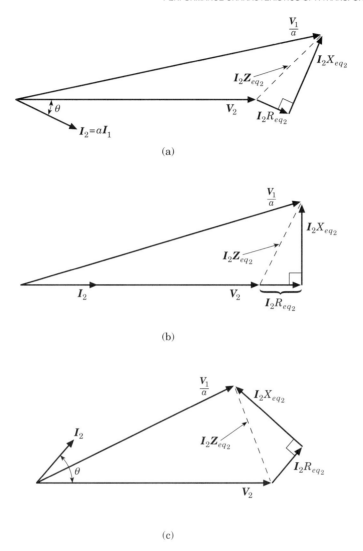

Figure 4.22   Phasor diagram of a transformer operating at: (a) lagging power factor; (b) unity power factor; (c) leading power factor. All quantities are referred to the secondary side of the transformer.

It possible to use an approximate value for the primary voltage by taking into account only the horizontal components in the phasor diagram, as shown in Figure 4.23. Therefore, when all quantities are referred to the secondary side of the transformer, the approximate value of the primary voltage is

$$\frac{V_1}{a} = V_2 + I_2 R_{eq_2} \cos\theta + I_2 X_{eq_2} \sin\theta \qquad (4.84)$$

Transformers used in power system applications are usually designed with taps on one winding in order to change its turns ratio over a small range. Such *tap changing* is frequently achieved automatically in large power transformers to maintain a reasonably constant secondary-side voltage[31] as the magnitude and power factor of the load connected to the secondary side terminals change. Tap changing is also used to compensate for the deviations in primary-side voltage as a result of feeder impedance. In distribution transformers, however, the tap changing is normally done manually.

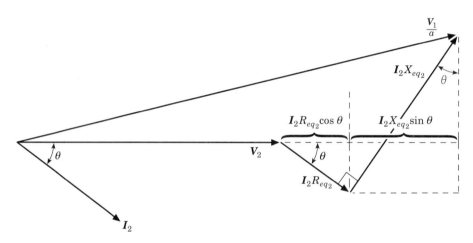

Figure 4.23   Phasor diagram showing the derivation of the approximate equation for V_1/a.

Example 4.7

A 75-kVA 2,400/240-V 60-Hz distribution transformer has equivalent resistance and reactance of 0.009318 Ω and 0.058462 Ω respectively, which are both referred to its secondary side. Use the exact equation for V_1 and determine the full-load voltage regulation:
(a) At 0.85 lagging power factor.
(b) At unity power factor.
(c) At 0.85 leading power factor.

SOLUTION:
(a) At 0.85 lagging power factor,

---

[31] Certain types of loads such as incandescent lamps and motors require rated voltage and frequency for their optimum operation. Otherwise, at voltages above the rated voltage, the lives of the incandescent lamps are shortened. Similarly, when providing a rated load at subnormal voltage, motors draw overcurrents, which in turn cause the motors to overheat. Therefore, such loads must be served by transformers that have a *small* voltage regulation. However, arc welding transformers require a *large* voltage regulation so that they can operate at almost constant current.

$$I_2 = \frac{S}{V_2}$$

$$= \frac{75,000 \text{ VA}}{240 \text{ V}} = 312.5 \text{ A}$$

and

$$\theta = \cos^{-1} PF$$

$$= \cos^{-1} 0.8 = 31.79°$$

Therefore,

$$\boldsymbol{I}_2 = 312.5\angle - 31.79° \text{ A}$$

Using equation (4.83),

$$\frac{\boldsymbol{V}_1}{a} = \boldsymbol{V}_2 + \boldsymbol{I}_2 R_{eq_2} + j\boldsymbol{I}_2 X_{eq_2}$$

$$= 240\angle 0° + (312.5\angle - 31.79°)(0.009318)$$

$$+ j(312.5\angle - 31.79°)(0.058462)$$

$$= 252.4873\angle 318° \text{ V}$$

Thus, the voltage regulation can be found by using equation (4.79) as

$$\% \, V \, Reg = \frac{\dfrac{V_1}{a} - V_{2, \, FL}}{V_{2, \, FL}} \times 100$$

$$= \frac{252.4873 - 240}{240} \times 100 = 5.2$$

or $V \, Reg = 5.2\%$

(b) At unity power factor,

$$PF = \cos \theta = 1.0 \text{ thus } \theta = 0°$$

so that

$$I_2 = 312,5\angle 0° \text{ A}$$

Thus,

$$\frac{V_1}{a} = 240 + (312.5\angle 0°)(0.0918) + j(312.5\angle 0°)(0.058462)$$

$$= 243.598\angle 4.3° \text{ V}$$

Therefore, the voltage regulation is

$$\% V \, Reg = \frac{243.598 - 240}{240} \times 100 \cong 1.5$$

or $V \, Reg = 1.5\%$

(c) At 0.85 leading power factor,

$$I_2 = 312.5\angle 31.79° \text{ A}$$

hence

$$\frac{V_1}{a} = 240 + (321.5\angle 31.79°)(0.009318) + j(312.5\angle 31.79°)(0.058462)$$

$$= 233.4755\angle 4.19° \text{ V}$$

Thus, the voltage regulation is

$$\% V \, Reg = \frac{233.4755 - 240}{240} \times 100 = -2.72$$

or $V \, Reg = -2.72 \%$

4.9.2 TRANSFORMER EFFICIENCY

The efficiency of any equipment can be defined as the ratio of output power to input power. Therefore, the efficiency η is

$$\eta = \frac{P_{out}}{P_{in}}$$

(4.85)

$$= \frac{P_{out}}{P_{in} + P_{loss}}$$

$$= \frac{P_{in} - P_{loss}}{P_{in}}$$

$$\text{(4.85)}$$

$$= 1 - \frac{P_{loss}}{P_{in}}$$

The power losses in a given transformer are the *core losses*, which can be considered constant for a given voltage and frequency, and the *copper losses*,[32] caused by the resistance of the windings. The *core losses* are the sum of hysteresis and eddy-current losses. Therefore, the input power can be expressed as

$$P_{in} = P_{out} + P_{core} + P_{cu} \qquad \text{(4.86)}$$

where

$$P_{out} = V_2 I_2 \cos \theta = S_{out} \cos \theta \qquad \text{(4.87)}$$

Here, the $\cos \theta$ is the load power factor. Therefore, the percent efficiency of the transformer is

$$\eta = \frac{V_2 I_2 \cos \theta}{V_2 I_2 \cos \theta + P_{core} + P_{cu}} \times 100 \qquad \text{(4.88)}$$

where

$$P_{cu} = I_1^2 R_1 + I_2^2 R_2$$

$$= I_1^2 R_{eq_1} \qquad \text{(4.89)}$$

$$= I_2^2 R_{eq_2} \cong P_{sc}$$

and

$$P_{core} = P_{oc} \qquad \text{(4.90)}$$

The current, voltage, and equivalent circuit parameters must be referred to the same side of the transformer. The *maximum efficiency* is achieved when the core loss is equal to the copper loss, that is

$$P_{core} = P_{cu} \qquad \text{(4.91)}$$

In general, the efficiency of transformers at a rated load is very high and increases with their ratings. For example, transformers as small as 1 kVA may

---

[32] The term *copper loss* is still used for the losses caused by the resistances of the windings regardless of whether they are copper or aluminum.

have an efficiency of 90 percent. Power transformer efficiencies vary from 95 percent to 99 percent. In a well-designed transformer, both core losses and copper losses are extremely small, so that efficiency is very high.[33] For example, efficiency for very large transformers is about 99 percent.

In contrast to power transformers, distribution transformers operate well below the rated power output most of the time. Therefore, their efficiency performance is approximately evaluated based on *all-day* (or *energy*) *efficiency*, which is defined as

$$\eta_{AD} = \frac{Energy\ output\ over\ 24\ hours}{Energy\ input\ over\ 24\ hours} \times 100 \tag{4.92}$$

or

$$\eta_{AD} = \frac{Energy\ output\ over\ 24\ hours}{Energy\ output\ over\ 24\ hours + losses\ over\ 24\ hours} \times 100 \tag{4.93}$$

Hence, if the load cycle of the transformer is known, the all-day efficiency can easily be found. Here, the load cycle is segmented into periods where the load is approximately constant, and the energy and losses for each period are calculated.

Example 4.8

Consider a 100-kVA 7,200/240-V 60-Hz transformer. Assume that the open-circuit and short-circuit tests were performed on the transformer and that the following data was obtained:

	Open-Circuit Test (on Primary)	Short-Circuit Test (on Primary)
Voltmeter	V_{oc} = 7,200 V	V_{sc} = 250 V
Ammeter	I_{oc} = 0.65 A	I_{sc} = 13.889 A
Wattmeter	P_{oc} = 425 W	P_{sc} = 1420 W

---

[33] The maximum allowable temperature that the transformer may be permitted to reach is imposed by the temperature rating of the insulation used for the coils. Because of this, the losses in the transformer must not be permitted to remain at excessively high temperatures for too long. The copper losses dictate a maximum allowable continuous current value and the iron losses set a maximum voltage value. (Because of the saturation of the iron core, operating a transformer above the rated voltage causes the no-load current to increase drastically. This also dictates a maximum allowable operating voltage.) The two limitations are independent of each other and the load power factor. Because of this, transformers are rated in kVA rather than kW.

Also assume that the transformer operates at full load with a 0.90 lagging power factor. If the given power factor belongs to the load, not to the transformer, determine the following:

(a) The equivalent impedance, resistance, and reactance of the transformer all referred to the primary side.
(b) Total losses, including the copper and core losses, at full load.
(c) The efficiency of the transformer.
(d) Percent voltage regulation of the transformer.
(e) The phasor diagram of the transformer.

SOLUTION:

(a) Referred to the primary side,

$$Z_{eq_1} = \frac{V_{sc}}{I_{sc}}$$

$$= \frac{250 \text{ V}}{13.889 \text{ A}} = 18 \,\Omega$$

and

$$R_{eq_1} = \frac{P_{sc}}{I_{sc}^2}$$

$$= \frac{1,420 \text{ W}}{13.889^2 \text{A}} = 7.36 \,\Omega$$

Therefore,

$$X_{eq_1} = (Z_{sc}^2 - R_{sc}^2)^{1/2}$$

$$= (18^2 - 7.36^2)^{1/2} = 16.43 \,\Omega$$

(b) The full load current is

$$I_1 = \frac{S}{V_1}$$

$$= \frac{100,000 \text{ VA}}{7,200 \text{ V}} = 13.89 \text{ A}$$

thus,

$$P_{cu} = I_1^2 R_{eq_1}$$

$$= (13.89 \text{ A})^2 (7.36 \,\Omega) = 1,419.98 \text{ W}$$

and

$$P_{core} = P_{oc} = 425 \text{ W}$$

Hence, the total loss at full load is

$$P_{loss} = P_{cu} + P_{core}$$

$$= 1,419.98 + 425 = 1,844.98 \text{ W}$$

(c) The output power is

$$P_{out} = S_{out} \cos \theta$$

$$= 90,000 + 1,844.98 = 91,844.98 \text{ W}$$

Hence, the efficiency of the transformer is

$$\eta = 1 - \frac{P_{loss}}{P_{in}}$$

$$= 1 - \frac{1,844.98}{91,844.98} = 0.9799 \text{ or } 97.99$$

(d) By using equation (4.81),

$$V_1 = aV_2 + I_1(R_{eq_1} + jX_{eq_1})$$

$$= 7,200\angle 0° + (13.89\angle -25.84°)(7.36 + j16.43)$$

$$= 7,393.19\angle 1.25° \text{ V}$$

Therefore, the percent voltage regulation is

$$\% \ V \ Reg = \frac{V_1 - aV_2}{aV_2} \times 100$$

$$= \frac{7,393.19 - 7,200}{7,200} \times 100 = 2.68$$

or $V \ Reg = 2.68 \ \%$

(e) The phasor diagram of the transformer is as shown in Figure 4.22(a). Note that $\theta = 25.84°$, $I_1 = 13.89\angle -25.84°$ A, $aV_2 = 7,200\angle 0°$ V, and $V_1 = 7,393.19\angle 1.25°$ V.

4.10 THREE-PHASE TRANSFORMERS

Today, for reasons of efficiency and economy, most electrical energy is generated, transmitted, and distributed using a three-phase system rather than a single phase system. Three-phase power may be transformed either by the use of a single three-phase transformer or three single-phase transformers, which are properly connected with each other for a three-phase operation. A three-phase transformer, in comparison to a bank of three single-phase transformers, weighs less, costs less, needs less floor space, and has a slightly higher efficiency. In the event of failure, however, the entire three-phase transformer must be replaced. On the other hand if three separate single-phase units (i.e., a *three-phase transformer bank*) are used, only one of them needs to be replaced.[34] Also, a stand-by three-phase transformer is more expensive than a single-phase spare transformer. Figure 4.24 shows the two versions of three-phase core construction that are normally used: core-type and shell-type.

In the *core-type* design, both the primary and secondary windings of each phase are placed only on one leg of each transformer, as shown in Figure 4.24(a). For balanced, three-phase sinusoidal voltages, the sum of the three-core fluxes at any given time must be zero. This is a requirement that does not have to be met in the *shell-type* construction. In the *core-type construction*, the magnetic reluctance of the flux path of the center phase is less than that of the outer two phases. Therefore, there is some imbalance in the magnetic circuits of the three phases of the transformer. This in turn results in unequal magnetizing currents that affect their harmonic composition. In essence, such a design prevents the existence of the third-harmonic flux and thus pretty much avoids third-harmonic voltages. For example, third-harmonic flux components (which are in time phase in this design) of three-phase core type-transformers are reduced to at least one-tenth of what shell-type or single-phase transformer cores have. In the case of wye-wye connected windings with isolated neutrals, no third-harmonic excitation current components are present.

A shell-type transformer is quite different in character from a core-type transformer. As shown in Figure 4.24(b), the flux in the outside paths of the core is reduced by 42 percent since in a *shell-type construction* the center phase windings are wound in the opposite direction of the other two phases. Since all yoke cross-sections are equal, not only is the amount of core requirement reduced, but also the manufacturing process involved is simplified. Furthermore, in a shell-type transformer, the no-load losses are less than those in a core-type transformer.

Figure 4.25 shows a 40-MVA, 110-kV ± 16%/21-kV three-phase core-type transformer. Notice that its primary-side voltage can be adjusted by ±16 percent. Figure 4.26 shows a 850/950/1,100-MVA, 415-kV ± 11%/27-kV, three-

---

[34] However, it is not possible to use transformers to convert a single-phase system to a three-phase system for a large amount of power. Relatively very small amounts of power can be developed from a single-phase system using R-C phase shift networks (or an induction phase converter) to produce two-phase power which in turn can be transformed into three-phase.

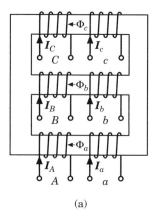

(a)

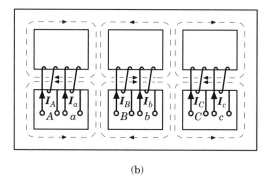

(b)

Figure 4.24 Three-phase, two-winding transformer core construction: (a) core type; (b) shell type.

phase shell-type transformer. Figure 4.27 shows a 10-MVA and a 40-MVA core-type three-phase transformer with *GEAFOL solid dielectric core*. Finally, Figure 4.28 shows a typical core and coil assembly of a three-phase core-type power transformer. Notice that its core and coil assembly are rigidly supported and clamped by heavy, fabricated clamping structures. The windings are concentrically placed on the core legs and laterally braced by inserting kiln-dried, maple wood dowels between the windings and the core. The complete assembly is rigidly braced to withstand the mechanical forces experienced under fault conditions and to resist vibration and shock forces encountered during shipment and installation. All high-voltage leads are brought to tap changers, terminal blocks, or bushings.

4.11 THREE-PHASE TRANSFORMER CONNECTIONS

As previously stated, in a three-phase power system, it is often necessary to step up or step down the voltage levels at various locations in the system. Such transformations can be achieved by means of transformer banks that

Figure 4.25   A 40-MVA, 110-kV ± 16%/21-kV, three-phase, core-type transformer, 5.2 m
high, 9.4 m long, 3 m wide, weighing 80 tons. (Courtesy of Siemens AG.)

have three identical single-phase transformer units, one for each phase, or by
the use of three-phase transformer units. In either case, each phase has one
primary and secondary winding associated with it. These primary and secondary windings may be connected independently in either **delta** (Δ) or **wye** (Y)
configurations.[35] There are four possible connections for a three-phase transformer, namely, **wye-wye** (Y-Y), **delta-delta** (Δ-Δ), **wye-delta** (Y-Δ), and
delta-wye (Δ-Y), as shown in Figure 4.29.

The primary benefit of using a wye-connected winding in a transformer is
that it provides a neutral point so that phase voltages are also available. In a
wye-wye connection, there is no phase displacement between the primary
and secondary line-to-line voltages, even though it is possible to shift the secondary voltages 180° by reversing all three secondary windings. The use of a
wye-wye connection creates no problem as long as it has *solidly grounded neutrals* (especially the neutral for the primary side[36]). Here, the addition of a
primary neutral connection makes each transformer independent of the other

---

[35] The delta and wye configurations are also known as the *mesh* and *star* configurations, respectively.

[36] In the event that the neutral point of the primary windings is connected to the neutral point of
the power source, there will be no difference between the behavior of a wye-wye connected core-
type or shell-type three-phase transformer and a three-phase transformer bank.

Figure 4.26   A 850/950/1100-MVA, 415-kV ± 11%/27-kV, three phase, shell-type transformer, 11.3 m high, 14 m long, 5.7 m wide, weighing (without cooling oil) 552 tons. (Courtesy of Siemens AG.)

two. Also, dissimilar transformers will not cause voltage unbalance under no-load conditions. Due to the neutrals, the additive third-harmonic components cause a current flow in the neutral rather than building up large voltages. If there is a delta connected *tertiary* winding, in addition to the primary and secondary windings, the third-harmonic voltages are suppressed by trapping third-harmonic (circulating) currents within the delta tertiary winding. However, *if such a neutral is not provided*, the phase voltages become drastically unbalanced when the load is unbalanced. This causes neutral instability that makes unbalanced loading impractical, even though the line-to-line voltages remain normal. There are also problems with third harmonics. In summary, any attempt to operate a wye-wye connection of transformers without the presence of a primary neutral connection will lead to difficulty and potential failure.

---

[37] In general, all single-phase transformers when excited at rated voltage produce a third harmonic. This is due to the fact that their cores saturate fast and, because of this, their magnetization currents become distorted. Therefore, when a perfectly sinusoidal voltage (for example, at 60 Hz) is applied to the primary of a transformer, it produces a magnetization current that has the fundamental component. Luckily, in single-phase transformers, the magnetization current is small in comparison to the load current. Therefore, the resulting distortion in the current waveform is negligible; whereas in three-phase transformers, the three fundamental magnetization currents are displaced by 120°. The third harmonic currents, however, are *in phase with respect*

Figure 4.27   10-MVA and 50-kVA, core-type three-phase transformers with GEAFOL solid dielectric cores. (Courtesy of Siemens AG.)

In fact, trouble occurs even under no-load conditions. Therefore, such a wye-wye connection is seldom used in practice.[37]

In the **delta-delta connection**, under balanced conditions, the line currents are $\sqrt{3}$ times the currents in the windings when the third harmonics in the excitation current are ignored. There is no phase shift and no problem with unbalanced loads or harmonics. If a center tap is available on one transformer secondary, the bank may be used to supply a *three-phase four-wire delta system*.[38] Also, if one transformer fails in service, the remaining two transformers in the bank can be operated as an **open-delta** (or V-V) connection at about 58 percent of the original capacity of the bank. However, in a complete delta-delta bank, transformers tend to share the load inversely to their internal impedances, and therefore identical transformers have to be used.

---

to each other (as are the 6th, 9th, 12th, etc. harmonics). Therefore, in an ungrounded wye-wye connection, such a tripled third-harmonic component induces a secondary voltage waveform in each winding that has a large third harmonic voltage. Therefore, the output voltage waveforms are distorted. To prevent this, a neutral line to ground at either primary or secondary (or both) must be provided. However, if the connection is wye-delta, delta-delta, or delta-wye, the third harmonics circulate within the delta and thus the harmonic voltage is suppressed and no secondary voltage distortion is taking place.

[38] For further information, see Gönen [33].

Figure 4.28   A typical core and coil assembly of a three-phase, core-type, power transformer. (Courtesy of North American Transformer.)

In the **wye-delta connection**, there is no problem with third-harmonic components in its voltages, since they are absorbed in a circulating current on the delta side. This connection can be used with unbalanced loads. In high-voltage transmission systems, the high-voltage side is connected in delta and the low-voltage-side is connected in wye. Due to the delta connection, the secondary voltage is shifted 30° with respect to the primary voltage. *In the United States, it is standard practice to make the secondary voltage* (i.e., *the lower voltage) lag the primary voltage* (i.e., *the higher voltage) by* 30°. This connection is basically used to step down a high voltage to a lower voltage.

In the **delta-wye connection**, there is also no problem with third-harmonic components in its voltages. It has the same advantages and the same phase shift as the wye-delta connection. *The secondary voltage lags the primary voltage by* 30°, as is the case for the wye-delta connection. This connection is basically used to step up a low voltage to a high voltage. In general, when a wye-delta or delta-wye connection is used, the wye is preferably on the high-voltage side, and the neutral is grounded. Thus, the transformer insulation can be manufactured to withstand $1/\sqrt{3}$ times the line voltage, instead of the total line voltage.

Example 4.9

Consider a three-phase 15-MVA 138/13.8-kV distribution substation transformer that is being used as a step down transformer. Determine the ratings

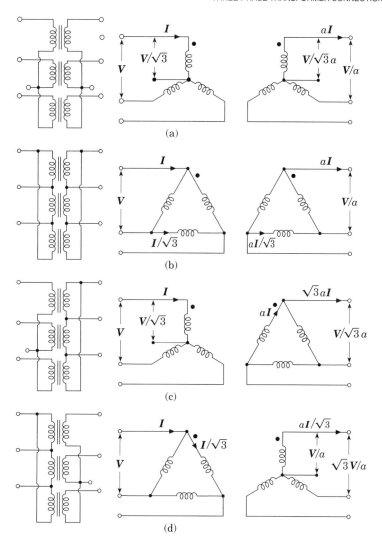

Figure 4.29 Basic three-phase transformer connections: (a) wye-wye; (b) delta-delta; (c) wye-delta; (d) delta-wye.

and turn ratios of the transformer, if it is connected in:

(a) Wye-delta.
(b) Delta-wye.
(c) Delta-delta.
(d) Wye-wye.

SOLUTION:

The rated primary line current is

$$I_{L1} = \frac{S_{3\phi}}{\sqrt{3}V_{L1}}$$

$$= \frac{15 \times 10^6 \text{ VA}}{\sqrt{3}(138,000 \text{ V})} = 62.7555 \text{ A}$$

The rated secondary line current is

$$I_{L2} = \frac{S_{3\phi}}{\sqrt{3}V_{L2}}$$

$$= \frac{15 \times 10^6 \text{ VA}}{\sqrt{3}(13,800 \text{ V})} = 627.555 \text{ A}$$

(a) If the transformer is connected in wye-delta:

$$Rated\ total\ kVA = S_{3\phi} = \frac{15 \times 10^6 \text{ VA}}{1,000} = 15,000 \text{ kVA}$$

$$Rated\ kVA\ per\ phase = \frac{S_{3\phi}}{3}$$

$$= \frac{15,000 \text{ kVA}}{3} = 5,000 \text{ kVA}$$

$$Rated\ I_1 = I_{L1} = 62.7775 \text{ A}$$

$$Rated\ I_2 = \frac{I_{L2}}{\sqrt{3}} = \frac{627.555 \text{ A}}{\sqrt{3}} = 362.3188 \text{ A}$$

$$Rated\ V_{L1} = 138 \text{ kV}$$

$$Rated\ V_{L2} = 13.8 \text{ kV}$$

$$Rated\ V_1 = \frac{V_{L1}}{\sqrt{3}} = \frac{138,000 \text{ V}}{\sqrt{3}} = 79,674.3 \text{ V}$$

$$Rated\ V_2 = V_{L2} = 13,800 \text{ V}$$

$$Turns\ ratio = a = \frac{V_1}{V_2} = \frac{79,674.3 \text{ V}}{13,800 \text{ V}} = 5.7735$$

(b) If the transformer is connected in delta-wye:

$$Rated\ total\ kVA = \frac{S_{3\phi}}{1,000} = 15,000\ \text{kVA}$$

$$Rated\ kVA\ per\ phase = 5,000\ \text{kVA}$$

$$Rated\ I_1 = \frac{I_{L1}}{\sqrt{3}} = \frac{62.7555\ \text{A}}{\sqrt{3}} = 36.2319\ \text{A}$$

$$Rated\ I_2 = I_{L2} = 627.555\ \text{A}$$

$$Rated\ V_{L1} = 138\ \text{kV and } Rated\ V_{L2} = 13.8\ \text{kV}$$

$$Rated\ V_1 = V_{L1} = 138\ \text{kV}$$

$$Rated\ V_2 = \frac{V_{L2}}{\sqrt{3}} = \frac{13,800\ \text{V}}{\sqrt{3}} = 7,967.4337\ \text{V}$$

$$Turns\ ratio = a = \frac{V_1}{V_2} = \frac{138,000\ \text{V}}{7,967.4337\ \text{V}} = 17.3205$$

(c) If the transformer is connected in delta-delta:

$$Rated\ total\ kVA = \frac{S_{3\phi}}{1,000} = 15,000\ \text{kVA}$$

$$Rated\ kVA\ per\ phase = 5,000\ \text{kVA}$$

$$Rated\ I_1 = \frac{I_{L1}}{\sqrt{3}} = \frac{62.7555\ \text{A}}{\sqrt{3}} = 36.2319\ \text{A}$$

$$Rated\ I_2 = \frac{I_{L2}}{\sqrt{3}} = \frac{627.555\ \text{A}}{\sqrt{3}} = 362.319\ \text{A}$$

$$Rated\ V_{L1} = 138\ \text{kV and } Rated\ V_{L2} = 13.8\ \text{kV}$$

$$Rated\ V_1 = V_{L1} = 138\ \text{kV}$$

$$Rated\ V_2 = V_{L2} = 13.8\ \text{kV}$$

$$Turns\ ratio = a = \frac{V_1}{V_2} = \frac{138\ \text{kV}}{13.8\ \text{kV}} = 10$$

(d) If the transformer is connected in wye-wye:

$$Rated\ total\ kVA = \frac{S_{3\phi}}{1,000} = 15,000\ kVA$$

$$Rated\ kVA\ per\ phase = 5,000\ kVA$$

$$Rated\ I_1 = I_{L1} = 62.7555\ A$$

$$Rated\ I_2 = I_{L2} = 627.555\ A$$

$$Rated\ V_{L1} = 138\ kV\ and\ Rated\ V_{L2} = 13.8\ kV$$

$$Rated\ V_1 = \frac{138,000\ V}{\sqrt{3}} = 79,674.3\ V$$

$$Rated\ V_2 = \frac{V_{L2}}{\sqrt{3}} = \frac{13,800\ V}{\sqrt{3}} = 7,967.4337\ V$$

$$Turns\ ratio = a = \frac{V_1}{V_2} = \frac{138\ kV}{13.8\ kV} = 10$$

or

$$a = \frac{V_{L1}}{V_{L2}} = \frac{138\ kV}{13.8\ kV} = 10$$

4.12 AUTOTRANSFORMERS

The two windings in the usual two-winding transformer are not connected to each other; in other words, they are electrically isolated from each other. Therefore, power is transferred inductively from one side to the other. However, an autotransformer has a single winding, part of which is common to both the primary and the secondary simultaneously. Thus, *in an autotransformer, there is no electrical isolation between the input side and the output side.* As a result, the power is transferred from the primary to the secondary through both *induction* and *conduction*. As shown in Figure 4.30, an autotransformer can be used as a step-down or step-up transformer. Consider the step-down connection shown in Figure 4.30(a). The *common winding* is the winding between the low-voltage terminals while the remainder of the winding belonging exclusively to the high-voltage circuit is called the *series winding*. This combined with the common winding forms the *series-common winding* between the high-voltage terminals. In a sense, an autotransformer is just a normal two-winding transformer connected in a special way. The only structural difference is that the *series winding* must have extra insulation in order to be just as strong as the one on the common winding. In a *variable autotransformer*, the tap is movable. Autotransformers are increasingly used to interconnect two high-voltage transmission lines operating at different voltages, as shown in Figure 4.2. They can be used as *step-down* or *step-up* transformers.

Consider the equivalent circuit of an ideal transformer (neglecting losses) shown in Figure 4.30(a). The output voltage V_2 is related to the input voltage as it is in a two-winding transformer. Therefore,

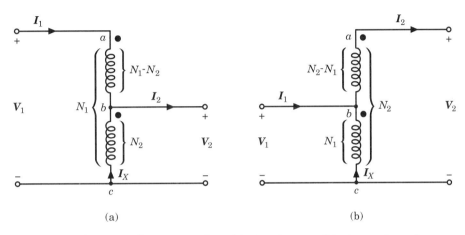

(a)   (b)

Figure 4.30   Autotransformers used as: (a) step-down or (b) step-up transformers.

$$\frac{V_1}{V_2} = \frac{N_1}{N_2} = a \qquad (4.94)$$

where $a > 1$ for a *step-down transformer*, since $N_1 > N_2$. Also, since an ideal transformer is assumed,

$$V_1 I_1 = V_2 I_2 \qquad (4.95)$$

or

$$\frac{V_1}{V_2} = \frac{I_2}{I_1} = a \qquad (4.96)$$

Since the excitation current is neglected, then I_1 and I_2 are in phase, and the current in the common section of the winding is

$$I_x = I_2 - I_1 \qquad (4.97)$$

Also, the mmfs of the two windings are equal. Thus, according to the mmf balance equation

$$N_2 I_x = (N_1 - N_2) I_1 \qquad (4.98)$$

or

$$I_x = \frac{N_1 - N_2}{N_2} = (a-1)I_1 = I_2 - I_1 \qquad (4.99)$$

Hence

$$I_1 = \frac{I_2}{a} \qquad (4.100a)$$

$$= \frac{N_2}{N_1}I_2 \qquad (4.100b)$$

or

$$\frac{I_2}{I_1} = a \qquad (4.101)$$

Since

$$a = \frac{N_1}{N_2} = \frac{N_c + N_s}{N_c} \qquad (4.102)$$

then

$$\frac{I_2}{I_1} = \frac{N_c + N_s}{N_c} \qquad (4.103)$$

where N_c = number of turns in common winding
$\quad = N_2$
$\quad N_s$ = number of turns in series winding
$\quad = N_1 - N_2$

Similarly, it can be shown that

$$\frac{V_2}{V_1} = \frac{N_c}{N_c + N_s} \qquad (4.104)$$

The apparent power delivered to the load is S_{out} and can be expressed as

$$S_{out} = V_2 I_2 \qquad (4.105)$$

From equation (4.97),

$$I_2 = I_1 + I_x = I_1 + (I_2 - I_1) \qquad (4.106)$$

Substituting equation (4.106) into equation (4.105),

$$S_{out} = V_2 I_1 + V_2 (I_2 - I_1) \qquad (4.107a)$$

$$= S_c + S_i \qquad (4.107b)$$

where S_c = conductively transferred power to the load through N_2 winding

$$= V_2 I_1 \qquad (4.108)$$

S_i = inductively transferred power to the load through $N_1 - N_2$ winding

$$= V_2 (I_2 - I_1) \qquad (4.109)$$

The S_c and S_i are related to S_{out} by

$$\frac{S_i}{S_{out}} = \frac{I_2 - I_1}{I_2} = \frac{a-1}{a} \qquad (4.110a)$$

$$= \frac{N_s}{N_c + N_s} \qquad (4.110b)$$

$$= \frac{N_1 - N_2}{N_1} \qquad (4.110c)$$

and

$$\frac{S_c}{S_{out}} = \frac{I_1}{I_2} = \frac{1}{a} \qquad (4.111a)$$

$$= \frac{N_2}{N_1} \qquad (4.111b)$$

where $a > 1$ for a *step-down transformer*.

Similarly, for a *step-up transformer*, as shown in Figure 4.30(b), the current in the common section of the winding is

$$I_x = I_2 - I_1 \qquad (4.97)$$

so that

$$I_1 = I_2 - I_x = I_2 - (I_2 - I_1)$$

$$= I_2 + (I_1 - I_2) \qquad (4.112)$$

Substituting equation (4.112) into equation (4.95),

$$S_{out} = V_2 I_2 = V_1 I_1 = S_{in} \qquad (4.113a)$$

$$= V_1 I_2 + V_1 (I_1 - I_2) \qquad (4.113b)$$

$$= S_c + S_i \qquad (4.113c)$$

where S_c = conductively transferred power to the load through N_2 winding

$$= V_1 I_2 \qquad (4.114)$$

S_i = inductively transferred power to the load through $N_2 - N_1$ winding

$$= V_1 (I_1 - I_2) \qquad (4.115)$$

The S_c and S_i are related to S_{out} by

$$\frac{S_i}{S_{out}} = \frac{I_1 - I_2}{I_1} = 1 - a \qquad (4.116)$$

and

$$\frac{S_c}{S_{out}} = \frac{I_2}{I_1} = a \qquad (4.117)$$

where

$$a = \frac{N_c}{N_c + N_s} \qquad (4.118)$$

and a < 1 for a *step-up transformer*.

The advantages of autotransformers include lower leakage reactances, lower losses, and smaller excitation current requirements. Most of all, an autotransformer is cheaper than the equivalent two-winding transformer (especially when the voltage ratio does not vary too greatly from 1 to 1). The disadvantages of autotransformers are that there is no electrical isolation between the primary and secondary and that there is a greater short-circuit current than for the two-winding transformer.

Three-phase autotransformer banks generally have wye-connected main windings with the neutral normally connected solidly to ground. In addition, it is common practice to include a third winding connected in delta, called the *tertiary winding*.

Example 4.10

Assume that a single-phase 100-kVA 2,400/240-V two-winding transformer is connected as an autotransformer to step down the voltage from 2,640 V to 2,400 V. The transformer connection is as shown in Figure 4.30(a), with 240-V and 2,400-V windings for sections *ab* and *bc*, respectively. Compare the kVA rating of the autotransformer with that of the original two-winding transformer, and determine all three currents as well as S_{out}, S_i, and S_c.

SOLUTION:

The rated current in the 240-V winding (or in the ab section) is

$$I_1 = \frac{100,000 \text{ VA}}{240 \text{ V}} = 416.6667 \text{ A}$$

Similarly, the rated current in the 2,400-V winding (or in the bc section) is

$$I_x = I_2 - I_1 = \frac{100,000 \text{ VA}}{2,400 \text{ V}} = 41.6667 \text{ A}$$

Therefore, the load current is

$$I_2 = I_1 + I_x = 416.6667 + 41.6667 = 458.3334 \text{ A}$$

Alternatively, by first calculating the turns ratio as

$$a = \frac{2,640 \text{ V}}{2,400 \text{ V}} = 1.10$$

then

$$I_2 = aI_1 = \frac{2,640 \text{ V}}{2,400 \text{ V}}(416.6667 \text{ A}) = 458.3334 \text{ A}$$

as before. The kVA (or output) rating of the autotransformer is

$$S_{auto} = V_1 I_1 = V_2 I_2$$

$$= \frac{2,640 \times 458.3334}{1,000} = 1,210 \text{ kVA}$$

Notice that the two-winding transformer rating was 100 kVA. Therefore, the ratio of the autotransformer capacity to the two-winding transformer capacity is

$$\frac{S_{auto}}{S_{two\,wdg}} = \frac{1,210 \text{ kVA}}{100 \text{ kVA}} = 12.1$$

In other words, the kVA capacity of the transformer increased 12.1 times when it was connected as an autotransformer. Here S_{auto} and S_{out} are the same rating.

Also, the *inductively supplied power* to the load is

$$S_i = V_2(I_2 - I_1)$$

$$= 2,640(458.3334 - 416.66667) = 110 \, \text{kVA}$$

or

$$S_i = \frac{a-1}{a} \times S_{auto}$$

$$= \frac{1.10 - 1}{1.10} \times 1,210 = 110 \, \text{kVA}$$

The *conductively supplied power* to the load is

$$S_c = \frac{S_{auto}}{a}$$

$$= \frac{1,210 \, \text{kVA}}{1.10} = 1,100 \, \text{kVA}$$

4.13 THREE-WINDING TRANSFORMERS

Figure 4.31(a) shows a single-phase three-winding transformer. Three-winding transformers are usually used in bulk power (transmission) substations to lower the transmission voltage to the subtransmission voltage level. They are also frequently used at distribution substations. If excitation impedance is neglected, the equivalent circuit of a three-winding transformer can be represented by a wye of impedances, as shown in Figure 4.31(b), where the primary, secondary, and tertiary windings are denoted by 1, 2, and 3, respectively. Note that the common point 0 is fictitious and is not related to the neutral of the system. While the primaries and secondaries are usually connected in wye-wye, the tertiary windings of a three-phase and three-winding transformer bank are connected in delta. The tertiaries are used for: (1) providing a path for the third harmonics and their multiples in the excitation and the zero-sequence currents (the zero-sequence currents are trapped and circulate in the delta connection); (2) in-plant power distribution; and (3) the application of power factor correcting capacitors or reactors. If the three-winding transformer can be considered an ideal transformer, then

$$\frac{V_2}{V_1} = \frac{N_2}{N_1} \tag{4.119}$$

$$\frac{V_3}{V_1} = \frac{N_3}{N_1} \tag{4.120}$$

$$N_1 I_1 = N_2 I_2 + N_3 I_3 \tag{4.121}$$

where V_1, V_2, and V_3 are the primary, secondary, and tertiary terminal voltages, respectively, and N_1, N_2, and N_3 are the turns in the respective windings. Also, I_1, I_2, and I_3 are the currents in the three windings.

The impedance of any of the branches shown in Figure 4.31(b) can be determined by considering the short-circuit impedance between pairs of windings with the third winding open. Therefore,

$$Z_{12} = Z_1 + Z_2 \tag{4.122}$$

$$Z_{13} = Z_1 + Z_3 \tag{4.123)}$$

$$Z_{23} = Z_2 + Z_3 \tag{4.124}$$

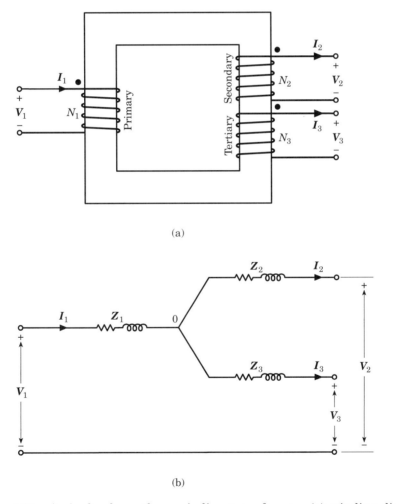

(a)

(b)

Figure 4.31   A single-phase three-winding transformer: (a) winding diagram; (b) equivalent circuit.

If the leakage impedances Z_1, Z_2, and Z_3 are referred to the primary, they are then expressed as

$$Z_1 = \frac{1}{2}(Z_{12} + Z_{13} - Z_{23}) \tag{4.125}$$

$$Z_2 = \frac{1}{2}(Z_{23} + Z_{12} - Z_{13}) \tag{4.126}$$

$$Z_3 = \frac{1}{2}(Z_{13} + Z_{23} - Z_{12}) \tag{4.127}$$

where Z_{12} = leakage impedance measured in primary with secondary short-circuited and tertiary open.

Z_{13} = leakage impedance measured in primary with tertiary short-circuited and secondary open.

Z_{23} = leakage impedance measured in secondary with tertiary short-circuited and primary open.

Z_1 = leakage impedance of primary winding.

Z_2 = leakage impedance of secondary winding.

Z_3 = leakage impedance of tertiary winding.

In most large transformers the value of Z_2 is very small and can be negative. In contrast to the situation with a two-winding transformer, the kVA ratings of the three windings of a three-winding transformer bank are not usually equal. Therefore, all impedances as defined above should be expressed based on the same kVA base. Figure 4.32 shows a completely assembled 910-MVA 20.5/500-kV three-phase step-up transformer.

4.14 INSTRUMENT TRANSFORMERS

In general, instrument transformers are of two types: *current transformers* (CTs) and *voltage transformers*[39] (VTs). They are used in ac power circuits to provide safety for the operator and equipment from high voltage; they permit proper insulation levels and current-carrying capacity in relays, meters, and other instruments. In the United States, the standard instruments and relays are rated at 5 A and/or 120 V, 60 Hz. Regardless of the type of instrument transformer in use, the external load applied to its secondary is referred to as its *burden*. The burden usually describes the impedance connected to the transformer's secondary winding, but may specify the voltamperes supplied to the load.[40] For example, a transformer supplying 5 A to a resistive burden of 0.5 Ω may also be said to have a burden of 12.5 VA at 5 A.

---

[39] Voltage transformers were formerly called *potential transformers* (PTs).
[40] For further information, see Gönen [35].

Figure 4.32   A completely assembled 910-MVA, 20.5/500-kV three-phase step-up transformer, 39 ft high, 36 ft long, 20 ft wide, weighing 562 tons. (Courtesy of ABB Power T&D Company.)

Current transformers are connected in series with the line, as shown in Figure 4.33(a). They are used to step down the current at a rated value of 5 A for ammeters, wattmeters, and relays. As shown in the figure, frequently the primary is not an integral part of the transformer itself, but is part of the line in which current is being measured.

It is very important to note that current transformers can be very dangerous. During the operation of a current transformer, **its secondary terminals must never be open-circuited**! Unlike other types of transformers, the number of primary ampere-turns is constant for any given primary current. When the secondary is open-circuited, the primary mmf is not balanced by a corresponding secondary mmf. (In other words, there will be no secondary mmf to oppose the primary mmf.) Therefore, all of the primary current becomes excitation current. Consequently, a very high flux density is produced in the core, causing a very high voltage to be induced in the secondary. In addition to endangering the user, it may damage the transformer insulation and also cause overheating due to excessive core losses. Furthermore, if such high magnetizing forces are suddenly removed from the core, they may leave behind substantial amounts of residual magnetism, causing the turns ratio to be different from the one that existed before. As shown in Figure 4.33(a), if the ammeter needs to be removed, the proper procedure is to **close the shorting switch first**.

The **voltage transformer** primary is connected across the potential difference to be measured, as shown in Figure 4.33(b). The secondary is connected to the voltmeter, wattmeter, or relay potential winding. The voltage transformers are specially designed to be very accurate step-down transformers. The rated output of a voltage transformer seldom exceeds a few hundred voltamperes. As shown in Figure 4.33(b), for safety reasons the secondary side of a VT is always grounded and well insulated from the high-voltage side.

4.15 CURRENT INRUSH

Occasionally upon energizing a power transformer, a transient phenomenon (due to magnetizing current characteristics) takes place even if there is no load connected to its secondary. As a result, its magnetizing current peak may be several times (about eight to ten times) the rated transformer current, or it may be practically unnoticeable. Because of losses in the excited winding and magnetic circuit, this current ultimately decreases to the normal value of the excitation current (i.e., to about 5 percent or less of the rated transformer current). Such a transient event is known as the **current inrush** phenomenon. It may cause (1) a momentary dip in the voltage if the impedance of the excitation source is significant; (2) undue stress in the transformer windings; or (3) improper operation of protective devices (e.g., tripping overload or common differential relays[41]). The magnitude of such a current inrush depends on the magnitude, polarity, and rate of change in applied voltage at the time of switching. For example, assume that the applied voltage, at $t = 0$, happens to be

$$v_1(t) = \sqrt{2}V_1 \sin \omega t = \frac{d\lambda_1}{dt} = N_1 \frac{d\phi}{dt} \tag{4.128}$$

The resultant flux is

$$\Phi = \frac{\sqrt{2}V_1}{N_1} \int_0^t \sin \omega t dt + \Phi(0) \tag{4.129}$$

where $\Phi(0) = \Phi_r$ (i.e., the residual flux). Therefore,

$$\Phi = \frac{\sqrt{2}V_1}{\omega N_1}(1 - \cos \omega t) + \Phi_r \tag{4.130}$$

or

$$\Phi = -\Phi_m \cos \omega t + \Phi_m + \Phi_r \tag{4.131}$$

---

[41] For further information, see Blume [8].

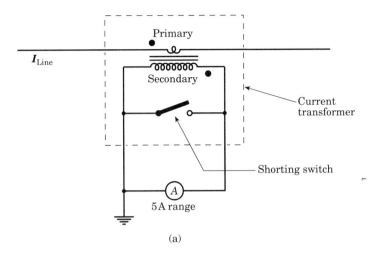

(a)

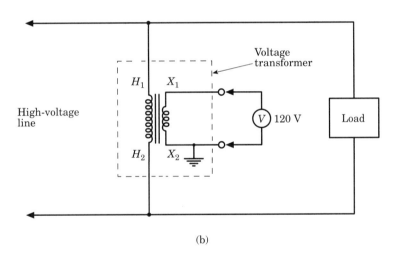

(b)

Figure 4.33   Instrument transformer connections: (a) current transformer connection; (b) voltage transformer connection.

Assuming that the *dc* component flux $\Phi_m + \Phi_r$ are constant, at $\omega t = \pi$, the instantaneous flux is

$$\Phi = 2\Phi_m + \Phi_r \tag{4.132}$$

That is, the maximum value of the flux may be more than twice the maximum of the normal flux, since there is often residual magnetism in the core when it is initially energized. Obviously, such doubling of the maximum flux in the core causes a tremendously large magnetization current. As shown in Figure 4.34, as time progresses (i.e., in about a few cycles), there will be a fast decay in the inrush current. Luckily, the probability of the occurrence of this theoretical maximum inrush current is relatively small.

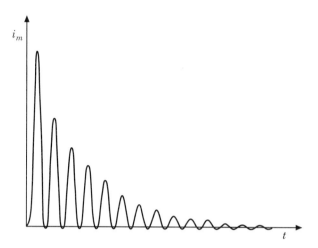

Figure 4.34   Current inrush phenomenon in a power transformer.

PROBLEMS

4.1   Assume that an ideal transformer is used to step down 13.8 kV to 2.4 kV and that it is fully loaded when it delivers 100 kVA. Determine the following:

(a)   Its turns ratio.
(b)   The rated currents for each winding.
(c)   The load impedance, referred to the high-voltage side, corresponding to full load.
(d)   The load impedance referred to the low-voltage side, corresponding to full load.

4.2   A single-phase, 2,500/250-V, two-winding ideal transformer has a load of $10\angle40°\ \Omega$ connected to its secondary. If the primary of the transformer is connected to a 2,400-V line, determine the following:

(a)   The secondary current.
(b)   The primary current.
(c)   The input impedance as seen from the line.
(d)   The output power of the transformer in kVA and in kW.
(e)   The input power of the transformer in kVA and in kW.

4.3   Assume that a single-phase two-winding transformer is rated 50 kVA, 345/138 kV, and has 10 percent impedance. When the transformer has a current of 0.8 per unit in its high-voltage winding, determine the following:

(a) The corresponding primary and secondary currents in per unit and in amperes.

(b) The internal impedance of the transformer, in ohms, referred to the high- and the low-voltage windings, respectively.

(c) If the low-voltage terminals of the transformer are short-circuited and a 0.3 pu voltage is applied to the high-voltage winding, the resulting high-voltage side and low-voltage side currents in amperes and in per units.

(d) If a current of 1.0 pu flows in the high-voltage winding, the resultant (internal) IZ voltage drop in the transformer in volts and in per units.

4.4 A 60-Hz 50-kVA 2,400/240-V single-phase ideal transformer has 50 turns on its secondary winding. Determine the following:
(a) The values of its primary and secondary currents.
(b) The number of turns on its primary windings.
(c) The maximum flux Φ_m in the core.

4.5
4.6 A 60-Hz 75-kVA 2,400/240-V single-phase transformer is used to step down the voltage of a distribution system. If the low voltage is to be kept constant at 240V, determine the following:
(a) The value of the load impedance connected to the low-voltage side that will cause the transformer to be fully loaded.
(b) The value of such a load impedance referred to the high-voltage side.
(c) The values of the load current referred to the low-voltage and high-voltage sides.

4.7 An audio-frequency transformer is employed to couple a 100-Ω resistive load to an electronic source that can be represented by a constant voltage of 6 V in series with an internal resistance of 4,000 Ω. Assume that the transformer is an ideal transformer and determine:
(a) The turns ratio needed to provide maximum power transfer by matching the load and source impedances.
(b) The values of the current, voltage, and power at the load under such conditions.

4.8 Repeat Example 4.3 but assume that the impedance of the power line is $0.5 + j1.6$ Ω and that the impedance of the load is $4 + j7$ Ω.

4.9 Repeat Example 4.3, but assume that the impedances of the power line and the load are $4 + j9$ Ω and $3 + j10$ Ω, respectively. Also assume that the generator bus voltage V_G is $220\angle0°$ V.

4.10 Use the data given in Problem 4.8 and repeat Example 4.4. The impedances of the transformers T_1 and T_2 are $Z_{T_1} = j0.10$ Ω and $Z_{T_2} = j0.10$ Ω, respectively. Both Z_{T_1} and Z_{T_2} are referred to the high voltage side of each transformer, respectively.

4.11   Use the data given in Problem 4.9 and repeat Example 4.4. The imped-ances of the transformers T_1 and T_2 are $\mathbf{Z}_{T_1} = j0.10\ \Omega$ and $\mathbf{Z}_{T_2} = j0.10\ \Omega$, respectively. Both $\mathbf{Z}_{T_1}$ and $\mathbf{Z}_{T_2}$ are referred to the high-voltage side of each transformer, respectively.

4.12   Repeat Example 4.3 but assume that the generator has an internal impedance as shown in Figure P4.12 and that $\mathbf{Z}_G = jX_d = j9.5\ \Omega$.

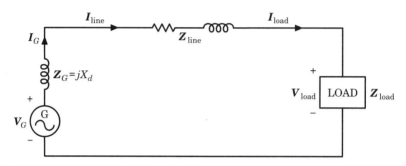

Figure P4.12   Figure for Problem 4.12.

4.13   Repeat Example 4.4 but assume that the generator has an internal impedance of $\mathbf{Z}_G = jX_d = j9.5\ \Omega$.

4.14   Repeat Example 4.3 but assume that the generator has an internal impedance of $\mathbf{Z}_G = jX_d = j6.4\ \Omega$. Also assume that the impedance of the power line is $0.5 + j1.6\ \Omega$ and that the impedance of the load is $4 + j7\ \Omega$.

4.15   Repeat Example 4.4 but assume that the generator has an internal impedance of $\mathbf{Z}_G = jX_d = j6.4\ \Omega$. The impedance of the power line is $0.5 + j1.6\ \Omega$ and that the impedance of the load is $4 + j7\ \Omega$. The impedances of the transformers T_1 and T_2 are $\mathbf{Z}_{T_1} = j0.10\ \Omega$ and $\mathbf{Z}_{T_2} = j0.10\ \Omega$, respectively. Both $\mathbf{Z}_{T_1}$ and $\mathbf{Z}_{T_2}$ are referred to the high-voltage side of each trans-former, respectively.

4.16   Consider a 50-kVA 2,400/240-V 60-Hz distribution transformer. Assume that the open-circuit and short-circuit tests were performed on the pri-mary side of the transformer and that the following data were obtained:

	Open-Circuit Test (on Primary)	Short-Circuit Test (on Primary)
Voltmeter	$V_{oc} = 2400$ V	$V_{sc} = 52$ V
Ammeter	$I_{oc} = 0.2083$ A	$I_{sc} = 20.8333$ A
Wattmeter	$P_{oc} = 185$ W	$P_{sc} = 615$ W

Determine the impedances of the approximate equivalent circuit referred to the primary side.

4.17 Consider a 100-kVA 7,200/240-V 60-Hz distribution transformer. Assume that the open-circuit and short-circuit tests were performed on the primary side of the transformer and that the following data were obtained:

	Open-Circuit Test (on Primary)	Short-Circuit Test (on Primary)
Voltmeter	$V_{oc} = 7200$ V	$V_{sc} = 250$ V
Ammeter	$I_{oc} = 0.45$ A	$I_{sc} = 13.8889$ A
Wattmeter	$P_{oc} = 355$ W	$P_{sc} = 1275$ W

Determine the impedances of the approximate equivalent circuit referred to the primary side.

4.18 Consider a 75-kVA 7,500/480-V 60-Hz distribution transformer. Resolve Example 4.6 using the following data:

	Open-Circuit Test (on Primary)	Short-Circuit Test (on Primary)
Voltmeter	$V_{oc} = 7500$ V	$V_{sc} = 499$ V
Ammeter	$I_{oc} = 0.35$ A	$I_{sc} = 10$ A
Wattmeter	$P_{oc} = 473$ W	$P_{sc} = 1050$ W

4.19 Consider a 25-kVA 2,400/240-V 60-Hz distribution transformer. Resolve Example 4.6 using the following data:

	Open-Circuit Test (on Primary)	Short-Circuit Test (on Primary)
Voltmeter	$V_{oc} = 2400$ V	$V_{sc} = 214$ V
Ammeter	$I_{oc} = 0.125$ A	$I_{sc} = 10.417$ A
Wattmeter	$P_{oc} = 72$ W	$P_{sc} = 422$ W

4.20 Consider a 37.5-kVA 7,200/240-V 60-Hz distribution transformer. Resolve Example 4.6 using the following data:

	Open-Circuit Test (on Primary)	Short-Circuit Test (on Primary)
Voltmeter	V_{oc} = 7200 V	V_{sc} = 255 V
Ammeter	I_{oc} = 0.55 A	I_{sc} = 5.2083 A
Wattmeter	P_{oc} = 555 W	P_{sc} = 263 W

4.21   A 50-kVA 2,400/240-V 60-Hz transformer is to be tested to find out its excitation branch components and its series impedances. The following test data has been taken from the primary side of the transformer:

	Open-Circuit Test (on Primary)	Short-Circuit Test (on Primary)
Voltmeter	V_{oc} = 2400 V	V_{sc} = 45 V
Ammeter	I_{oc} = 0.65 A	I_{sc} = 20.8333 A
Wattmeter	P_{oc} = 65 W	P_{sc} = 300 W

(a)   Find the values of R_c and X_m of the shunt (excitation) branch.
(b)   Find the equivalent impedance of the transformer, referred to the primary side.
(c)   Find the equivalent resistance and reactance of the transformer, referred to the primary side.
(d)   Draw the equivalent circuit of the transformer, referred to the high-voltage side.

4.22   A 25-kVA 2,400/240-V 60-Hz transformer is to be tested to find out its excitation branch components and its series impedances. The following test data has been taken from the primary side of the transformer:

	Open-Circuit Test (on Primary)	Short-Circuit Test (on Primary)
Voltmeter	V_{oc} = 2400 V	V_{sc} = 70 V
Ammeter	I_{oc} = 0.4 A	I_{sc} = 10.41 A
Wattmeter	P_{oc} = 65 W	P_{sc} = 250 W

(a)   Find the values of R_c and X_m of the shunt (excitation) branch.
(b)   Find the equivalent impedance of the transformer, referred to the primary side.
(c)   Find the equivalent resistance and reactance of the transformer,

referred to the primary side.

(d) Draw the equivalent circuit of the transformer, referred to the high-voltage side.

4.23 A 75-kVA 2,400/240-V 60-Hz transformer is to be tested to determine its excitation branch components and its series impedances. The following test data has been taken from the primary side of the transformer:

	Open-Circuit Test (on Primary)	Short-Circuit Test (on Primary)
Voltmeter	$V_{oc} = 2400$ V	$V_{sc} = 185$ V
Ammeter	$I_{oc} = 0.7$ A	$I_{sc} = 31.25$ A
Wattmeter	$P_{oc} = 285$ W	$P_{sc} = 910$ W

(a) Find the equivalent circuit parameters of this transformer, referred to the high-voltage side.

(b) Find the equivalent circuit parameters of this transformer, referred to the low-voltage side.

4.24 A 50-kVA 2,400/240-V 60-Hz transformer is to be tested to determine its excitation branch components and its series impedances. The following test data has been taken from the primary side of the transformer:

	Open-Circuit Test (on Primary)	Short-Circuit Test (on Primary)
Voltmeter	$V_{oc} = 2400$ V	$V_{sc} = 195$ V
Ammeter	$I_{oc} = 0.9$ A	$I_{sc} = 20.8333$ A
Wattmeter	$P_{oc} = 395$ W	$P_{sc} = 950$ W

(a) Find the equivalent circuit parameters of this transformer, referred to the high-voltage side.

(b) Find the equivalent circuit parameters of this transformer, referred to the low-voltage side.

4.25 Consider the power system given in Example 4.4 and determine the input impedance (i.e., the Thévenin's equivalent impedance) of the system looking into the system from bus 2. (Hint: Refer all impedances to the bus 2 side and then find the equivalent impedance.)

4.26 Consider the power system given in Problem 4.10 and determine the input impedance (i.e., the Thévenin's equivalent impedance) of the system looking into the system from bus 2. (Hint: Refer all impedances to the bus 2 side and then find the equivalent impedance.)

4.27   Resolve Example 4.5 but assume that the transformer delivers the rated load current I_L at 0.9 leading PF.

4.28   Resolve Example 4.5 by using the following data: A 100-kVA 2,400/240-V 60-Hz distribution transformer with $Z_1 = 0.595 + j1.150\ \Omega$, $Z_2 = 0.0059 + j011\ \Omega$, and $Y_{e_1} = 5.55 \times 10^{-4} \angle -82°$ S. Assume that the transformer delivers rated I_L at 0.85 lagging PF.

4.29   Resolve Example 4.7 but use the approximate equation, i.e., Equation (4.84), for V_1.

4.30   A 50-kVA 2,400/240-V 60-Hz distribution transformer has equivalent resistance and equivalent reactance, both referred to its secondary side, of 0.021888 Ω and 0.09101 Ω, respectively. Use the exact equation for V_1 and determine the full-load voltage regulation:
   (a) At a 0.8 lagging power factor.
   (b) At unity power factor.
   (c) At a 0.8 leading power factor.

4.31   Solve Problem 4.30 using the approximate equation for V_1.

4.32   Consider a 75-kVA 2,400/240-V 60-Hz transformer. Assume that the open-circuit and short-circuit tests were performed on the transformer and that the following data was obtained:

	Open-Circuit Test (on Primary)	Short-Circuit Test (on Primary)
Voltmeter	$V_{oc} = 2400$ V	$V_{sc} = 200$ V
Ammeter	$I_{oc} = 0.50$ A	$I_{sc} = 31.25$ A
Wattmeter	$P_{oc} = 75$ W	$P_{sc} = 800$ W

Also assume that the transformer operated at full load with a 0.92 lagging power factor. Note that the given power factor belongs to the load, not to the transformer. Determine the following:
   (a) The equivalent impedance, resistance, and reactance of the transformer, all referred to its primary side.
   (b) Total loss, including the copper and core losses, at full load.
   (c) The efficiency of the transformer.
   (d) The percent voltage regulation of the transformer.

4.33   Assume that a single-phase, two-winding transformer has a core loss of 1,500 W. The secondary (i.e., output) voltage is 480 V and the output

power is 40 kW at a 0.9 lagging power factor. The current in the primary winding is 9.2593 A. If its secondary and primary winding resistances are 0.03 Ω and 3 Ω, respectively, determine:
(a) The secondary current.
(b) The complex impedance of the load.
(c) The copper losses of the primary and secondary windings.
(d) The input power.
(e) The transformer efficiency.

4.34   Consider the data given in Problem 4.24 and determine the following:
(a) The efficiency of the transformer at full load, operating at a 0.8 lagging power factor.
(b) The efficiency of the transformer at full load, operating at the same power factor.
(c) The amount of load at which the transformer operates at its maximum efficiency.

4.35   A 2,400/2,000-V autotransformer supplies a load of 100-kW at a power factor of 0.8. Find the current in each winding section and the kVA rating of the autotransformer.

4.36   A single-phase transformer has a core loss of 600 W. Its copper loss is 700 W, when the full-load secondary current is 20 A.
(a) Determine the amount of the secondary current at which the transformer has its maximum efficiency.
(b) If the transformer output is given in hp (Good grief!) as 20 hp, find its efficiency at full load.

4.37   Consider the data given in Problem 4.23 and the results of Example 4.7, and suppose that the transformer has 0.009318 Ω equivalent resistance referred to its secondary side. Determine the efficiency of the transformer at full load with these power factors:
(a) PF = 0.85 lagging.
(b) PF = unity.
(c) PF = 0.85 leading.

4.38   Consider the data given in Problem 4.24 and determine the efficiency of the transformer at full load with these power factors:
(a) PF = 0.90 lagging.
(b) PF = unity.
(c) PF = 0.90 leading.

4.39   Consider a 100-kVA 7,200/240-V 60-Hz transformer. Assume that the open-circuit and short-circuit tests were performed on the transformer and that the following data was obtained:

	Open-Circuit Test (on Primary)	Short-Circuit Test (on Primary)
Voltmeter	V_{oc} = 7200 V	V_{sc} = 250 V
Ammeter	I_{oc} = 0.45 A	I_{sc} = 13.889 A
Wattmeter	P_{oc} = 355 W	P_{sc} = 1275 W

Also assume that the transformer operated at full load with a 0.85 lagging power factor. Note that the given power factor belongs to the load, not to the transformer. Determine the following:
(a) The equivalent impedance, resistance, and reactance of the transformer, all referred to its primary side.
(b) Total loss, including the copper and core losses, at full load.
(c) The efficiency of the transformer.
(d) The percent voltage regulation of the transformer.
(e) The phasor diagram of the transformer.

4.40 Consider a 75-kVA 7,200/240-V 60-Hz transformer. Assume that the open-circuit and short-circuit tests were performed on the transformer and that the following data was obtained:

	Open-Circuit Test (on Primary)	Short-Circuit Test (on Primary)
Voltmeter	V_{oc} = 7200 V	V_{sc} = 250 V
Ammeter	I_{oc} = 0.55 A	I_{sc} = 10.4167 A
Wattmeter	P_{oc} = 399 W	P_{sc} = 775 W

Also assume that the transformer operated at full load with a 0.90 lagging power factor. Note that the given power factor belongs to the load, not to the transformer. Determine the following:
(a) The equivalent impedance, resistance, and reactance of the transformer, all referred to its primary side.
(b) Total loss, including the copper and core losses, at full load.
(c) The efficiency of the transformer.
(d) Percent voltage regulation of the transformer.

4.41 Consider a 25-kVA 2,400/240-V 60-Hz transformer with Z_1 = 2.533 + $j2.995$ Ω and Z_2 = (2.5333 + $j2.995$) × 10^{-2} Ω, referred to the primary and secondary sides, respectively.
(a) Find V_1, aV_2, I_1, I_2, R_{eq_1}, X_{eq_1}.
(b) The transformer is connected at the receiving end of a feeder that has an impedance of 0.3 + $j1.8$ Ω. Let the sending end voltage mag-

nitude of the feeder be 2,400 V. Also there is a load connected to the secondary side of the transformer that draws rated current from the transformer at a 0.85 lagging power factor. Neglect the excitation current of the transformer. Determine the secondary-side voltage of the transformer under such conditions.

(c)   Draw the associated phasor diagram.

4.42   Consider a 75-kVA 2,400/240-V 60-Hz transformer with $Z_1 = 0.52 + j3.85$ Ω and $Z_2 = (0.52 + j3.85) \times 10^{-2}\,\Omega$, referred to the primary and secondary sides, respectively.

(a)   Find $V_1, aV_2, I_1, I_2, R_{eq_1}, X_{eq_1}$.

(b)   The transformer is connected at the receiving end of a feeder that has an impedance of $0.5 + j2.3\,\Omega$. Let the sending end voltage magnitude of the feeder be 2,400 V. Also there is a load connected to the secondary side of the transformer that draws rated current from the transformer at a 0.92 lagging power factor. Neglect the excitation current of the transformer. Determine the secondary-side voltage of the transformer under such conditions.

(c)   Draw the associated phasor diagram.

4.43   Repeat Example 4.9 assuming that the three-phase transformer is rated 45 MVA, 345/34.5 kV.

4.44   A three-phase 150-kVA 12,470/208-V distribution transformer bank supplies 150 kVA to a balanced three-phase load connected to its secondary-side terminals. Find the kVA rating and turns ratio of each single-phase transformer unit as well as their primary- and secondary-side voltages and currents, if the transformer bank is connected in:

(a)   Wye-wye.

(b)   Delta-delta.

4.45   A three-phase 300-kVA 480/4,170-V distribution transformer bank supplies 300 kVA to a balanced three-phase load connected to its secondary-side terminals. Find the kVA rating and turns ratio of each single-phase transformer unit as well as their primary- and secondary-side voltages and currents, if the transformer bank is connected in:

(a)   Wye-wye.

(b)   Delta-delta.

4.46   Repeat Example 4.10 for a 2,400/2,640-V step-up connection, as shown in Figure 4.26(b).

CHAPTER 5

ELECTRO-MECHANICAL ENERGY CONVERSION PRINCIPLES

Use all of our resources with maximum efficiency and effectiveness.
Lawrence J. Peter

Truth is the most valuable thing we have. Let us economize it.
Mark Twain, *Henry Ward Beecher's Farm*, 1885

If a man said 'all men are liars', would you believe him?
Author Unknown

5.1 INTRODUCTION

According to the **energy conversion principle**, *energy is neither created nor destroyed: it is simply changed in form*. The role of electromagnetic (or electromechanical) machines is to transmit energy or convert it from one type of energy to another. For example, the transformer transmits electrical energy, changing only the potential difference and current at which it exists. However, it also converts a small amount of electrical energy to heat. This is an unwanted result that needs to be minimized at the design stage. However, a rotational or translational electromagnetic machine converts energy from mechanical to electrical form, or vice versa; that is, it operates as a **generator** or **motor**. In the process,[1] it also converts some electrical or mechanical energy to heat, which is

---

[1] Such a process occurs through the medium of the electric or magnetic field of the conversion device. In general, electromechanical devices can be classified as follows: (1) **transducers**, which

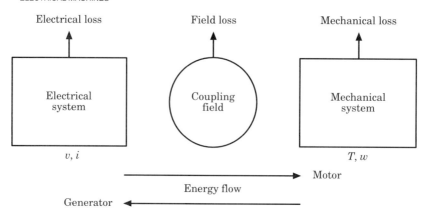

Figure 5.1 A representation of electromechanical energy conversion.

also unwanted. In general, electric generators and motors of all kinds can be defined as **electromechanical energy converters**. Their main components are an electrical system, a mechanical system, and a coupling field, as shown in Figure 5.1.

5.2 FUNDAMENTAL CONCEPTS

In this section, some of the basic concepts involving electrical rotating machines are reviewed. Such concepts include angular velocity, angular acceleration, mechanical work, power, and torque. As explained in Chapter 2, most electrical machines rotate around an axis known as the shaft of the machine. As can be seen in Figures 5.1 and 5.2, the input to a generator and the output of the motor are mechanical in nature. If the shaft rotates in a *counterclockwise* (CCW) direction rather than in a *clockwise* (CW) direction, the resultant rotational angle and direction are, by definition, considered positive; otherwise, they are considered negative. The *angular position* θ of the shaft is the angle at which it is positioned measured from some arbitrarily selected reference point. Note that the angular position concept conforms to the linear distance concept along a line, and that it is measured in radians or degrees. However, the *angular velocity* (or *speed*) represents the rate of change in the angular position with respect to time. Thus, angular velocity can be expressed as

$$\omega = \frac{d\theta}{dt} \quad \text{rad/s} \tag{5.1}$$

---

are the devices used for measurement and control, such as torque motors, loudspeakers, and microphones; (2) **force producing devices**, such as relays, electromagnets, and solenoid actuators; and (3) **continuous energy-conversion apparatus**, such as generators and motors.

as long as the angular position θ is measured in radians. Usually, the rotational speed n is given in revolutions per minute (i.e., rpm); so that

$$n = \left(\frac{60}{2\pi}\right)\omega \quad \text{rev/min} \tag{5.2}$$

The rate of change in angular velocity with respect to time is defined as the angular acceleration and is expressed as

$$\alpha = \frac{d\omega}{dt} \quad \text{rad/s}^2 \tag{5.3}$$

In rotational mechanics, the **torque** (or twisting action on the cylinder), T, is defined as the tangential force times the radial distance at which it is applied, measured from the axis of rotation. In other words, the torque is a function of the magnitude of the applied force F, and the distance between the axis of rotation and the line of action of the force. Hence, as illustrated in Figure 5.3, the rotational torque T can be expressed as

$$T = (applied\ force)(perpendicular\ distance)$$

$$= (F)(r\sin\theta) \tag{5.4}$$

$$= Fr\sin\theta$$

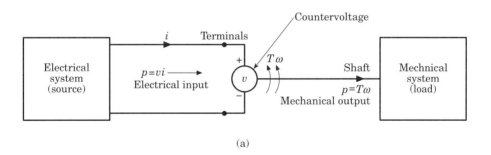

(a)

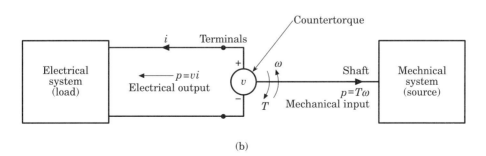

(b)

Figure 5.2   Illustration of some of the fundamental concepts associated with the operation of an electrical rotating machine: (a) motor action; (b) generator action.

In the SI system[2] of units, force is given in newtons (N) and distance is given in meters (m).

For linear motion, according to Isaac Newton, mechanical work W is defined as the *integral of force over distance*. Therefore,

$$W = f \int dx \quad \text{J} \tag{5.5}$$

The units of work are joules (J) on the SI system and foot-pounds in the English system. However, for rotational motion, work is defined as the *integral of torque through an angle*. Thus,

$$W = T \int d\theta \quad \text{J} \tag{5.6}$$

If the torque is unchanging,

$$W = T\theta \quad \text{J} \tag{5.7}$$

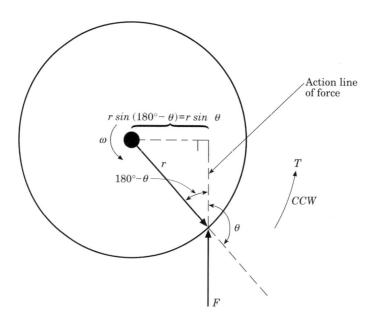

Figure 5.3   Illustration of the relationship between force and torque.

Power can be defined as the rate of doing work. Hence,

$$P = \frac{dW}{dt} = F\frac{dx}{dt} \quad \text{W} \tag{5.8}$$

[2] It is interesting to note that since 1954 the metric system known as the **International System of Units** (or **Système International**) (**SI**) has been in use all over the world with the exception of the United States. Today, the SI system is used even in England and Canada.

However, for rotational motion having constant torque, power[3] can be expressed as

$$P = \frac{dW}{dt} = T\frac{d\theta}{dt} \quad \text{W} \tag{5.9}$$

or

$$P = T\omega \quad \text{W} \tag{5.10}$$

Also, from equation (5.10) torque can be found as

$$T = \frac{P}{\omega} \quad \text{N} \cdot \text{m} \tag{5.11}$$

Since in the United States the English system of units is still in use, knowing the following conversion formulas may be useful:

$$T = \frac{7.04\,P(\text{watts})}{n} \quad \text{lb} \cdot \text{ft} \tag{5.12}$$

or

$$T = \frac{5252\,P(\text{horsepower})}{n} \quad \text{lb} \cdot \text{ft} \tag{5.13}$$

where

$$n = \left(\frac{60}{2\pi}\right)\omega \quad \text{rev/min} \tag{5.2}$$

$$P(horsepower) = \frac{P(watts)}{746} \tag{5.14}$$

$$T(\text{lb} \cdot \text{ft}) = 0.738T \ (\text{N} \cdot \text{m}) \tag{5.15}$$

Some of the fundamental concepts associated with the operation of an electrical rotating machine are illustrated in Figure 5.2. The magnetic field in such a machine establishes the necessary link between the electrical and mechanical systems, producing mechanical torque as well as inducing voltages in the coils. The magnetic field itself is developed by the current flowing through these coils.

Consider the **motor action** of the machine as illustrated in Figure 5.2(a). The instantaneous power input to the motor is

---

[3] In the SI system of units, the work is in joules (since 1 J = 1 W/s) if the force is in newtons and the distance, x, is in meters. Thus, one watt equals one joule per second (i.e., 1 W = 1 J/s).

$$p = v \times i \tag{5.16}$$

where v and i are the terminal voltage and current, respectively, as shown in Figure 5.2(a). The magnetic field develops the output torque and induces a countervoltage (also called **counter emf** since it opposes the current flow) which makes it possible for the machine to receive power from the electrical source and convert it into mechanical output. Here, the torque and the angular velocity are in the same direction.

In the generator action as shown in Figure 5.2(b), the magnetic field induces the generated voltage and develops a countertorque (since it opposes the torque of the mechanical source) which makes it possible for the rotating machine to receive power from the mechanical source in order to convert it into electrical output. Note that the countertorque and the angular velocity are in the opposite direction.

Some large electromechanical machines that are used in different applications are shown in Figures 5.4 through 5.11. The dc motors, similar to the one shown in Figure 5.6, are typically used for continuous high-impact load duty on blooming and slabbing mills. The ac generator shown in Figure 5.10 has a wound stator. Figure 5.11 shows the artist's rendition of one of the three ac generators that are in use at the Grand Coulee Dam. It is a water-cooled, low-speed hydroelectric generator with a combined thrust and guide bearing below the rotor. The stator's outside diameter is 23.9 meters. The generator's complete weight is 2500 metric tons; its rotor weight is 1400 metric tons. The stator windings are water cooled. This is one of the world's most powerful hydroelectric generators.

Figure 5.4 The stator of a 9-MW, 250 rpm, 1,650-V mill motor. (Courtesy of Siemens.)

Figure 5.5   The rotor of a 5-MW, 60/120 rpm, 1,750-V converter mill motor. (Courtesy of Siemens.)

Figure 5.6   A dc 8-MW, 50/100 rpm dc reversing mill motor operating at 899 V. (Courtesy of Siemens).

Figure 5.7   A large synchronous motor used in the mining industry. (Courtesy of General Electric Canada, Inc.)

Figure 5.8   A vertical synchronous generator used by a large industrial firm. (Courtesy of General Electric Canada, Inc.)

Figure 5.9   Two large dc motors used in the metals industry. (Courtesy of General Electric Canada, Inc.)

Figure 5.10   Installation of a 500-MVA, 200 rpm, ac generator rotor used at Churchill Falls hydroelectric plant. (Courtesy of General Electric Canada, Inc.)

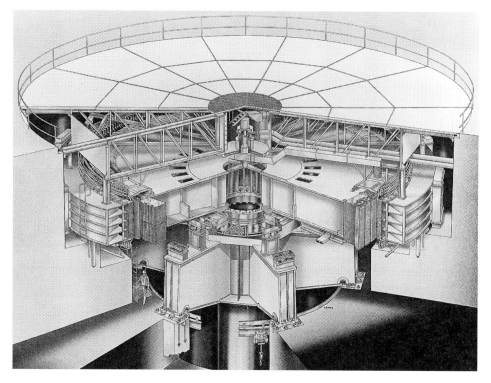

Figure 5.11 Cutaway view of an ac 718-MVA, 86 rpm, Grand Coulee generator. (Courtesy of General Electric Canada, Inc.)

Example 5.1

The rotational speed of a motor (i.e., its shaft speed) is 1800 rev/min. Determine its angular velocity (i.e., its shaft speed) in rad/s.

SOLUTION:
From equation (5.2),

$$n = \left(\frac{60}{2\pi}\right)\omega \quad \text{rev/min}$$

or

$$\omega = \left(\frac{2\pi}{60}\right)n$$

$$= \left(\frac{2\pi}{60}\right)(1800 \text{ rev/min}) = 188.5 \text{ rad/s}$$

Example 5.2

Consider Figure 5.3 and assume that the rotational torque is 200 N · m and the radius of the rotor is 0.25 m. Determine the applied force in N:
(a) If the angle θ is 15°.
(b) If the angle θ is 45°.
(c) If the angle θ is 90°.
(d) If the angle θ is 120°.

SOLUTION:
(a) From equation (5.4),

$$T = \mathscr{F} r \sin \theta$$

from which

$$\mathscr{F} = \frac{T}{r \sin \theta}$$

Thus, at $\theta = 15°$,

$$\mathscr{F} = \frac{200 \, \text{N} \cdot \text{m}}{(0.25 \, \text{m}) \sin 15°} = 3{,}091. \, \text{N}$$

(b) At $\theta = 15°$,

$$\mathscr{F} = \frac{200 \, \text{N} \cdot \text{m}}{(0.25 \, \text{m}) \sin 45°} = 1{,}131.4 \, \text{N}$$

(c) At $\theta = 90°$,

$$\mathscr{F} = \frac{200 \, \text{N} \cdot \text{m}}{(0.25 \, \text{m}) \sin 90°} = 800 \, \text{N}$$

(d) At $\theta = 10°$,

$$\mathscr{F} = \frac{200 \, \text{N} \cdot \text{m}}{(0.25 \, \text{m}) \sin 120°} = 923.8 \, \text{N}$$

Example 5.3

A coil, having a sectional area of 0.3 m^2 with $N = 20$ turns, is rotating around its horizontal axis with a constant speed of 3600 rpm in a uniform and vertical magnetic field of flux density $B = 0.8$ T. If the total magnetic flux passing through the coil is given by $\mathscr{F} = A \, B \cos \omega t$ in Wb, where A is the sectional area of the coil, determine the maximum and effective values of the induced voltage in the coil.

SOLUTION:

Since the total magnetic flux passing through the coil is given as

$$\Phi = A\,B\,\cos\,\omega t$$

According to Faraday's Law, the induced voltage is

$$v = N\frac{d\Phi}{dt}$$

$$= N\frac{d(AB\cos\,\omega t)}{dt}$$

$$= -N\omega AB\sin\,\omega t$$

Thus, its maximum voltage is

$$v_{\text{max}} = N\omega AB$$

where

$$\omega = \left(\frac{2\pi}{60}\right)(3600\text{ rpm})$$

$$= 377\text{ rad/s}$$

Hence,

$$v_{\text{max}} = (20\text{ turns})(377\text{ rad/s})(0.3\text{ m}^2)(0.8\text{ T})$$

$$= 1,809.6\text{ V}$$

and its effective value is

$$V_{rms} = \frac{V_{max}}{\sqrt{2}}$$

$$= \frac{1,809.6\text{ V}}{\sqrt{2}} = 1,279.6\text{ V}$$

5.3 ELECTROMECHANICAL ENERGY CONVERSION

The energy conservation principle with regard to electromechanical systems can be expressed in various forms. For example, as shown in Figure 5.1, for a *sink* of electrical energy such as an electric motor, it can be expressed as

$$\begin{pmatrix} Electrical \\ energy\ input \\ from\ source \end{pmatrix} = \begin{pmatrix} Mechanical \\ energy\ output \\ to\ load \end{pmatrix} + \begin{pmatrix} Increase \\ in\ stored\ energy \\ in\ coupling\ field \end{pmatrix} + \begin{pmatrix} Energy\ loss \\ converted \\ to\ heat \end{pmatrix} \quad (5.17)$$

The last term of equation (5.17), can be expressed as

$$\begin{pmatrix} Energy\ loss \\ converted \\ to\ heat \end{pmatrix} = \begin{pmatrix} Resistance \\ loss\ of \\ winding \end{pmatrix} + \begin{pmatrix} Friction \\ and \\ windage \\ losses \end{pmatrix} + \begin{pmatrix} Field \\ losses \end{pmatrix} \quad (5.18)$$

The resistance loss is the i^2R loss in the resistance (R) of the winding. The friction and windage losses are associated with motion. Since the coupling field is the magnetic field, the field losses[4] are due to hysteresis and eddy-current losses; that is, the *core losses* are due to the changing magnetic field in the magnetic core. If the energy losses[5] which are given in equation (5.18) are substituted into equation (5.17), the energy balance equation can be expressed as

$$\begin{pmatrix} Electrical \\ energy\ input \\ from\ source \\ minus \\ resistance \\ losses \end{pmatrix} = \begin{pmatrix} Mechanical \\ energy\ output \\ to\ load\ plus \\ friction\ and \\ windage\ losses \end{pmatrix} + \begin{pmatrix} Increase \\ in\ stored\ energy \\ in\ coupling\ field \\ plus\ core\ losses \end{pmatrix} \quad (5.19)$$

Assume a differential time interval dt during which an increment of electrical energy dW_e (without including the i^2R loss) flows to the system. Then the net electrical input W_e can be equated to the increase in energy W_m so that, in incremental form:

$$dW_e = dW_m + dW_f \quad \text{J} \quad (5.20)$$

where dW_e = differential electrical energy input[6]

   dW_m = differential mechanical energy output[7]

   dW_f = differential increase in energy stored in the magnetic field

---

[4] They are also known as the **iron losses**, as previously stated.

[5] Furthermore, there are additional losses that arise from the nonuniform current distribution in the conductors and the core losses generated in the iron due to the distortion of the magnetic flux distribution from the load currents. Such losses are known as the **stray-load losses** and are very hard to determine precisely. Because of this, estimates that are based on tests, experience, and judgment are used. Typically, such stray losses range from 0.5 percent of the output in large machines to 5 percent of the output in medium-sized machines. In dc machines, they are usually estimated to be about one percent of the output.

[6] It is also denoted by dW_i.

[7] It is also called the **differential developed energy** and is denoted by dW_d.

The equation (5.20) is also known as the **incremental** (or **differential**) **energy-balance equation**. It provides a basis for the analysis of the operation of electromechanical machines. Since in time dt,

$$dW_e = v \times i \times dt \tag{5.21}$$

where v is the (reaction) voltage induced in the electric terminals by the changing magnetic stored energy. Therefore,

$$dW_e = v \times i \times dt = dW_m + dW_f \tag{5.22}$$

According to Faraday's Law, the induced voltage v based on the flux linkages λ can be expressed as

$$v = \frac{d\lambda}{dt} \tag{5.23}$$

Then the net differential electrical energy input in time dt can be expressed as

$$dW_e = v \times i \times dt$$

$$= \left(\frac{d\lambda}{dt}\right) i \times dt \tag{5.24}$$

$$= i \times d\lambda$$

The differential mechanical energy output for a **virtual displacement** (i.e., **linear motion**) dx when the force is $\mathscr{F}_f$ can be expressed as

$$dW_m = \mathscr{F}_f \, dx \tag{5.25}$$

Substituting equations (5.24) and (5.25) into equation (5.22),

$$dW_f = i \, d\lambda - \mathscr{F}_f \, dx \tag{5.26}$$

If the differential mechanical energy output is for a rotary motion, the force $\mathscr{F}_f$ is replaced by torque[8] T_f and the linear (differential) displacement dx is replaced by the angular (differential) displacement $d\theta$ so that

$$dW_m = T_f \, d\theta \tag{5.26}$$

and therefore

$$dW_f = i d\lambda - T_f d\theta \tag{5.28}$$

---

[8] It is also known as the **developed torque** and is denoted by T_d.

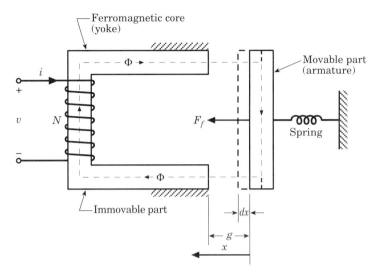

Figure 5.12   A simple electromechanical system.

5.3.1 FIELD ENERGY

Suppose that the electromechanical system shown in Figure 5.12 has a moveable part (i.e., **armature**) that can be kept in static equilibrium by a spring. If the movable part is kept stationary at some air gap and the current is increased from zero to a value i, a flux ϕ will be maintained in the electromagnetic system. Since no mechanical output can be produced,

$$dW_m = 0 \tag{5.29}$$

and substituting equation (5.29) into equation (5.20),

$$dW_e = dW_f \tag{5.30}$$

Thus, if core loss is ignored, all the incremental electric energy input must be stored in the magnetic field. Since, from equation (5.24),

$$dW_e = i\, d\lambda \tag{5.31}$$

then

$$dW_f = i\, d\lambda \tag{5.32}$$

or

$$dW_f = dW_e = v \times i\, dt = i\, d\lambda \tag{5.33}$$

Figure 5.13(a) shows the relationship between coil flux linkage λ and current i for a particular air-gap length. Since core loss is being ignored, the curve will be a single-valued curve passing through the origin. The incremental field energy dW_f is shown as the crosshatched area in Figure 5.13(a). If the applied terminal voltage v is increased, causing a change in current from i_1 to i_2, there will be a matching change in flux linkage from λ_1 to λ_2. Therefore, the corresponding increase in stored energy is

$$dW_f = \int_{\lambda_1}^{\lambda_2} i \, d\lambda \qquad (5.34)$$

as shown in Figure 5.13(a). When the coil current and flux linkage are zero, the field energy is zero. Thus, if the flux linkage is increased from zero to l, the total energy stored in the field is

$$W_f = \int_0^{\lambda} i \, d\lambda \qquad (5.35)$$

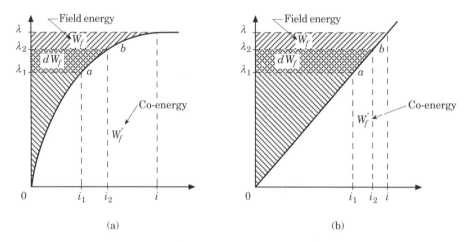

(a)   (b)

Figure 5.13 λ–i characteristic of: (a) a magnetic system; (b) an idealized magnetic system.

Such an integral represents the shaded area W_f between the λ ordinate and the λ–i characteristic, as shown in Figure 5.13(a). Equation (5.35) can be used for any lossless electromagnetic system. Also, if the leakage flux is negligibly small, then all flux Φ in the magnetic circuit links all N turns of the coil. Therefore,

$$\lambda = N\Phi \qquad (5.36)$$

so that from equations (5.33) and (5.36),

$$dW_f = i\,d\lambda = Ni\,d\Phi = \mathscr{F}\,d\Phi \qquad (5.37)$$

where

$$\mathscr{F} = Ni \quad \text{A} \cdot \text{turns} \tag{5.38}$$

Thus, if the characteristic shown in Figure 5.13(a) is rescaled to show the relationship between Φ and $\mathscr{F}$ (so that the ordinate represents the Φ rather than the λ and the axis represents the $\mathscr{F}$ rather than the i), the shaded area again represents the stored energy.

Also, if the reluctance of the air gap makes up a considerably larger portion of the total reluctance of the magnetic circuit, then that of the magnetic material used may be ignored. The resultant λ-i characteristic is represented by a straight-line through the origin. Figure 5.13(b) shows such a characteristic of an idealized magnetic circuit. Hence, for this idealized system,

$$\lambda = Li \tag{5.39}$$

where the inductance of the coil is given by L. By substituting equation (5.23) into equation (5.35), the total energy stored in the field can be expressed as

$$
\begin{aligned}
W_f &= \int_0^\lambda \frac{\lambda}{L} d\lambda \\
&= \frac{\lambda^2}{2L} \\
&= \frac{Li^2}{2} \\
&= \frac{i\lambda}{2}
\end{aligned}
\tag{5.40}
$$

On the other hand, if the reluctance of the magnetic system (i.e., of the air gap) as viewed from the coil is $\mathscr{R}$, then

$$\mathscr{F} = \mathscr{R}\Phi \quad \text{A} \cdot \text{turns} \tag{5.41}$$

and from equation (5.37), the total energy stored in the field can be expressed as

$$W_f = \int_0^\Phi \mathscr{F} d\Phi = \mathscr{R} \frac{\Phi^2}{2} = \mathscr{F}^2 / 2\mathscr{R} \tag{5.42}$$

Furthermore, if it is assumed that there is no fringing at the air gaps, and that the total field energy is distributed uniformly, the total energy stored in the field can be expressed as

$$W_f = \frac{i\lambda}{2} = \frac{i(N\Phi)}{2} = \frac{(Ni)\Phi}{2} = \mathscr{F}\frac{\Phi}{2} \quad (5.43)$$

or

$$W_f = \frac{HB\ell A}{2}$$
$$= \frac{B^2(vol)}{2\mu_0} \quad (5.44)$$

where $\ell = 2g$ = total length of the air gap in a flux path
$vol = \ell A$ = total air gap volume
A = cross-sectional area of the core
B = flux density in the air gaps
$\mu_0 = \dfrac{B}{H}$
= permeability of free space, H/m
$= 4\pi \times 10^{-7}$

Since ℓA is the total gap volume, the energy density w_f in the air gaps can be expressed as

$$w_f = \frac{W_f}{\ell A}$$
$$= \frac{BH}{2}$$
$$= \frac{\mu_0 H^2}{2} \quad (5.45)$$
$$= \frac{B^2}{2\mu_0}$$

The unit of the energy density is J/m^3.

Example 5.4

Consider Example 3.6 and determine the following:
(a) The mmf required by the air gap.
(b) The total mmf required.
(c) The mmf required by the ferromagnetic core.
(d) The energy density in the air gap.
(e) The energy stored in the air gap.

(f) The total energy stored in the magnetic system.
(g) The energy stored in the ferromagnetic core.
(h) The energy density in the ferromagnetic core.

SOLUTION:
(a) The mmf required by the air gap is found from

$$\mathscr{F}_g = \Phi_g \mathscr{R}_g = \Phi_4 \mathscr{R}_4 = \Phi_2 \mathscr{R}_4$$

as

$$\mathscr{F}_g = (0.001197\ \text{Wb})(31{,}870.9886\ \text{A} \cdot \text{turns/Wb}) = 38.1017\ \text{A} \cdot \text{turns}$$

(b) The total mmf required is

$$\mathscr{F}_{tot} = \Phi_{tot} \mathscr{R}_{tot}$$

$$= (0.0015652\ \text{Wb})(255{,}565.4983\ \text{A} \cdot \text{turns/Wb}) = 400\ \text{A} \cdot \text{turns}$$

(c) The mmf required by the ferromagnetic core is

$$\mathscr{F}_c = \mathscr{F}_{tot} - \mathscr{F}_g$$

$$= 400 - 38.1017 = 361.8983\ \text{A} \cdot \text{turns}$$

(d) Since

$$B_g = B_2$$

$$= 0.2394\ \text{Wb/m}^2$$

From equation (5.45) the energy density in the air gap is

$$w_g = \frac{B_g^2}{2\mu_0}$$

$$= \frac{(0.2394\ \text{Wb/m}^2)^2}{2(4\pi \times 10^{-7})} = 0.0023 \times 10^7 \text{J/m}^3$$

(e) The total air gap volume is

$$vol_g = \ell_g A_g$$

$$= (0.0002\ \text{m})(0.005\ \text{m}^2) = 1 \times 10^{-6} \text{m}^3$$

Thus, the energy stored in the air gap is

$$W_g = w_g(vol_g)W_g = w_g(vol_g)$$

$$= (0.0023 \times 10^7 \, \text{J/m}^3)(1 \times 10^{-6} \, \text{m}^3) = 0.023 \, \text{J}$$

Alternatively, using equation (5.42),

$$W_g = \mathscr{R}_g \frac{\Phi_g^2}{2}$$

$$= \frac{(31,830.9886 \, \text{A} \cdot \text{turns/Wb})(0.001197 \, \text{Wb})^2}{2} = 0.023 \, \text{J}$$

(f) The total energy stored in the magnetic system is

$$W_f = \mathscr{F}_{tot} \frac{\Phi_{tot}}{2}$$

$$= \frac{(400 \, \text{A} \cdot \text{turns/Wb})(0.0015652 \, \text{Wb})^2}{2} = 0.313 \, \text{J}$$

(g) Since

$$W_f = W_c + W_g$$

then the energy stored in the ferromagnetic core is

$$W_c = W_f - W_g$$

$$= 0.313 - 0.023 = 0.29 \, \text{J}$$

(h) The volume of the ferromagnetic core is

$$vol_c = 2(0.05 \times 0.25)0.05 + (0.05 \times 0.4)0.05 + (0.15 - 0.0002)(0.10)0.05$$

$$= 4 \times 10^{-3} \, \text{m}^3$$

Since the energy stored in the ferromagnetic core is

$$W_c = w_c \, (vol_c)$$

then the energy density in the ferromagnetic core is

$$w_c = \frac{W_c}{vol_c}$$

$$= \frac{0.29 \, \text{J}}{4 \times 10^{-3} \, \text{m}^3} = 72.5 \, \text{J/m}^3$$

5.3.2 MAGNETIC FORCE

The magnetic flux that crosses an air gap in a magnetic material produces a force $\mathscr{F}_f$ of attraction between the faces of the air gap, as shown in Figure 5.12. The core shown in the figure has an air gap of variable length g as dictated by the position of the movable part (i.e., the armature), which in turn is determined by the magnetic pulling force $\mathscr{F}_f$ and the spring. Note that, in Figure 5.12, the differential displacement can also be expressed as

$$dx = dg \qquad (5.46)$$

Based on the symmetry involved (considering only one pole of the magnetic circuit), the differential change in volume can be found from

$$d(vol) = A\,dg \qquad (5.47)$$

Ignoring leakage and fringing of the flux at the gaps

$$dWf = \frac{BH\,d(vol)}{2}$$

$$= \frac{BH\,A\,dg}{2} \qquad (5.48)$$

$$= \frac{B^2 A\,dg}{2\mu_0}$$

From the definition of work,

$$dW_m = \mathscr{F}_f\,dx = \mathscr{F}_f\,dg \qquad (5.49)$$

When a magnetic pulling force is applied to the movable part (i.e., the armature), an energy dW equal to the magnetic energy dW_f stored in the magnetic field is expended. Therefore, at the equilibrium,

$$dW_f = dW_m \qquad (5.50)$$

or substituting equations (5.48) and (5.49) into equation (5.50)

$$\frac{B^2 A\,dg}{2\mu_0} = \mathscr{F}_f\,dg \qquad (5.51)$$

from which the *magnetic pulling force per pole* on the movable part can be found as

$$\mathscr{F}_f = \frac{B^2 A}{2\mu_0} \quad \text{N} \qquad (5.52)$$

Thus, the total magnetic pulling force on the movable part can be expressed as

$$\mathscr{F}_{f,\,total} = 2\frac{B^2 A}{2\mu_0} = \frac{B^2 A}{\mu_0} \quad \text{N} \tag{5.53}$$

It is important to understand that since the electrical input makes no contribution to the energy in the air gaps, due to the constant air-gap flux, the mechanical energy must be obtained from the stored energy in the air-gap fields (i.e., $id\ell = 0$). In other words, the air gaps give off energy by virtue of their decreased volume.

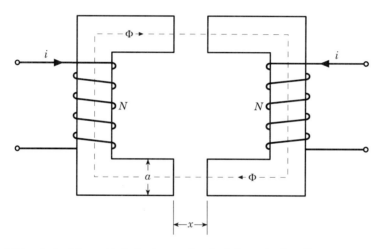

Figure 5.14 A doubly excited electromechanical (translational) system.

Example 5.5

Consider the linear electromechanical system shown in Figure 5.14. Assume that only the coil shown on the left is energized and that the core on the right acts as an armature (i.e., the movable part). The cross-sectional area of each air gap is $25 \times 10^{-6}\,\text{m}^2$. If the flux density is 1.1 Wb/m^2, determine:
(a) The magnetic pulling force per pole.
(b) The total magnetic pulling force.

SOLUTION:
(a) The magnetic pulling force per pole can be found from equation (5.52) as,

$$\mathscr{F}_f = \frac{B^2 A}{2\mu_0} = \frac{(1.1\,\text{Wb}/\,\text{m}^2)^2(25 \times 10^{-6}\,\text{m}^2)}{2(4\pi \times 10^{-7})} = 12.04\ \text{N}$$

(b) The total magnetic pulling force can be found from equation (5.53) as,

$$\mathscr{F}_{f,\,total} = \frac{B^2 A}{\mu_0} = \frac{(1.1\ \text{Wb}/\text{m}^2)^2 (25 \times 10^{-6}\ \text{m}^2)}{4\pi \times 10^{-7}} = 24.08\ \text{N}$$

Alternatively,

$$\mathscr{F}_{f,\,total} = 2\mathscr{F}_f = 2(12.04\ \text{N}) = 24.08\ \text{N}$$

5.3.3 ENERGY AND COENERGY

As previously stated, the shaded area in Figure 5.13(a) represents the total energy stored in a coil (which magnetizes the field) from zero to i A. Such energy can be determined by using equation (5.35); that is, from

$$W_f = \int_0^\lambda i\, d\lambda$$

In Figure 5.13(a), the area between the i axis and $\lambda–i$ characteristic is defined as the **coenergy** and can be determined from

$$W_f' = \int_0^i \lambda\, di \tag{5.54}$$

Such a magnetic coenergy has no physical meaning. However, it can be useful in determining force (or torque) developed in an electromagnetic system. From Figure 5.13(a), for a coil current i and the resultant flux linkage λ

$$W_f + W_f' = \lambda i \tag{5.55}$$

or

$$energy + coenergy = \lambda i \tag{5.56}$$

Notice that W_f' is greater than W_f, if the $\lambda–i$ characteristic is nonlinear and that W_f' is equal to W_f, if the $\lambda–i$ characteristic is linear,[9] as shown in Figure 5.13(b).

5.3.4 MAGNETIC FORCE IN A SATURABLE SYSTEM

Consider the electromechanical system shown in Figure 5.12 and assume that it is made up of saturable ferromagnetic material. It shows that when the air gap is large, the resultant $\lambda–i$ characteristic is almost a straight line; while,

---

[9] In other words, if the magnetic core has a constant permeability, for example as in the air, the energy and coenergy are equal.

when the air gap is very small, the characteristic is almost a straight line for small values of flux linkage. However, as flux linkage is increased, the curvature of the characteristic starts to appear because of the saturation of the magnetic core.

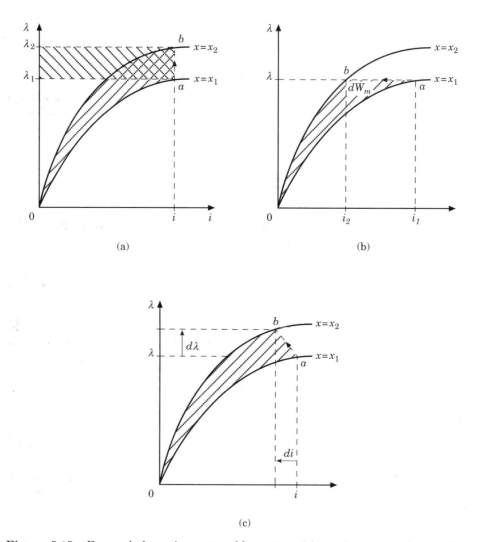

Figure 5.15   Energy balance in a saturable system: (a) constant current operation; (b) constant flux operation; (c) a general case.

Assume that λ is a function of x and i and that there is a differential movement of the operating point corresponding to a differential displacement of dx of the armature (i.e., the movable part) made at a low speed (i.e., at a constant current), as shown in Figure 5.15(a). In other words, the armature of Figure 5.12 moves from the operating point a (where $x = x_1$) to a new operating point b (where $x = x_2$) so that at the end of the movement the air gap decreases. If the armature has moved slowly, the current i has stayed constant during the motion, causing the operating point to move upward from point a to b, as shown in Figure 5.15(a). For this displacement,[10] during which the flux linkage changes, neither the emf nor the dW_e is zero. Therefore, from equation (5.20),

$$dW_m = dW_e - dW_f = d(W_e - W_f) \tag{5.57}$$

Hence for the displacement,

$$\mathscr{F}_f = \left. \frac{\partial(W_e - W_f)}{\partial x} \right|_{i=constant} \tag{5.58}$$

Since the motion has taken place under constant-current conditions, the mechanical work performed is depicted by the shaded area in Figure 5.15(a). That area also represents the increase in the coenergy. Thus,

$$dW_m = dW'_f \tag{5.59}$$

Substituting equation (5.25) into equation (5.59),

$$\mathscr{F}_f dx = dW'_f \tag{5.60}$$

The force on the armature is then

$$\mathscr{F}_f = \left. \frac{\partial W'_f(i, x)}{\partial x} \right|_{i=contant} \tag{5.61}$$

Since for any armature position,

$$W'_f(i, x) = \int_0^i \lambda di \tag{5.62}$$

and

$$\lambda = N\Phi \tag{5.63}$$

$$i = \frac{\mathscr{F}}{N} \tag{5.64}$$

Substituting equations (5.63) and (5.64) into (5.62) gives the coenergy as a function of the mmf and displacement as

---

[10] Notice the increase in coenergy during the move from position a to b, as shown in Figure 5.15(a).

$$W'_f(\mathscr{F}, x) = \int_0^F \Phi \, d\mathscr{F} \tag{5.65}$$

and the force on the armature is then

$$\mathscr{F}_f = \left. \frac{\partial W'_f(F, x)}{\partial x} \right|_{F=constant} \tag{5.66}$$

Figure 5.15(b) illustrates a differential movement of the operating point in the λ–i diagram, corresponding to a differential displacement dx of the armature made at high speed, that is, at constant flux linkage. Here,

$$W_f = W_f(\lambda, x) \tag{5.67}$$

The electrical energy input for the movement is zero, since λ does not change and the emf is zero. The mechanical work done during the motion is represented by the shaded area, which depicts the decrease in the field energy. Since

$$dW_m = dW_f(\lambda, x) \tag{5.68}$$

and

$$\mathscr{F}_f dx = dW_m = - dW_f \tag{5.69}$$

Therefore

$$\mathscr{F}_f = \frac{dW_m}{dx} = - \left. \frac{\partial W_f(\lambda, x)}{\partial x} \right|_{\lambda=constant} \tag{5.70}$$

It is interesting to see that at high-speed motion the electrical input is zero (i.e., $id\lambda = 0$) because the flux linkage has stayed constant and the mechanical output energy has been provided totally by the field energy. In the discussions so far, either i or λ has been kept constant. In reality, however, neither condition is true. It is more likely that the change from position a to b follows a path such as the one shown in Figure 5.15(c).

Also notice that for the linear case (i.e., when flux Φ is proportional to mmf $\mathscr{F}$), the energy and coenergy are equal. Thus,

$$W'_f = W_f \tag{5.71}$$

and

$$\mathscr{F}_f = \frac{\partial W'_f(F, x)}{\partial x} = \frac{\partial W_f(F, x)}{\partial x} \tag{5.72}$$

It is easier to use the inductance L of the excitation coil because L is independent of the current. Therefore,

$$W_f = \frac{Li^2}{2} \tag{5.73}$$

so that

$$\mathcal{F}_f = \frac{d}{dx}\left(\frac{Li^2}{2}\right) = \frac{i^2}{2}\frac{dL}{dx} \tag{5.74}$$

5.4 THE STUDY OF ROTATING MACHINES

In previous sections, the development of **translation motion** in an electromagnetic system has been reviewed extensively. However, most of the energy converters, especially the ones with higher power, develop **rotational motion**. Such a rotating electromagnetic system is made up of a fixed part known as the **stator** and a moving part known as the **rotor**, as previously explained in Section 2.2. In the following sections, singly excited and multiply excited rotating systems will be studied.

5.5 SINGLY EXCITED ROTATING SYSTEMS

To illustrate the application of electromechanical energy conversion principles to rotating systems, consider an elementary, singly excited two-pole rotating system, as shown in Figure 5.16. Such a system represents an elementary reluctance machine. Note that the **stator (pole) axis** is called the **direct axis** or simply the **d-axis** and that its **interpole axis** is also called the **quadrature axis** or simply the **q-axis**. Assume that a sinusoidal excitation is supplied to the stator winding, while the rotor is free to rotate on its shaft. The variables are torque T and angle θ and the differential mechanical energy output is $Td\theta$ when the torque and angle are assumed positive in the same direction (i.e., **motor action**). The developed torque can be expressed as

$$
\begin{aligned}
T_d = T_f &= \frac{\partial W'_f}{\partial \theta} \\
&= \frac{\partial W'_f(i,\theta)}{\partial \theta}
\end{aligned}
\tag{5.75}
$$

For each revolution of the rotor there are two cycles of reluctance, since the reluctance varies sinusoidally. Figure 5.16 shows the variation of inductance with rotor angular position θ as the rotor rotates with a uniform speed ω_m in a reluctance machine. Because the inductance is a periodic function of 2θ, it can be represented by a Fourier series as

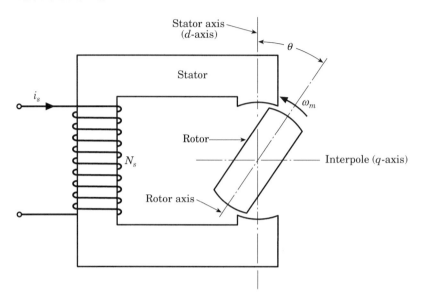

Figure 5.16   A singly excited rotating system.

$$L(\theta) = L_0 + L_2 \cos 2\theta \qquad (5.76)$$

ignoring the higher order terms. The variables used are defined as in Figure 5.17. The stator excitation current is

$$i = I \sin \omega_s t \qquad (5.77)$$

which is a sinusoidal excitation whose angular frequency is ω_s. Since the air-gap region is linear, the coenergy in the magnetic field of the air-gap region can be expressed as

$$W_f' = W_f = \frac{1}{2} L(\theta) i^2 \qquad (5.78)$$

and therefore the developed torque can be expressed as

$$T_d = \frac{\partial W_f'(i,\theta)}{\partial \theta}$$

$$= \frac{1}{2} i^2 \frac{\partial L(\theta)}{\partial \theta} \qquad (5.79)$$

which in terms of current and inductance variations can be reexpressed as

$$T_d = -I^2 L_2 \sin 2\theta \sin^2 \omega_s t \qquad (5.80)$$

Here, it is assumed that the rotor rotates at an angular velocity ω_m; therefore, at any given time

$$\theta = \omega_m t - \delta \qquad (5.81)$$

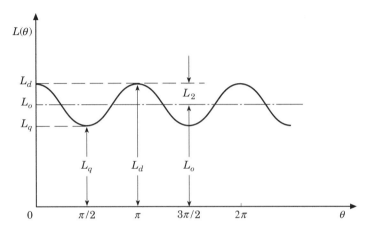

Figure 5.17   Variation of inductance with rotor angular position θ as the rotor rotates in a reluctance machine.

Note that at $t = 0$ the current i is zero, thus the angular rotor position becomes

$$\theta = -\delta \qquad (5.82)$$

The instantaneous torque expression given by equation (5.80) can be expressed in terms of ω_m and ω_s by using the following trigonometric equations:

$$\sin^2 A = \frac{1 - \cos 2A}{2} \qquad (5.83)$$

and

$$\sin A \cos B = \frac{1}{2} \sin (A + B) + \frac{1}{2} \sin (A - B) \qquad (5.84)$$

Hence, the instantaneous (electromagnetic) developed torque becomes

$$T_d = -\frac{I^2 L_2}{2} \{ \sin 2(\omega_m t - \delta) - \frac{1}{2} \sin 2[(\omega_m + \omega_s)t - \delta] \\ - \frac{1}{2} \sin 2[(\omega_m - \omega_s)t - \delta] \} \qquad (5.85)$$

As can be seen from equation (5.85) the torque equation is made up of the sum of sinusoids of various frequencies. Therefore, in most cases, the average torque over a period of time is zero, since the value of each term integrated over a period is zero. Under such conditions the machine cannot operate as a motor to provide a load torque[11] to its shaft. The only case in which the average **(load) torque**[12] is nonzero is when

$$\omega_m = \omega_s \quad \text{rad/s} \tag{5.86}$$

so that

$$T_{d(ave)} = -\frac{I^2 L_2}{4}\sin 2\delta \tag{5.87}$$

Also, it can be seen in Figure 5.17 that

$$L_2 = \frac{L_d - L_q}{2} \tag{5.88}$$

where L_d and L_q are defined as the **direct-axis inductance** and **quadrature-axis inductance**, representing the maximum and minimum values of inductance, respectively. Therefore, substituting equation (5.88) into equation (5.87), the average developed torque[13] can be expressed as

$$T_{d(ave)} = -\frac{I^2(L_d - L_q)}{8}\sin 2\delta \tag{5.89}$$

Based on the previous review, the following summary and conclusions can be made:

1. Only at a certain speed, given by equation (5.86), can such a machine develop an average torque in either rotational direction. This speed is defined as the **synchronous speed**, at which the speed of mechanical rotation in radians per second is equal to the angular frequency of the electrical source.

---

[11] **Load torque** is defined as a torque in opposition to the rotor motion. Therefore, the total torque is equal to the difference between the magnetic torque and the load torque in the forward direction.
[12] It is also interesting to note that the total torque as a function of time has pulsating components even when $\omega_m = \omega_s$. However, because of the typical heavy steel rotor of a synchronous machine, it cannot significantly react to such pulsating components. Therefore, they cannot affect the average torque. Succinctly put, the rotor's mass functions as a **low-pass filter**.
[13] In general, the basic difference between various rotating machines is based on how the stator and rotor mmfs are kept displaced with respect to each other at all times so that they incline to align continuously and develop an average torque. This phenomenon is known as the **alignment principle**.

2. Because the torque is a function of the reluctance variation with rotor position, such an apparatus is called a **synchronous reluctance machine**. Therefore, if there is no inductance or reluctance variation with rotor position (i.e., if $L_d = L_q$), the torque becomes zero. This can easily be concluded from equation (5.89).

3. As can be concluded from equation (5.89), the developed torque is a function of the angle δ, which is called the **torque angle**. The torque varies sinusoidally with the angle δ. Therefore, the angle δ can be used as a measure of the torque.

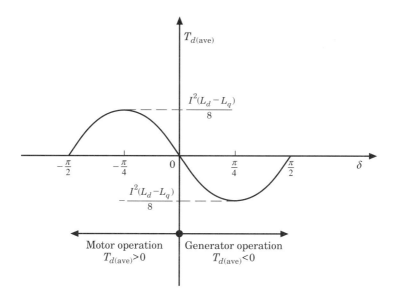

Figure 5.18   Variation of developed torque by a synchronous reluctance machine.

4. When $\delta < 0$ *and* $T_{d(ave)} > 0$, the developed torque is in the direction of rotation, and the machine operates as a *motor*, as can be seen in Figure 5.18. This torque maintains the speed of the rotor against friction, windage, and any external load torque applied to the rotor shaft.

5. Ignoring the effects of friction and windage, the load torque can determine the angle δ. For example, a greater load torque can cause the rotor to operate at a larger negative δ. Since power and torque are proportional to constant speed, there is definitely a power limit and at loads beyond the crest of the curve shown in Figure 5.18 the motor will stall. The maximum torque for motor operation takes place at $\delta = -\pi/4$, and is called the **pull-out torque**. As previously stated, any load that requires a torque greater than the maximum torque causes an unstable operation of the machine; the machine pulls out of synchronism and comes to a standstill.

6. If the shaft of the same machine is driven by a prime mover, the angle δ will advance and the machine will absorb torque and power, and will supply

electrical power as a generator. In other words, when $\delta > 0$ and $T_{d(ave)} < 0$, the developed torque resists the rotation. Therefore, an external driving torque has to be applied to the rotor shaft to maintain the rotor at a synchronous speed. The mechanical energy supplied to the system after meeting the friction and windage losses is converted into electrical energy, i.e., the machine operates as a generator. But this can happen only if the stator winding is already connected to an ac source, which acts as a sink when the external driving torque is applied and the machine begins to generate. As shown in Figure 5.18, the maximum torque for generator operation takes place at $\delta = \pi/4$.

7. If the driving torque provided by the prime mover is greater than the sum of the developed torque and that due to friction and windage, then the machine is driven above synchronous speed. It may therefore run away unless the prime mover speed is controlled and the continuous energy conversion process is stopped. In summary, a given machine can develop only a certain maximum power and is limited to the rate of energy conversion.

8. It is interesting that a mechanical speed $\omega_m = \omega_s$ will also provide a non-zero average developed torque. Therefore, such a reluctance motor cannot start by itself, but will continue to run in the direction in which it is started.

Because of the variation of reluctance with rotor position, the induced voltage in the stator coil will have a third-harmonic component. Such an unwanted characteristic makes reluctance machines useless as practical generators and restricts their size as motors. However, small reluctance motors, when they are designed to develop starting torque, can be used to drive electric clocks, record players, and other devices, since they provide constant speed.

Example 5.6

Suppose that a two-pole reluctance motor operates at 60 Hz and 6 A. If its direct-axis inductance and quadrature-axis inductance are 0.8 H and 0.2 H, respectively, determine its maximum average developed torque.

SOLUTION:
From equation (5.89),

$$T_{d(ave)} = -\frac{I^2(L_d - L_q)}{8}\sin 2\delta$$

$$= -\frac{36(0.8 - 0.2)}{8}\sin 2\delta = -2.7\sin 2\delta$$

Since $\sin 2\delta = 1$ when $\delta = 45°$,

$$T_{d(ave)} = -2.7\sin 90° = -2.7 \text{ N} \cdot \text{m}$$

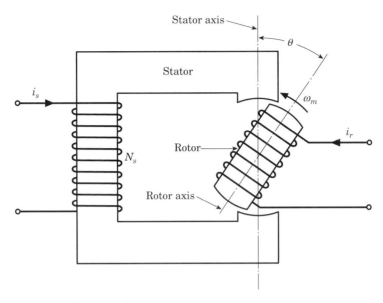

Figure 5.19   A doubly excited rotating system.

5.6 MULTIPLY EXCITED ROTATING SYSTEMS

The general principles that were developed in the previous section also apply to multiply excited (i.e., multicoil) rotating systems. As an example, consider the doubly excited rotating system shown in Figure 5.19. Notice that this system is the same as the one shown in Figure 5.16 except that the rotor also has a coil which is connected to its electrical source through *fixed (carbon) brushes* and rotor-mounted *slip rings* (or *collector rings*). The flux linkages of the stator and rotor windings, respectively, can be written as

$$\lambda_s = L_{ss}i_s + L_{sr}i_r \tag{5.90}$$

$$\lambda_r = L_{rs}i_s + L_{rr}i_r \tag{5.91}$$

where L_{ss} = the self-inductance of the stator winding
$\quad\quad L_{rr}$ = the self-inductance of the rotor winding
$\quad\quad L_{sr} = L_{rs}$ = the mutual inductances between stator and rotor windings

Note that all these inductances depend on the position θ of the rotor (which is the angle between the magnetic axes of the stator and the rotor windings). Since for a linear magnetic system $L_{sr} = L_{rs}$, equations (5.90) – (5.91) can be expressed in the matrix form as

$$\begin{bmatrix} \lambda_s \\ \lambda_r \end{bmatrix} = \begin{bmatrix} L_{ss} & L_{sr} \\ L_{sr} & L_{rr} \end{bmatrix} \begin{bmatrix} i_s \\ i_r \end{bmatrix} \tag{5.92}$$

If the system's rotor is prevented from rotating so that there is no mechanical output from its shaft, then the stored field energy W_f of the system can be found by establishing the currents i_s and i_r in its stator and rotor windings, respectively. Therefore,

$$dW_f = v_s i_s dt + v_r i_r dt$$

$$= i_s d\lambda_s + i_r d\lambda_r$$

(5.93)

Thus, for such a linear system, the differential field energy can be found by substituting equations (5.90) and (5.91) into equation (5.93) so that

$$dW_f = i_s d\lambda_s + i_r d\lambda_r$$

$$= i_s d(L_{ss} i_s + L_{sr} i_r) + i_r d(L_{sr} i_s + L_{rr} i_r)$$

$$= L_{ss} i_s di_s + L_{sr} d(i_s i_r) + L_{rr} i_r di_r$$

(5.94)

The total (stored) field energy can be determined by integrating equation (5.94) as

$$W_f = L_{ss} \int_0^{i_s} i_s di_s + L_{sr} \int_0^{i_s, i_r} d(i_s i_r) + L_{rr} \int_0^{i_r} i_r di_r$$

$$= \frac{1}{2} L_{ss} i_s^2 + L_{sr} i_s i_r + \frac{1}{2} L_{rr} i_r^2$$

(5.95)

The developed torque can be determined from

$$T_d = \left. \frac{\partial W_f'(i,\theta)}{\partial \theta} \right|_{i=constant}$$

(5.96)

Since in a linear magnetic system, energy and coenergy are equal, that is

$$W_f = W_f'$$

(5.71)

the instantaneous (electromagnetic) developed torque can be expressed as

$$T_d = \frac{i_s^2}{2} \frac{dL_{ss}}{d\theta} + i_r i_r \frac{dL_{sr}}{d\theta} + \frac{i_r^2}{2} \frac{dL_{rr}}{d\theta}$$

(5.97)

Note that the first and third terms on the right-hand side of equation (5.97) depict torques developed in the rotating machine due to variations of self-inductances as a function of rotor position. They represent the reluctance torque components of the torque; however the second term represents the torque developed by the variations of the mutual inductance between the stator and rotor windings. Furthermore, multiply excited rotating systems, having more than two coils, are treated in a similar manner.

Consider the doubly excited rotating system shown in Figure 5.19 and assume that R_s and R_r are the resistances of the stator and rotor windings, respectively. The voltage-current relationships for the stator and rotor circuits can be written as

$$v_s = i_s R_s + \frac{d\lambda_s}{dt} \tag{5.98}$$

and

$$v_r = i_r R_r + \frac{d\lambda_r}{dt} \tag{5.99}$$

In general, the inductances L_{ss}, L_{rr}, and L_{sr} are functions of the angular position θ of the rotor, and the currents are time functions. Therefore, for the stator

$$\frac{d\lambda_s}{dt} = \frac{d}{dt}[L_{ss}(\theta)i_s(t) + L_{rs}(\theta)i_r(t)]$$

or

$$\frac{d\lambda_s}{dt} = L_{ss}\frac{di_s}{dt} + i_s\frac{dL_{ss}}{dt}\frac{d\theta}{dt} + L_{rr}\frac{dir}{dt} + i_r\frac{dL_{rs}}{d\theta}\frac{d\theta}{dt} \tag{5.100}$$

Similarly, for the rotor

$$\frac{d\lambda_r}{dt} = L_{rs}\frac{di_r}{dt} + i_s\frac{dL_{rs}}{d\theta}\frac{d\theta}{dt} + L_{rr}\frac{di_r}{dt} + i_r\frac{dL_{rr}}{d\theta}\frac{d\theta}{dt} \tag{5.101}$$

By substituting equations (5.100) and (5.101) into equations (5.98) and (5.99), respectively,

$$v_s = \left[i_s R_s + L_{ss}\frac{di_s}{dt}\right] + \left[\left(i_s\frac{dL_{ss}}{d\theta} + i_r\frac{dL_{rs}}{d\theta}\right)\frac{d\theta}{dt}\right] + \left[L_{rs}\frac{di_r}{dt}\right] \tag{5.102}$$

and

$$v_r = \left[i_r R_r + L_{rr}\frac{di_r}{dt}\right] + \left[\left(i_s\frac{dL_{rs}}{d\theta} + i_r\frac{dL_{rr}}{d\theta}\right)\frac{d\theta}{dt}\right] + \left[L_{rs}\frac{di_r}{dt}\right] \tag{5.103}$$

In equations (5.102) and (5.103), the first terms on the right sides of the equations represent the **self impedance voltage** v_z, the second terms represent the **speed voltage** or **motional voltage** v_m, and the third terms represent the **transformer voltage** v_t. Therefore, the voltage equations for the stator and rotor can be expressed in the form

$$v = v_z + v_m + v_t \tag{5.104}$$

Note that in many cases, the self-inductances L_{ss} and L_{rr} are not dependent on the angular position of the rotor. Thus, equations (5.97), (5.102), and (5.103) reduce to

$$T_d = i_s i_r \frac{dL_{sr}}{dt} \tag{5.105}$$

$$v_s = \left(i_s R_s + L_{ss} \frac{di_s}{dt} \right) + \left(i_r \frac{d\theta}{dt} \right) \frac{dL_{sr}}{d\theta} + L_{rs} \frac{di_r}{dt} \tag{5.106}$$

$$v_r = \left(i_r R_r + L_{rr} \frac{di_r}{dt} \right) + \left(i_s \frac{d\theta}{dt} \right) \frac{dL_{sr}}{d\theta} + L_{rs} \frac{di_s}{dt} \tag{5.107}$$

If the resistances of the stator and rotor are negligible, then equations (5.106) and (5.107) further reduce to

$$v_s = L_{ss} \frac{di_s}{dt} + \left(i_r \frac{d\theta}{dt} \right) \frac{dL_{sr}}{d\theta} + L_{rs} \frac{di_r}{dt} \tag{5.108}$$

$$v_r = L_{rr} \frac{di_r}{dt} + \left(i_s \frac{d\theta}{dt} \right) \frac{dL_{sr}}{d\theta} + L_{rs} \frac{di_s}{dt} \tag{5.109}$$

In matrix notation, the total (stored) field energy, given by equation (5.95) can be expressed as

$$W_f = \frac{1}{2} [\,i\,]^t [L][\,i\,] \tag{5.110}$$

where $[\,i\,]$ = column matrix
$[\,i\,]^t$ = transpose of matrix $[i]$, i.e., a row matrix
$[L]$ = inductance matrix of the system

Also, in matrix notation, the developed torque can be expressed as

$$T_d = \frac{1}{2} [\,i\,]^t \frac{\partial}{\partial \theta} ([L][\,i\,]) \tag{5.111}$$

and the voltage can be expressed as

$$[v] = [i][R] + \frac{d}{dt}([L][i]) \qquad (5.112)$$

Example 5.7

Consider the doubly excited rotating system shown in Figure 5.19. Assume that the self-inductances of the stator and rotor windings are 9 H and 1 H, respectively, and that the mutual inductance between its stator and rotor windings is 2 H. If its stator and rotor currents are 16 A and 8 A, respectively, determine the total stored magnetic field energy in the system.

SOLUTION:

From equation (5.95), the total stored magnetic field energy in the system can be found as

$$W_f = \frac{1}{2} L_{ss} i_s^2 + L_{sr} i_s + \frac{1}{2} L_{rr} i_r^2$$

$$= \frac{1}{2}(9\text{ H})(16\text{ A})^2 + 2(16\text{ H})(8\text{ A}) + \frac{1}{2}(1\text{ H})(8\text{ A})^2 = 1,440\text{ J}$$

5.7 CYLINDRICAL MACHINES

Figure 5.20 shows a cross-sectional view of a single-phase two-pole cylindrical rotating machine with a uniform air gap. Such machines are also called **smooth-air-gap machines**, **uniform-air-gap machines**, or **round-rotor machines**. Note that previous sections dealt with rotating machines with salient poles.[14] As shown in Figure 5.20, a cylindrical machine[15] has a cylindrical rotor in its cylindrical stator. The rotor is free to rotate and its instantaneous angular position θ is defined as the displacement of the rotor's magnetic axis with respect to the stator's magnetic axis. In a real rotating machine the windings are distributed over a number of slots so that their mmf waves can be approximated by space sinusoids. The structure shown in Figure 5.20 is called a *smooth-air-gap* machine because it can be accurately modeled mathematically by assuming that the reluctance of the magnetic path seen by each circuit is

---

[14] The rotating machines can be classified based on their structures: (1) Those machines with salient stator but nonsalient rotor (i.e., the rotor is round or cylindrical), for example, dc commutator machines. (2) Those machines with nonsalient stator but salient rotor, for example, small reluctance machines and low-speed synchronous machines. (3) Those machines with salient stator and salient rotor, for example, some special rotating machines. (4) Those machines with nonsalient stator and nonsalient rotor, for example, induction motors and high-speed synchronous machines.

[15] Most electrical machines are of the cylindrical type because they develop greater torques even though their construction is more complex.

independent of rotor position. Also, such a model ignores the effects of slots and teeth on the magnetic path as the angle is changed. Of course, in an actual machine the slots and teeth are relatively smaller than those shown in Figure 5.20. Furthermore, special construction techniques, such as skewing the slots of one member slightly with respect to a line parallel to the axis, substantially minimize these effects. As a result of such construction it can be assumed that the self-inductances L_{ss} and L_{rr} are constant and no reluctance torques are produced. The mutual inductance L_{sr} changes with rotor position. Therefore,

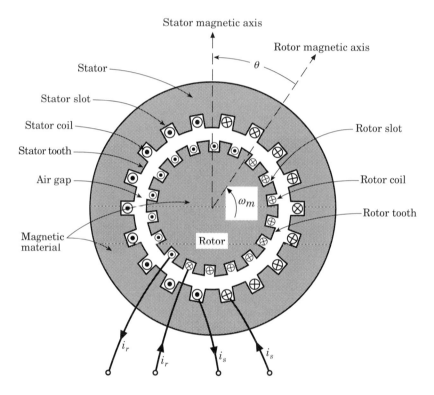

Figure 5.20   Cross-sectional view of a single-phase smooth-air-gap machine having a cylindrical rotor in a cylindrical stator.

$$L_{sr} = M\cos\theta \tag{5.113}$$

where θ = the angle between the magnetic axis of the stator and rotor windings

   M = the peak value of the mutual inductance L_{sr}

Thus,

$$\lambda_s = L_{ss}i_s + M\cos\theta i_r \tag{5.114}$$

$$\lambda_r = M\cos\theta i_s + L_{rr} i_r \tag{5.115}$$

where

$$i_s = I_s \cos\omega_s t \tag{5.116}$$

$$i_r = I_r \cos(\omega_r t + \alpha) \tag{5.117}$$

The torque developed in the cylindrical machine is

$$T_d = i_s i_r \frac{dL_{sr}}{d\theta} \tag{5.118}$$

The position of the rotor at any instant is given as

$$\theta = \omega_m t + \delta \tag{5.119}$$

where ω_m is the angular velocity of the rotor in radians per second, and δ is the rotor position at $t = 0$. Hence, by substituting equations (5.113), (5.116), and (5.117) into equation (5.118), the instantaneous electromagnetic torque developed by the machine can be expressed as

$$T_d = -I_s I_r M \cos\omega_s t \cos(\omega_r t + \alpha)\sin(\omega_m t + \delta) \tag{5.120}$$

Further, by using the trigonometric identities,

$$T_d = -\frac{I_s I_r M}{4}\{\{\sin[\omega_m + (\omega_s + \omega_r)]t + \alpha + \delta\}$$

$$+ \sin\{[\omega_m - (\omega_s + \omega_r)]t - \alpha + \delta\}$$

$$+ \sin\{[\omega_m + (\omega_s - \omega_r)]t - \alpha + \delta\} \tag{5.121}$$

$$+ \sin\{[\omega_m - (\omega_s - \omega_r)]t + \alpha + \delta\}\}$$

Thus, the torque changes sinusoidally with time. As a result, the average value of each of the sinusoidal terms in equation (5.121) is zero, except when the coefficient t is zero. Hence, the average developed torque will be nonzero if

$$\omega_m = \pm(\omega_s \pm \omega_r) \tag{5.122}$$

which may also be expressed as

$$|\omega_m| = |\omega_s \pm \omega_r| \tag{5.123}$$

In other words, the machine will develop average torque if it rotates in either direction, at a speed that is equal to the sum or difference of the angular speeds of the stator and rotor currents.

5.7.1 THE SINGLE-PHASE SYNCHRONOUS MACHINE

Assume that $\omega_r = 0$, $\alpha = 0$ and $\omega_m = \omega_s$. Here, the rotor excitation current is a direct current I_r and the machine rotates at synchronous speed. Therefore, from equation (5.121) the developed torque can be expressed as

$$T_d = -\frac{I_s I_r M}{2}[\sin(2\omega_s t + \delta) + \sin \delta] \tag{5.124}$$

This torque is a pulsating instantaneous torque. Thus, the average developed torque is

$$T_{d(ave)} = -\frac{I_s I_r M}{2}\sin \delta \tag{5.125}$$

The machine operates as an idealized **single-phase synchronous machine** and has an average (unidirectional) developed torque. It has dc excitation in the rotor and ac excitation in the stator. It is important to point out that when $\omega_m = 0$, the machine cannot develop an average torque and hence is not self-starting. Note that the pulsating torque can cause noise, speed fluctuation, and vibration; and therefore is waste of energy. Such pulsating torque can be avoided in a polyphase machine. All large synchronous machines are polyphase machines.

5.7.2 THE SINGLE-PHASE INDUCTION MACHINE

Assume that $\omega_m = \omega_s - \omega_r$ and that ω_s and ω_r are two different angular frequencies. Therefore, both stator and rotor windings have ac currents but at different frequencies. The motor operates at an **asynchronous speed** (i.e., $\omega_m \neq \omega_r$ or $\omega_m \neq \omega_r$). From equation (5.121) the instantaneous developed torque can be expressed as

$$T_d = -\frac{I_s I_r M}{4}[\sin(2\omega_s t + \alpha - \delta) + \sin(-2\omega_r t - \alpha + \delta) \\ + \sin(2\omega_s t - 2\omega_r t - \alpha + \delta) + \sin(\alpha + \delta)] \tag{5.126}$$

This instantaneous torque is a pulsating torque. The average developed torque is

$$T_{d(ave)} = -\frac{I_s I_r M}{4}\sin(\alpha + \delta) \tag{5.127}$$

The machine operates as a **single-phase induction machine**. Its stator winding is excited by an ac current and an ac current is induced in the rotor winding. Such a single-phase induction machine *cannot self-start*, since when $\omega_m = 0$ no average unidirectional torque is developed. The machine has to be brought up to the speed of $\omega_m = \omega_s - \omega_r$ to achieve an average developed torque. In order to avoid pulsating torque, polyphase induction machines are used in most applications.

Example 5.8

Consider a two-pole cylindrical rotating machine as shown in Figure 5.20. If it operates with a speed of $\omega_s = \omega_r = \omega_m = 0$ and $\alpha = 0$, determine the following:
(a) The instantaneous developed torque.
(b) The average developed torque.

SOLUTION:
(a) Since $\omega_s = \omega_r = \omega_m = 0$ and $\alpha = 0$, the excitations are direct currents I_s and I_r. Therefore, from equation (5.121), the instantaneous developed torque can be found as

$$T_d = -I_s I_r M \sin \delta$$

which is a constant.

(b) Thus, the average developed torque is

$$T_{d(ave)} = -I_s I_r M \sin \delta$$

Such a machine operates as a **dc rotary actuator**, developing a constant torque against any displacement δ caused by an external torque placed on the rotor shaft.

5.8 THE FORCE PRODUCED ON A CONDUCTOR

According to field theory, the force on a differential length of conductor $d\boldsymbol{L}$, carrying i, and located in a field $\boldsymbol{B}$ can be expressed as

$$d\mathcal{F} = id\boldsymbol{L} \times \boldsymbol{B} \tag{5.128}$$

The direction of the force is determined from the cross product of the vectors $d\boldsymbol{L}$ and $\boldsymbol{B}$.

Assume that a current-carrying conductor, having a length of $\boldsymbol{L}$, is within a uniform magnetic field of flux density $\boldsymbol{B}$ as shown in Figures 5.21(a) and

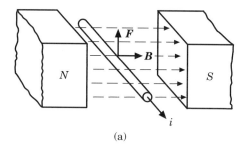

(a)

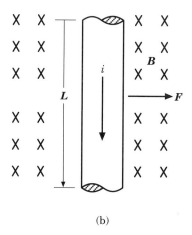

(b)

Figure 5.21   A current-carrying straight conductor in a uniform magnetic field.

5.21(b). Figure 5.21(b) shows the flux density B, pointing into the page. The developed force on the conductor will make the conductor move and the induced force can be expressed as

$$\mathscr{F} = i(L \times B) \quad \text{N} \tag{5.129}$$

where i = magnitude of the current in the conductor
   L = length of the conductor, given as a vector and with the same direction as the current flow
   B = magnetic flux density vector

The direction of the force produced on the conductor is found by **Flemming's right-hand rule**. Therefore, if the index finger of the right hand points in the direction of the vector L, and the middle finger points in the direction of the flux density vector B, then the thumb will point in the direction of the developed force on the conductor. The resulting electromechanical force[16] can be expressed as

---

[16] According to the electromagnetic force law the interaction between a magnetic field and a current-carrying conductor produces a mechanical force.

$$\mathscr{F} = B \times i \times L \times \sin\theta \qquad (5.130)$$

where θ is the angle between the conductor and the flux density vector. It is important to note that as the current-carrying conductor is placed in the field B, *the field itself will change due to the effect of current i.* Therefore, the field B in the above equation is the magnetic field that exists *before the presence of the current i.* The maximum value of the force takes place when $\theta = 90°$. Thus,

$$\mathscr{F}_{max} = B \times i \times L \qquad (5.131)$$

In summary, the induction of such a mechanical force caused by a current flowing through the conductor in a magnetic field produces **motor action.**

Example 5.9

Consider a current-carrying conductor that is within a uniform magnetic field, as shown in Figure 5.21(a). Assume that the magnetic flux density is 0.3 Wb/m^2, pointing into the page, and that the current flowing through the 2 m-long conductor is 3 A. Determine the magnitude and direction of the developed force on the conductor.

SOLUTION:

Based on the right-hand rule, the direction of the force is to the right, as shown in Figure 5.21(a). Its magnitude is

$$\mathscr{F} = B \times i \times L \times \sin\theta$$

$$= (0.3 \text{ Wb/m}^2)(3 \text{ A})(2 \text{ m})\sin 90° = 1.8 \text{ N}$$

Thus,

$$\mathscr{F} = 1.8 \text{ N} \quad \text{to the right}$$

5.9 THE INDUCED VOLTAGE ON A CONDUCTOR MOVING IN A MAGNETIC FIELD

Suppose that a straight-conductor moves with velocity in a uniform magnetic field, as shown in Figure 5.22(a). There will be an induced voltage[17] in the conductor that can be expressed as

$$e_{ind} = (\boldsymbol{v} \times \boldsymbol{B}) \cdot \boldsymbol{L} \qquad (5.132)$$

---

[17] In 1831, Faraday called this voltage an **induced voltage** because it occurred only when there was relative motion between the conductor and a magnetic field without any actual "physical" contact between them.

where v = velocity of the conductor
B = magnetic flux density
L = length of the conductor

Assume that the vector L is in the same direction as the conductor's positive end.[18] The voltage induced in the conductor builds up so that the positive end is in the direction of the vector $(v \times B)$, as shown in Figure 5.22(a). In summary, the induction of voltages in a conductor moving in a magnetic field causes **generator action**. Note that mathematically the vector cross product $v \times B$ has a magnitude that is equal to the product of the magnitudes of v and B and the sine of the angle between them. Its direction can be found from the right-hand rule, which states that when the thumb of the right hand points in the direction of v and the index finger points in the direction of B, $v \times B$ will be

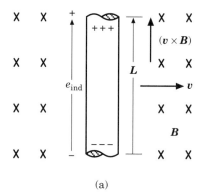

(a)

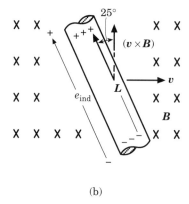

(b)

Figure 5.22   A straight conductor moving in a uniform magnetic field.

---

[18] The selection of the positive end is totally arbitrary, because if the selection is wrong, the resultant computed voltage value will be negative, indicating that a wrong assumption has been made in selecting the positive end.

parallel to L. If the conductor is not oriented on a vertical line, the direction of L must be selected to make the smallest possible angle with the direction of $v \times B$.

Example 5.10

Consider a 2 m-long conductor that is moving with a velocity of 4 m/s to the right, within a uniform magnetic field, as shown in Figure 5.22(a). The magnetic field density is 0.3 Wb/m², pointing into the page, and the conductor length is oriented from the bottom toward the top. Determine the following:
(a) The magnitude of the resulting induced voltage.
(b) The polarity of the resulting induced voltage.

SOLUTION:
(a) The velocity vector v is perpendicular to the magnetic field density vector B, and therefore $v \times B$ is parallel to the conductor length vector L. The magnitude of the resulting induced voltage is

$$e_{ind} = |(\,v \times B)\cdot L|$$

$$= (vB\sin\ 90°)L\cos\ 0°$$

$$= vBL$$

$$= (4\text{m/s})(0.3\ \text{Wb/m}^2)(2\ \text{m}) = 2.4\ \text{V}$$

(b) The polarity of the resulting induced voltage is positive at the top of the conductor and negative at the bottom of the conductor, as shown in Figure 5.22(a).

Example 5.11

Consider a 1.5 m-long conductor that is moving with a velocity of 5 m/s to the right within a uniform magnetic field, as shown in Figure 5.22(b). If the magnetic field density is 0.8 Wb/m², pointing into the page, and the conductor length is oriented from the bottom toward the top, determine the following:
(a) The magnitude of the resulting induced voltage.
(b) The polarity of the resulting induced voltage.

SOLUTION:
(a) The magnitude of the resulting induced voltage is

$$e_{ind} = |(\,v \times B)\cdot L|$$

$$= (vB\sin\ 90°)L\cos\ 25°$$

$$= (5\text{m/s})(0.8\ \text{Wb/m}^2)(\sin\ 90°)(1.5\ \text{m})\cos\ 25° = 5.44\ \text{V}$$

(b) The polarity of the resulting induced voltage is positive at the top of the conductor and negative at the bottom of the conductor, as shown in Figure 5.22(a).

PROBLEMS

5.1   The rotational speed of a motor (i.e., its shaft speed) is 3600 rev/min. Determine its angular velocity (i.e., its shaft speed) in rad/s.

5.2   If the motor in Problem 5.1 is operating at 50 Hz frequency, instead of 60 Hz, determine its new angular velocity in rad/s.

5.3   A special-purpose motor is operating at 25 Hz frequency. Determine the following:
(a) Its angular velocity (i.e., its shaft speed) in rad/s.
(b) Its rotational speed (i.e., its shaft speed) in rev/min.

5.4   If a motor is delivering 200 N•m of torque to its mechanical load at a shaft speed of 3600 rpm, determine the following:
(a) The power supplied to the load in watts.
(b) The power supplied to the load in horsepower.

5.5   Assume that a coil, having a sectional area of 0.25 m^2 with $N = 15$ turns, is rotating around its horizontal axis with a constant speed of 1800 rpm in a uniform and vertical magnetic field of flux density $B = 0.75$ T. If the total magnetic flux passing through the coil is given by $\phi = AB\cos \omega t$ Wb where A is the sectional area of the coil, determine the maximum and effective values of the induced voltage in the coil.

5.6   Consider the linear electromechanical system shown in Figure 5.14. Since the system is considered to be linear its core reluctance is negligibly small. The core depth is given as b. If the system is excited by two identical current sources, determine the following:
(a) The force of attraction between the poles in terms of current i and the geometry involved.
(b) The force between the poles, if the current is reversed in one coil.

5.7   Assume that there is a two-pole cylindrical rotating machine, as shown in Figure 5.20, and that it operates with $\omega_s = \omega_r$ and $\omega_m = 0$. Determine the following:
(a) The instantaneous developed torque.
(b) The average developed torque.

5.8   Consider the current-carrying conductor that is within a uniform magnetic field, as shown in Figure 5.21(a). Assume that the magnetic field density is 0.25 Wb/m^2, pointing out of the page, and that the current flowing through the 0.3 m-long conductor is 1.5 A. Determine the following:

(a) The magnitude of the developed force in N.

(b) The direction of the developed force, if the current is flowing from the top toward the bottom.

(c) The direction of the developed force, if the current is flowing from the bottom toward the top.

5.9 Consider a current-carrying conductor that is within a uniform magnetic field, as shown in Figure 5.21(a). Assume that the magnetic field density is 0. 5 Wb/m^2 and that the current flowing through the 0.6 m-long conductor (in the direction as shown, i.e., from the top toward the bottom) is 2 A. Determine the following:

(a) The magnitude of the developed force in N.

(b) The direction of the developed force, if the magnetic flux density vector is pointing into the page.

(c) The direction of the developed force, if the magnetic flux density vector is pointing out of the page.

(d) The direction of the developed force, if the magnetic flux density vector is pointing into the page and the direction of the current flow is reversed (i.e., from the bottom toward the top).

(e) The direction of the developed force, if the magnetic flux density is pointing out of the page and the direction of the current flow is reversed (i.e., from the bottom toward the top).

5.10 Consider a 0.5 m-long conductor that is moving with a velocity of 2-m/s to the right within a uniform magnetic field, as shown in Figure 5.22 (a). Assume that the magnetic field density is 0.25 Wb/m^2, pointing into the page, and that the conductor length is oriented from the bottom toward the top. Determine the following:

(a) The magnitude of the resulting induced voltage.

(b) The polarity of the resulting induced voltage.

(c) The polarity of the resulting induced voltage, if the conductor length is oriented from the top toward the bottom.

5.11 Consider a 0.5 m-long conductor that is moving with a velocity of 4-m/s to the right within a uniform magnetic field, as shown in Figure 5.22 (b). Assume that the magnetic field density is 0.75 Wb/m^2, pointing into the page, and that the conductor length is oriented from the bottom toward the top. Determine the following:

(a) The magnitude of the resulting induced voltage.

(b) The polarity of the resulting induced voltage.

(c) The polarity of the resulting induced voltage, if the magnetic flux density vector is pointing out of the page.

(d) The polarity of the resulting induced voltage, if the magnetic flux density vector is pointing into the page and the conductor length is oriented from the bottom toward the top.

(e) The polarity of the resulting induced voltage if the magnetic flux density vector is pointing out of the page, the conductor length is oriented from the top toward the bottom, and the conductor is moving to the left.

CHAPTER 6

INDUCTION MACHINES

Any man may make a mistake; none but a fool will stick to it.
M.T. Cicero, 51 B.C.

Time is the wisest counselor.
Pericles, 450 B.C.

When others agree with me, I wonder what is wrong!
Author Unknown

6.1 INTRODUCTION

Because of its relatively low cost, simple and rugged construction, minimal maintenance requirements, and good operating characteristics which satisfy a wide variety of loads, the induction motor is the most commonly used type of ac motor. Induction motors range in size from a few watts to about 40,000 hp. Small fractional-horsepower motors are usually single-phase and are used extensively for domestic appliances such as refrigerators, washers, dryers, and blenders. Large induction motors (usually above 5 horsepower) are always designed for three-phase operation to achieve a constant torque and balanced network loading. In particular, where very large machinery is to be operated, the three-phase induction motor[1] is the *workhorse* of the industry. In contrast to dc motors, induction motors can operate from supplies in excess of 10 kV. Figure 6.1(a) shows a typical three-phase induction motor, while Figure 6.1(b) shows a large three-phase induction motor. Figure 6.2 shows a totally enclosed, fan-cooled, three-phase induction motor. Figure 6.3 shows a totally enclosed, fan-cooled, explosion-proof, three-phase induction motor for use in hazardous environments.

---

[1] The whole concept of polyphase ac, including the induction motor, was developed by Nikola Tesla and patented in 1888. In 1895, the Niagara Falls hydroplant, using the Tesla polyphase ac system concept, went into operation. This was the first large-scale application of the polyphase ac system. However, the first paper written on the induction machine was authored by Galileo Ferraris, an Italian, who also developed a new per-phase equivalent circuit for the new motor. The circuit had a

(a)

(b)

Figure 6.1   (a) A small three-phase induction motor; (b) a large three-phase induction motor. (Courtesy of Reliance Electric Company.)

In typical induction motors, the stator winding (the *field winding*) is connected to the source, and the rotor winding (the *armature winding*) is short-circuited for many applications, or may be closed through external resistances. Thus, an induction motor is a *singly excited* motor (as opposed to a *doubly*

---

primary and secondary in much the same manner as the present per-phase circuit. Unfortunately he did not recognize the need for slip s in the circuit, instead he used R_2 rather than R_2/s as the secondary resistor. After some mathematical analysis, Ferraris concluded that this new motor was not practical since from the maximum power transfer theorem only a maximum efficiency of 50 percent can be attained when $R_2 = R_1$! Thus, Ferraris promptly gave up the development of the motor as impractical and went on to become famous in other areas, but not in electrical machines. Needless to say, Tesla, being an experimentalist, was never bothered with such niceties, but simply proceeded directly from the concept to the implementation and physically demonstrated that it worked.

Figure 6.2 A totally enclosed, fan-cooled, three-phase induction motor. (Courtesy of General Electric Canada, Inc.)

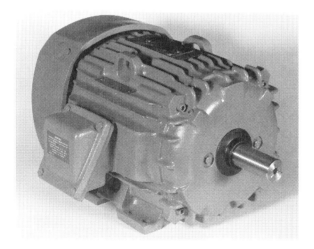

Figure 6.3 A totally enclosed, fan-cooled, explosion-proof three-phase induction motor for use in hazardous environments. (Courtesy of General Electric Company.)

excited synchronous motor). In such a motor, alternating current passing through a fixed stator winding sets up a rotating magnetic field. This moving field induces[2] currents in closed loops of wire mounted on the rotor. These

---

[2] An **induction motor** is so called because the driving force is provided by an electric current induced in a rotor, due to its interaction with a magnetic field.

currents set up magnetic fields around the wires and cause them to follow the main magnetic field as it rotates. Therefore, the operation of the induction motor depends on the rotating field passing through the loops on the rotor, which must always turn more slowly than the rotating field. Since no current has to be supplied to the rotor, the induction motor is simple to construct and reliable in operation. This class of rotating machines derives its name from the fact that the rotor current results from **induction**, rather than **conduction**.[3] A given induction machine can be operated in the motor region, generator region, or braking region, as shown in Figure 6.4. In the **motor mode**, its oper-

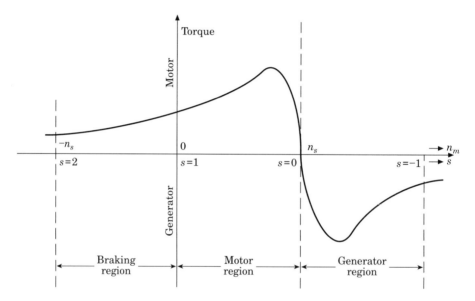

Figure 6.4   An induction machine's torque-speed characteristic curve showing braking, motor, and generator regions.

ating speed is slightly less than its synchronous speed;[4] but, in the **generator mode**, its operating speed is slightly greater than its synchronous speed and it needs magnetizing reactive power from the system that it is connected to in order to supply power. The full-load speed of three-phase induction motors is often within 7 percent of the synchronous speed, even though full-load speeds of 1 percent below the synchronous speed are not uncommon.

In the **braking mode** of operation, a three-phase induction motor running at a steady-state speed can be brought to a quick stop by interchanging

---

[3] Since an induction motor runs below synchronous speed, it is also known as an **asynchronous** (i.e., not synchronous) **machine**.

[4] As the winding loops of the rotor turn faster they try to catch up to the rotating magnetic field, and the difference between the two speeds gets smaller. The size of the induced currents, and therefore the size of the driving force, also gets smaller. The rotor thus settles down to a steady speed, which is slower than that of the rotating magnetic field.

two of its stator leads. By doing this, the phase sequence, and therefore the direction of rotation of the magnetic field, is suddenly reversed; the motor comes to a stop under the influence of torque and is immediately disconnected from the line before it can start in the other direction. This is also known as the **plugging operation**.

Since the induction motor cannot produce its excitation, it needs reactive power; it draws a lagging current from the source and operates at a power factor that is less than unity (usually, above 0.85). However, it runs at lower lagging power factors when lightly loaded. To limit the reactive power, the magnetizing reactance has to be high, and thus the air gap is shorter than in synchronous motors of the same size and rating (with the exception of small motors). Also, the starting current of an induction motor is usually five to seven times its full-load (i.e., rated) current. In general, the speed of an induction motor is not easily controlled.

Even though the induction machine, with a wound rotor, can be used as a generator, its performance characteristics (especially in comparison to a synchronous generator) have not been found satisfactory for most applications. However, induction generators are occasionally used at hydroelectric power plants. For example, they are presently in use as generators at the Folsom Dam in Northern California. Also, the induction machine with a wound rotor can be used as a **frequency changer**.

6.2 CONSTRUCTION OF INDUCTION MOTORS

In general, the stator construction of a three-phase induction machine is the same as that for a synchronous machine.[5] However, the same cannot be said for their rotors. In fact, the three-phase induction motors are classified based on their rotor types as wound-rotor or squirrel-cage motors.

Figure 6.5(a) shows a cross-sectional view of the magnetic circuit of an induction motor that has a wound rotor. The rotor iron is laminated and slotted to contain the insulated windings. The wound-rotor motor has a three-phase symmetrical winding similar to that in the stator and is wound for the same number of poles as the stator winding. These rotor phase windings are wye-connected with the open end of each phase connected to a slip ring mounted on the rotor shaft. Figure 6.6 shows that three equal external variable resistors used for speed control are connected to the slip rings by carbon brushes.[6] Note that the total rotor circuit is wye-connected, providing an external neutral that is usually grounded. Figure 6.7 also illustrates the concept of a three-phase induction motor that has wound-rotor windings connected to external

---

[5] That is, the stator core is built of sheet-steel laminations that are supported in a stator frame of cast iron or fabricated steel plate. Its windings, quite similar to those of the revolving field synchronous machine, are spaced in the stator slots 120 electrical degrees apart. The stator phase windings can be either wye- or delta-connected. The stator windings constitute the armature windings.
[6] The rotor winding is not connected to a supply. The slip rings and brushes simply provide a means of connecting an external variable-control resistance (called a **slip-ring rheostat**) to the rotor circuit.

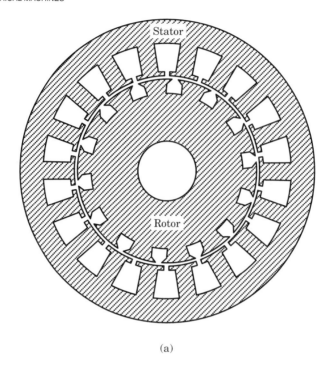

(a)

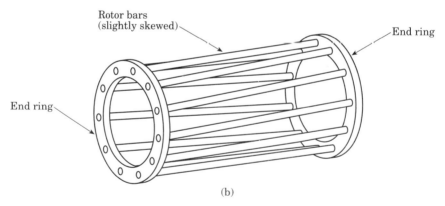

(b)

Figure 6.5   (a) A cross-sectional view of the magnetic circuit of an induction motor with a wound rotor; (b) the squirrel-cage winding of a cage rotor of an induction motor.

resistors. Wound-rotor motors are also called **slip-ring motors**, for obvious reasons.

It is important to know that the rotor winding need not be identical to the stator winding; however, the two have to be wound with an equal number of poles. The number of rotor and stator slots should not be equal, otherwise several slots may line up and cause a pulsating flux. Occasionally, if the slots do line up, the rotor may even lock up on starting and not turn.

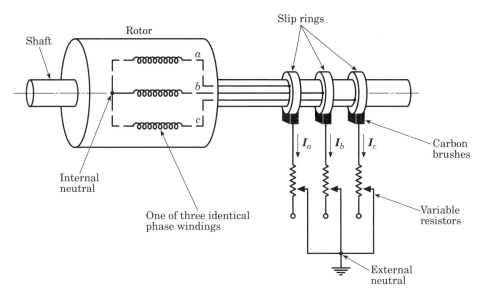

Figure 6.6   Illustration of a three-phase wound rotor winding with slip rings.

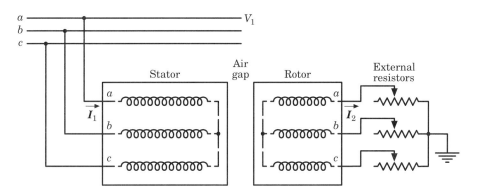

Figure 6.7   Illustration of a three-phase induction motor with its wound rotor windings connected to external resistors.

Figure 6.5(b) shows the *squirrel-cage winding* of a **cage rotor** of an induction motor. Instead of a winding, the slots in the **squirrel-cage**[7] rotor have bars of copper or aluminum, known as rotor bars, which are short-circuited with two *end rings* of the same material. There is one ring at each end of the stack of rotor laminations. The solid rotor bars are placed parallel, or approximately parallel, to the shaft and embedded in the surface of the core. The

---

[7] It is called this because of its appearance, since it resembles the exercise wheel used for hamsters and gerbils.

conductors are not insulated from the core, since the rotor currents naturally flow the path of least resistance through the rotor conductors. In large motors the rotor bars which may be copper alloy are driven into the slots and brazed to the end rings. In small motors, the rotors usually have die cast aluminum bars and their core laminations are stacked in a mold which is then filled with molten aluminum, as shown in Figure 6.8. Note that in such a rotor, the rotor bars, end rings, and cooling-fan blades are cast at the same time, for economical reasons.

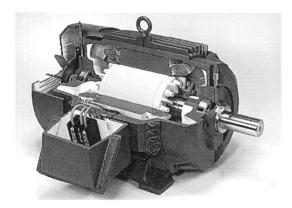

Figure 6.8   A cutaway view of a squirrel-cage three-phase induction motor (Omega XL). (Courtesy of Reliance Electric Company.)

Squirrel-cage rotor bars are not always placed parallel to the motor shaft, but are sometimes skewed as shown in Figure 6.4(b). This provides a more uniform torque and also reduces the magnetic humming noise and mechanical vibrations when the motor is running.

The induction motor is basically a fixed drive. Therefore, in order to function efficiently, its rotor has to rotate at a speed near the synchronous speed. The synchronous speed itself is a function of the frequency of the applied stator voltages and the number of poles of the motor. Thus, efficient variable-speed operation basically requires changing the frequency of the power supply. Recent developments in solid-state technology have resulted in more efficient variable-frequency power sources and have therefore substantially increased the possible applications of induction motors.

6.3 THE ROTATING MAGNETIC FIELD CONCEPT

When the three-phase stator windings of an induction motor are supplied by three-phase voltages, currents will flow in each phase. These currents are time-displaced from each other by 120 electrical degrees in a two-pole machine, as shown in Figure 6.9. An induction motor's operation depends on a rotating magnetic field established by the stator currents in the air gap of the motor. Because of the spacing of the windings and the phase difference of the currents in the windings, the pulsating (sinusoidally distributed) mmf wave produced by each phase combines to form a resultant mmf $\mathcal{F}$ which moves around the inner circumference of the stator surface (i.e., in the air gap) at a constant speed. The resultant flux is called the **rotating magnetic field**. If balanced three-phase excitation is applied with abc phase sequence, then the currents can be expressed as

$$i_a = I_m\cos \omega t \tag{6.1}$$

$$i_b = I_m\cos (\omega t - 120°) \tag{6.2}$$

$$i_c = I_m\cos (\omega t - 240°) \tag{6.3}$$

where I_m is the maximum value of the current and the time origin is arbitrarily assumed to be the instant when the phase-a current is at its positive maximum. Figure 6.9 shows such instantaneous currents. The resultant mmf wave is a function of the three component mmf waves caused by these currents. It can be determined either graphically or analytically.

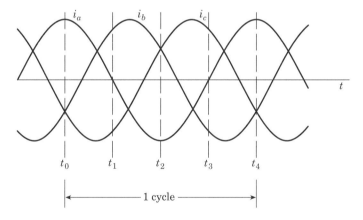

Figure 6.9   Balanced three-phase alternating currents applied to three-phase windings.

6.3.1 GRAPHICAL METHOD

Since the rotating magnetic field is produced by the mmf contribution of space-displaced phase windings with appropriate time-displaced currents, one has to take into account various instants of time and determine the magnitude and direction of the resultant mmf wave. For example, consider the instant of time (indicated in Figure 6.9) $t = t_0$ and notice that the currents in the phase windings a, b and c, respectively, are

$$i_a = I_m \tag{6.4}$$

$$i_b = -\frac{I_m}{2} \tag{6.5}$$

$$i_c = -\frac{I_m}{2} \tag{6.6}$$

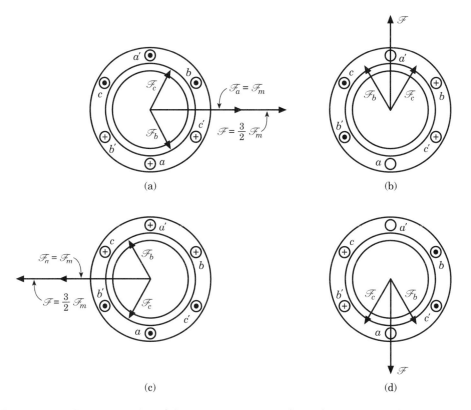

Figure 6.10 Representation of the rotating magnetic field of the stator at four different instants of time (indicated in Figure 6.9): (a) time $t = t_0 = t_4$; (b) time $t = t_1$; (c) time $t = t_2$; (d) time $t = t_3$.

Note that each phase in Figure 6.10(a), for the sake of convenience and simplicity, is represented by a single coil. For example, coil a-a' represents the entire phase-a winding (normally distributed over 60 electrical degrees), with its mmf axis directed along the horizontal. The right-hand rule readily confirms this statement. Similarly, the mmf axis of the phase-b winding is 120 electrical degrees apart from phase-a, and that of phase-c is 120 electrical degrees displaced from phase-b. Obviously, the unprimed and primed letters refer to the beginning and end terminals of each phase, respectively. Also, notice that the current directions in the corresponding coils are indicated by dots and crosses, as shown in Figure 6.10. The current in the phase-a winding is at its maximum at $t = t_0$, and is represented by a phasor $\mathscr{F}_a = \mathscr{F}_m$ along the axis of phase a, as shown in Figure 6.10(a). The mmfs of phases b and c are represented by phasors $\mathscr{F}_b$ and $\mathscr{F}_c$, respectively, each with a magnitude of $\mathscr{F}_m/2$ and located in the negative direction along their corresponding axes. The sum of the three phasors is a phasor $\mathscr{F} = 1.5\mathscr{F}_m$ affecting in the positive direction along the phase-a axis, as shown in Figure 6.10. Figure 6.11 shows the corresponding component mmf waves and the resultant mmf wave at the instant $t = t_0$.

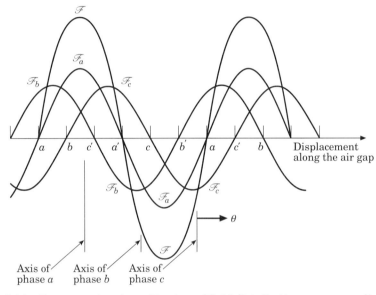

Figure 6.11 Component and resultant mmf field distributions corresponding to time $t = t_0$ in Figure 6.10(a).

Now consider a later instant of time t_1, as shown in Figure 6.10(b). The currents and mmf associated with the phase winding can be expressed as

$$i_a = 0 \qquad \text{and} \qquad \mathscr{F}_a = 0 \tag{6.7}$$

$$i_b = \frac{\sqrt{3}}{2} I_m \quad \text{and} \quad \mathscr{F}_b = \frac{\sqrt{3}}{2} \mathscr{F}_{max} \tag{6.8}$$

$$i_c = -\frac{\sqrt{3}}{2} I_m \quad \text{and} \quad \mathscr{F}_c = -\frac{\sqrt{3}}{2} \mathscr{F}_{max} \tag{6.9}$$

The figure shows the current directions, the component mmfs, and the resultant mmf at $t = t_1$. Note that the resultant mmf has now rotated counterclockwise 90 electrical degrees in space.

Similarly, Figures 6.10(c) and 6.10(d) show the corresponding current directions, component mmfs, and resultant mmfs at the other instants $t = t_2$ and $t = t_3$, respectively. It is obvious that as time passes, the resultant mmf wave keeps its sinusoidal form and amplitude, but shifts forward around the air gap. In one full cycle of the current variation, the resultant mmf wave comes back to the position shown in Figure 6.10(a). Thus, the resultant mmf wave completes one revolution per cycle of the current variation in a two-pole machine. Hence, in a p-pole machine, the mmf wave rotates by $2/p$ revolutions.

6.3.2 ANALYTICAL METHOD

Assume again that the two-pole machine has three-phase windings on its stator, so that the resultant stator mmf at any given instant is composed of the contributions of each phase. Each phase winding makes a contribution that changes with time along a fixed-space axis. Figure 6.12 shows a simplified two-pole three-phase stator winding arrangement. The resultant mmf wave, at any point in the air gap, can be defined by an angle θ. Notice the origin of the axis of phase-a, as shown in Figure 6.12. The resultant mmf along θ can be expressed as

$$\mathscr{F}(\theta) = \mathscr{F}_a(\theta) + \mathscr{F}_b(\theta) + \mathscr{F}_c(\theta) \qquad (6.10)$$

where each term on the right side of equation (6.10) represents the instantaneous contributions of the alternating mmfs of each phase. Hence, each phase winding produces a sinusoidally distributed mmf wave with its peak along the

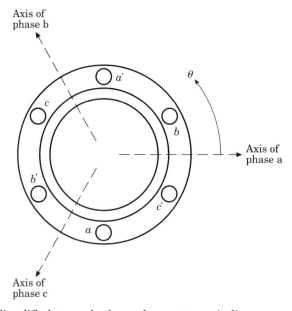

Figure 6.12   Simplified two-pole three-phase stator winding arrangement.

axis of the phase winding and its amplitude proportional to the instantaneous value of the phase current. For example, the contribution from phase-a along θ can be expressed as

$$\mathscr{F}_a(\theta) = \mathscr{F}_m \cos \theta \qquad (6.11)$$

where $\mathscr{F}_m$ is the maximum instantaneous value of the phase-a mmf wave. Therefore, equation (6.11) can be rewritten as

$$\mathscr{F}_a(\theta) = Ni_a \cos \theta \qquad (6.12)$$

where N = the effective number of turns in the phase-a winding
$\quad\quad i_a$ = the instantaneous value of the current in the phase-a winding

Since the phase axes shown in Figure 6.12 are shifted from each other by 120 electrical degrees, the mmf contributions from phase-b and phase-c can be expressed respectively as

$$\mathscr{F}_b(\theta) = \mathscr{F}_m \cos (\theta - 120°) \qquad (6.13)$$

$$\mathscr{F}_c(\theta) = \mathscr{F}_m \cos (\theta - 240°) \qquad (6.14)$$

or

$$\mathscr{F}_b(\theta) = Ni_b \cos (\theta - 120°) \qquad (6.15)$$

$$\mathscr{F}_c(\theta) = Ni_c \cos (\theta - 240°) \qquad (6.16)$$

Hence, the resultant mmf at point θ is

$$\mathscr{F}(\theta) = \mathscr{F}_m \cos \theta + \mathscr{F}_m \cos (\theta - 120°) + \mathscr{F}_m \cos (\theta - 240°) \qquad (6.17)$$

or, alternatively,

$$\mathscr{F}(\theta) = Ni_a \cos \theta + Ni_b \cos (\theta - 120°) + Ni_c \cos (\theta - 240°) \qquad (6.18)$$

N'_a

However, the instantaneous currents i_a, i_b, and i_c are functions of time and are expressed as

$$i_a = I_m \cos \omega t \qquad (6.19)$$

$$i_b = I_m \cos (\omega t - 120°) \qquad (6.20)$$

$$i_c = I_m \cos (\omega t - 240°) \qquad (6.21)$$

where I_m is the maximum value of the current and the time origin is arbitrarily taken as the instant when the phase-a current is at its positive maximum.

The quantity ω is the angular frequency of oscillation of the stator currents, which by definition is

$$\omega = 2\pi f \quad electrical\ radians\ per\ second \tag{6.22}$$

where f is the frequency of the stator currents in hertz. Therefore, equation (6.18) can be expressed as

$$\mathscr{F}(\theta, t) = NI_m \cos \omega t \cos \theta$$

$$+ NI_m \cos (\omega t - 120°)\cos (\theta - 120°) \tag{6.23}$$

$$+ NI_m \cos (\omega t - 240°)\cos (\theta - 240°)$$

By use of the identity

$$\cos x \cos y = \frac{1}{2} \cos (x + y) + \frac{1}{2} \cos (x - y) \tag{6.24}$$

each term on the right side of equation (6.23) can be rewritten as the sum of two cosine functions, one involving the difference and the other the sum of the two angles. The resultant mmf of the total three-phase winding can be expressed as

$$\mathscr{F}(\theta, t) = \frac{1}{2} NI_m \cos (\omega t - \theta) + \frac{1}{2} NI_m \cos (\omega t + \theta)$$

$$+ \frac{1}{2} NI_m \cos (\omega t - \theta) + \frac{1}{2} NI_m \cos (\omega t + \theta - 240°) \tag{6.25}$$

$$+ \frac{1}{2} NI_m \cos (\omega t - \theta) + \frac{1}{2} NI_m \cos (\omega t + \theta + 240°)$$

$$\underbrace{\hspace{4cm}}_{\substack{\text{Forward-rotating} \\ \text{components}}} \quad \underbrace{\hspace{4cm}}_{\substack{\text{Backward-rotating} \\ \text{components}}}$$

However, this expression defines a *space* field. Therefore, the second, fourth, and sixth terms, being equal in amplitude and 120° apart, yield a net value of zero. Thus, equation (6.25) simplifies to

$$\mathscr{F}(\theta, t) = \frac{3}{2} NI_m \cos (\omega t - \theta) \tag{6.26}$$

or

$$\mathscr{F}(\theta, t) = \frac{3}{2} \mathscr{F}_m \cos (\omega t - \theta) \tag{6.27}$$

which represents the resultant field mmf wave rotating counterclockwise with an angular velocity of ω radians per second in the air gap. The speed of such a revolving field is usually denoted by ω_s and is referred to as synchronous speed ($\omega_s = \omega$). Suppose that at a given time t_1, the resultant mmf wave is distributed sinusoidally around the air gap with its positive peak occurring along $\theta = \omega t_1$. If, at a later time t_2, the positive peak of the sinusoidally distributed wave is along $\theta = \omega t_2$, then the resultant mmf wave has moved by $\omega(t_2 - t_1)$ around the air gap. Therefore, polyphase currents cause a rotating magnetic field[8] to develop in the air gap as if there were a physically rotating permanent magnet present within the stator of the machine.

6.4 INDUCED VOLTAGES

Assume that the rotor winding is wound-type, wye-connected and open-circuited. Since the rotor winding is open-circuited, no torque can develop. This represents the **standstill operation** of a three-phase induction motor. The application of a three-phase voltage to the three-phase stator winding results in a rotating magnetic field that "cuts" both the stator and rotor windings at the supply frequency f_1. Hence, the rms value of the induced voltage per phase of the rotor winding can be expressed as

$$E_2 = \frac{2\pi}{\sqrt{2}} f_1 N_2 \phi k_{\omega 2} \tag{6.28}$$

or

$$E_2 = 4.44 f_1 N_2 \phi k_{\omega 2} \tag{6.29}$$

where the subscripts 1 and 2 are used to denote stator- and rotor-winding quantities, respectively. Since the rotor is at standstill, the stator frequency f_1 is used in equations (6.28) and (6.29). Here, the flux ϕ is the mutual flux per pole involving both the stator and rotor windings. Similarly, the rms value of the induced voltage per phase of the stator winding can be expressed as

$$E_1 = 4.44 f_1 N_1 \phi k_{\omega 1} \tag{6.30}$$

Thus, it can be shown that

---

[8] It is interesting to note that a reversal of the phase sequence of the currents on the stator windings causes the rotating mmf (as well as the shaft of the motor) to rotate in the opposite direction. For example, if current i_a flows through the phase-a winding as before, but the currents i_b and i_c now flow through the phase-c and phase-b windings, respectively, the rotating mmf (as well as the shaft of the motor) will rotate in a clockwise direction. In summary, the direction of the rotation of a three-phase motor may be reversed by interchanging any of the three motor supply lines.

$$\frac{E_1}{E_2} = \frac{N_1 k_{\omega 1}}{N_2 k_{\omega 2}} \qquad (6.31)$$

where $k_{\omega 1}$ and $k_{\omega 2}$ are the winding factors for the stator and rotor windings, respectively. Since usually they are the same, turn ratio a can be found from

$$\frac{E_1}{E_2} \cong \frac{N_1}{N_2} = a \qquad (6.32)$$

Notice the similarities between the induction motor at standstill and a transformer. Also note that the stator and rotor windings are represented by the primary and secondary, respectively.[9]

6.5 THE CONCEPT OF ROTOR SLIP

In the event that the stator windings are connected to a three-phase supply and the rotor circuit is closed, the induced voltages in the rotor windings produce three-phase rotor currents. These currents in turn cause another rotating magnetic field to develop in the air gap. This induced rotor magnetic field also rotates at the same synchronous speed, n_s. In other words, the stator magnetic field and the rotor magnetic field are stationary with respect to each other. As a result the rotor develops a torque according to the principle of alignment of magnetic fields. Thus, the rotor starts to rotate in the direction of the rotating field of the stator, due to Lenz's law. Here, the stator magnetic field can be considered as dragging the rotor magnetic field. The torque is maintained as long as the rotating magnetic field and the induced rotor currents exist. Also, the voltage induced in the rotor windings depends on the speed of the rotor *relative* to the magnetic fields. At steady-state operation, the rotor's shaft speed[10] n_m is less than the synchronous speed n_s at which the stator rotating field rotates in the air gap. The synchronous speed is determined by the applied stator frequency[11] f_1, in hertz, and the number of poles, p, of the stator winding. Therefore,

$$n_s = \frac{120 f_1}{p} \quad \text{rev/min} \qquad (6.33)$$

Of course, at $n_m = n_s$, there would be no induced voltages or currents in the rotor windings and, therefore, no torque. Thus, *the shaft speed of the rotor can never be equal to the synchronous speed*, but has to be at some value below that speed.

---

[9] Because of such similarities, the induction motor has also been called a "rotating transformer."
[10] It is also called the **mechanical shaft speed of the rotor**.
[11] In other words, the frequency of the applied three-phase supply system.

The **slip speed** (also called the slip rpm) is defined as the difference between synchronous speed and rotor speed and indicates how much the rotor slips[12] behind the synchronous speed. Hence,

$$n_{slip} = n_s - n_m \qquad (6.34)$$

where n_{slip} = slip speed of motor in rpm
$\qquad n_s$ = synchronous speed (i.e., speed of magnetic fields) in rpm
$\qquad n_m$ = mechanical shaft speed of rotor in rpm

Therefore, the term **slip** describes this relative motion in per unit or in percent. Thus the slip in per unit is

$$s = \frac{n_s - n_m}{n_s} \qquad (6.35)$$

and the slip in percent is

$$s = \frac{n_s - n_m}{n_s} \times 100 \qquad (6.36)$$

Alternatively, the slip can be defined in terms of angular velocity ω (radians per second) as

$$s = \frac{\omega_s - \omega_m}{\omega_s} \times 100 \qquad (6.37)$$

By closely inspecting equation (6.35) and Figure 6.4 and simply applying *deductive reasoning*,[13] one can observe the following:

1. If $s = 0$, it means that $n_m = n_s$, that is, the rotor turns at synchronous speed. (In practice, it can only occur if the direct current is injected into the rotor winding.)
2. If $s = 1$, it indicates that $n_m = 0$, that is, the rotor is stationary. In other words, the rotor is at standstill.
3. If $1 > s > 0$, it signals that the rotor turns at a speed somewhere between standstill and synchronous speed. In other words, the motor runs at an asynchronous speed as it should, as illustrated in Figure 6.13(a).
4. If $s > 1$, it signifies that the rotor rotates in a direction opposite of the stator rotating field, as shown in Figure 6.13(c). Therefore, in addition to electrical power, mechanical power (i.e., shaft power) must be provided.

---

[12] The term **slip** is used because it describes what an observer riding with the stator field sees when looking at the rotor; it appears to be slipping backward.
[13] Thanks to Sherlock Holmes, who once said: "You see, but you do not observe!" See the *Adventures of Sherlock Holmes* by Arthur Conan Doyle, 1891.

Since power comes in from both sides, the copper losses of the rotor increase tremendously. The rotor develops a braking torque that forces the motor to stop. This mode of induction machine operation is called **braking** (or **plugging**) **mode**.

5. If $s < 0$, it means that the machine operates as a generator with a shaft speed that is greater than the synchronous speed, as shown in Figure 6.13(b). This mode of operation is called **generating mode**.

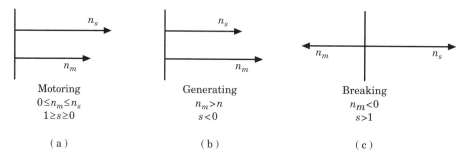

Motoring
$0 \leq n_m \leq n_s$
$1 \geq s \geq 0$

(a)

Generating
$n_m > n$
$s < 0$

(b)

Breaking
$n_m < 0$
$s > 1$

(c)

Figure 6.13   Three operation modes of an induction machine: (a) motoring; (b) generating; (c) plugging.

Also note that the mechanical shaft speed of the rotor can be obtained from the following two equations, which involve only slip and synchronous speed:

$$n_m = (1 - s)n_s \quad \text{rpm} \tag{6.38}$$

and

$$\omega_m = (1 - s)\omega_s \quad \text{rad/s} \tag{6.39}$$

6.6 THE EFFECTS OF SLIP ON THE FREQUENCY AND MAGNITUDE OF INDUCED VOLTAGE OF THE ROTOR

If the rotor of an induction motor is rotating, the frequency of the induced voltages (as well as the induced currents) in the rotor circuit is no longer the same as the frequency of its stator. Under such a running operation, the frequency of the induced voltages (and the currents) in the rotor is directly related to the slip rpm (i.e., the relative speed between the rotating field and the shaft speed of the rotor). Therefore

$$f_2 = \frac{p n_{slip}}{120} \tag{6.40a}$$

or

$$f_2 = \frac{p(n_s - n_m)}{120} \tag{6.40b}$$

where f_2 is the frequency of the voltage and current in the rotor winding. Using equation (6.35), equation (6.40) can be expressed as

$$f_2 = \frac{psn_s}{120} \tag{6.41a}$$

or

$$f_2 = s\frac{pn_s}{120} \tag{6.41b}$$

By substituting equation (6.33) into equation (6.41b),

$$f_2 = sf_1 \tag{6.42a}$$

or

$$f_r = sf_1 \tag{6.42b}$$

That is, the rotor frequency f_2 or f_r is found simply by multiplying the stator's frequency f_1 by the per unit value of the slip. Because of this,[14] f_2 or f_r is also called the slip frequency. Therefore, the voltage induced in the rotor circuit at a given slip s can be found from equation (6.29) simply by replacing f_1 with f_2 as

$$E_r = 4.44 f_2 N_2 \phi k_{\omega 2} \tag{6.43a}$$

or

$$E_r = 4.44 s f_1 N_2 \phi k_{\omega 2} \tag{6.43b}$$

or

$$E_r = s E_2 \tag{6.43c}$$

where E_2 is the induced voltage in the rotor circuit at standstill, that is, at the stator frequency f_1.

The induced currents in the three-phase rotor windings also develop a rotating field. The speed of this rotating magnetic field of the rotor with respect to rotor itself can be found from

$$n_2 = \frac{120 f_2}{p} \tag{6.44a}$$

or

[14] An induction machine with a wound rotor can also be used as a **frequency changer**.

$$n_2 = \frac{120 s f_1}{p} \qquad (6.44\text{b})$$

or

$$n_2 = s n_s \qquad (6.44\text{c})$$

or

$$n_r = s n_s \qquad (6.44\text{d})$$

where n_r is the speed of the rotating magnetic field of the rotor, and n_s is the speed of the rotating magnetic field of the stator. However, since the rotor itself is rotating at n_m, the developed rotor field rotates in the air gap at a speed of

$$n_m + n_2 = (1 - s)n_s + s n_s = n_s \qquad (6.45)$$

Thus, one can prove that both the stator field and the rotor field rotate in the air gap at the same synchronous speed n_s. In other words, the stator and rotor fields are stationary with respect to each other, producing a steady torque and maintaining rotation.

Example 6.1

A three-phase 60-Hz 25-hp wye-connected induction motor operates at a shaft speed of almost 1800 rpm at no load and 1650 rpm at full load. Determine the following:

(a) The number of poles of the motor.
(b) The per-unit and percent slip at full load.
(c) The slip frequency of the motor.
(d) The speed of the rotor field with respect to the rotor itself.
(e) The speed of the rotor field with respect to the stator.
(f) The speed of the rotor field with respect to the stator field.
(g) The output torque of the motor at the full load.

SOLUTION:
(a) From equation (6.33),

$$n_s = \frac{120 f_1}{p}$$

from which

$$p = \frac{120 f_1}{n_s}$$

$$= \frac{120 \times 60}{1800} = 4 \text{ poles}$$

(b) Since

$$n_m = n_s(1 - s)$$

then

$$s = \frac{n_s - n_m}{n_s}$$

$$= \frac{1800 - 1650}{1800}$$

$$= 0.08333 \text{ pu or } 8.33\%$$

(c) The slip frequency is

$$f_2 = sf_1 = 0.0833 \times 60 = 5 \text{ Hz}$$

(d) The speed of the rotor field with respect to the rotor itself can be determined from

$$n_2 = \frac{120 f_2}{p}$$

$$= \frac{120 \times 5}{4} = 150 \text{ rpm}$$

or

$$n_2 = s n_s$$

$$= 0.0833 \times 1800 = 150 \text{ rpm}$$

(e) The speed of the rotor field with respect to the stator can be found from

$$n_m + n_2 = 1650 + 150 = 1800 \text{ rpm}$$

or

$$n_m + n_2 = n_s = 1800 \text{ rpm}$$

(f) The speed of the rotor field with respect to the stator field can determined from

$$(n_m + n_2) - n_s = 1800 - 1800 = 0$$

or since

$$n_m + n_2 = n_s$$

then

$$n_s - n_s = 0$$

(g) The output torque of the motor at the full load can be determined from

$$T_{out} = T_{shaft} = \frac{P_{out}}{\omega_m}$$

$$= \frac{(25 \text{ hp})(746 \text{ W/hp})}{(1650 \text{ rev/min})(2\pi \text{ rad/rev})(1 \text{ min/60 s})} = 108 \text{ N} \cdot \text{m}$$

or in English units,

$$T_{out} = T_{shaft} = \frac{5252 P}{n}$$

$$= \frac{5252(25 \text{ hp})}{1650 \text{ rev/min}} = 79.6 \text{ lb} \cdot \text{ft}$$

6.7 THE EQUIVALENT CIRCUIT OF AN INDUCTION MOTOR

Assume that a three-phase wound rotor[15] induction motor has a balanced wye-connection, as shown in Figure 6.7 so that the currents are always line values and the voltages are always line-to-neutral values. If the currents flow in both the stator and rotor windings, there will be rotating magnetic fields in the air gap. Since these magnetic fields rotate at the same speed in the air gap, they will develop a resultant air-gap field rotating at synchronous speed. Because of this air gap-field, voltages will be induced in the stator windings at the supply frequency f_1 and in the rotor windings at the slip frequency f_2. As with a balanced polyphase transformer, only one phase of the circuit model need be considered.

6.7.1 THE STATOR CIRCUIT MODEL

Figure 6.14(a) shows the equivalent circuit of the stator. The stator terminal voltage differs from the induced voltage (i.e., the counter emf) in the stator winding because of the voltage drop in the stator leakage impedance. Therefore,

---

[15] In the case of a squirrel-cage rotor, the rotor circuit can be represented by an equivalent three-phase rotor winding.

$$V_1 = E_1 + I_1(R_1 + jX_1) \tag{6.46}$$

where V_1 = per-phase stator terminal voltage

$\quad E_1$ = per-phase induced voltage (counter emf) in the stator winding

$\quad I_1$ = stator current

$\quad R_1$ = per-phase stator winding resistance

$\quad X_1$ = per-phase stator leakage reactance

One can easily observe that the equivalent circuit of the stator winding is the same as the equivalent circuit of the transformer winding. As is the case in the transformer model, the stator current I_1 can be separated into two components, that is, a load component I_2 and an excitation component I_e. Here, the load component I_2 produces an mmf that exactly counteracts the mmf of the rotor current. The excitation component I_e is the extra stator current needed to create the resultant air-gap flux. In the shunt branch of the model, R_c and X_m represent per-phase stator core loss resistance and per-phase stator magnetizing reactance, respectively, as is the case in transformer theory. However, the magnitude of the parameters are considerably different. For example, I_e is much larger in the induction machine due to the air gap. It can be as high as 30 to 50 percent of the rated current in an induction machine versus 1 to 5 percent in a transformer. Due to the air gap, the value of magnetizing reactance X_m is relatively small in comparison to that of a transformer; but the leakage reactance X_1 is greater than the magnetizing reactance than in transformers. Another reason for this is that the stator and rotor windings are distributed along the periphery of the air gap instead of being stacked on a core as they are in transformers.

6.7.2 THE ROTOR CIRCUIT MODEL

Figure 6.14(b) shows the actual rotor circuit of an induction motor operating under load at a slip s. The rotor current per phase can be expressed as

$$I_2 = \frac{sE_2}{R_2 + jsX_2} \tag{6.47}$$

where E_2 = per-phase induced voltage in the rotor at standstill (i.e., at stator frequency f_1)

$\quad R_2$ = per-phase rotor circuit resistance

$\quad X_2$ = per-phase rotor leakage inductance *ive* reactance

The figure illustrates that I_2 is a slip-frequency current produced by the slip-frequency induced emf sE_2 acting in a rotor circuit with an impedance per phase of $R_2 + jsX_2$. Therefore, the total rotor copper loss can be expressed as

$$P_{2,cu} = 3I_2^2 R_2 \tag{6.48}$$

which represents the amount of real power involved in the rotor circuit. Equation (6.47) can be rewritten by dividing both the numerator and the denominator by the slip s so that

$$I_2 = \frac{E_2}{(R_2/s) + jX_2} \tag{6.49}$$

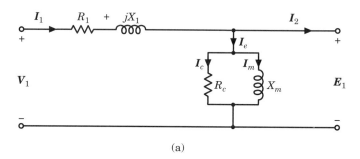

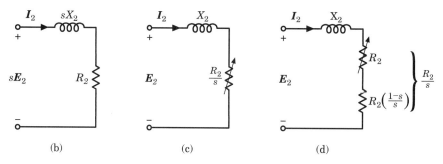

Figure 6.14   Development of the per-phase stator and rotor-equivalent circuits of an induction motor: (a) stator-equivalent circuit; (b) actual rotor circuit; (c) rotor-equivalent circuit; (d) modified equivalent rotor circuit.

This equation suggests the rotor equivalent circuit shown in Figure 6.14(c). Of course, the magnitude and phase angle of I_2 remain unchanged by this process, but there is a significant difference between these two equations and the circuits they represent. The current I_2 given by equation (6.47) is at slip frequency f_2, whereas I_2 given by equation (6.49) is at line frequency f_1. Also in equation (6.47) the rotor leakage reactance sX_2 changes with speed, but the resistance R_2 remains unchanged; whereas in equation (6.49) the resistance R_2/s changes with speed, but the leakage reactance X_2 remains unchanged. The total rotor copper loss associated with the equivalent rotor circuit shown in Figure 6.14(c) is

$$P = 3I_2^2 \frac{R_2}{s} = \frac{P_{2,cu}}{s} \tag{6.50}$$

Since induction machines are run at low slips the power associated with Figure 6.14(c) is substantially greater. The equivalent circuit given in Figure 6.14(c) is at the stator frequency and therefore is the rotor-equivalent circuit as seen from the stator. Thus, the power determined by using equation (6.50) is the power transferred across the air gap (i.e., P_g) from the stator to the rotor which includes the rotor copper loss as well as the developed mechanical power. Here, the equation can be expressed in a manner that stresses this fact. Therefore,

$$P = P_g = 3I_2^2 \frac{R_2}{s} = 3I_2^2 [R_2 + \frac{R_2}{s}(1-s)] \tag{6.51}$$

The corresponding equivalent circuit is shown in Figure 6.14(d). The speed-dependent resistance[16] $R_2(1-s)/s$ represents the mechanical power developed by the induction machine to overcome the mechanical shaft load. Therefore, the total developed mechanical power can be found from

$$P_d = P_{mech} = 3I_2^2 \frac{R_2}{s}(1-s) \tag{6.52a}$$

or

$$P_d = P_{mech} = (1-s)P_g \tag{6.52b}$$

or

$$P_d = P_{mech} = \frac{1-s}{s} P_{2,cu} \tag{6.52c}$$

where

$$P_{2,cu} = 3I_2^2 R_2 = sP_g \tag{6.53}$$

A small portion of the developed mechanical power is also lost due to windage and friction. The rest of the mechanical power is defined as the *output shaft power*.

6.7.3 THE COMPLETE EQUIVALENT CIRCUIT

If the stator-equivalent circuit shown in Figures 6.14(a) and 6.14(c) and the rotor-equivalent circuit shown in Figure 5.14(d) are at the same line frequency f_1, they can be joined together. However, if the turns in the stator winding and the rotor winding are different, then E_1 and E_2 can be different, as shown in Figure 6.15(a). Because of this, the turns ratio $(a = N_1/N_2)$ needs to be taken

---

[16] It is known as the **dynamic resistance** or **load resistance**. Note that in the braking mode of the operation, this resistance is negative and represents a source of energy.

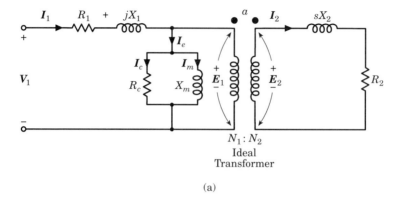

(a)

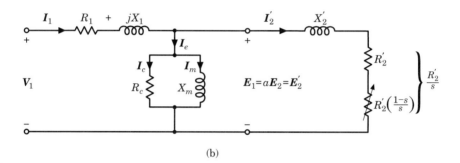

(b)

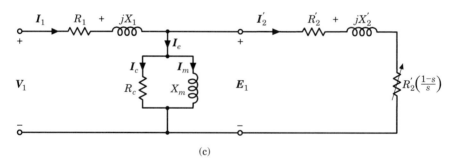

(c)

Figure 6.15   The per-phase equivalent circuit of an induction motor: (a) transformer model of an induction motor; (b) exact equivalent circuit; (c) alternative form of the equivalent circuit.

into account. Figure 6.15(c) shows the resultant equivalent circuit of the induction machine. Notice that such an equivalent circuit form is identical to that of the two-winding transformer. Also note that the prime notation is used to denote *stator-referred* rotor quantities. Therefore,

$$I'_2 = \frac{I_2}{a} \tag{6.54}$$

$$E_1 = aE_2 = E'_2 \tag{6.55}$$

$$R'_2 = a^2 R_2 \tag{6.56}$$

$$X'_2 = a^2 X_2 \tag{6.57}$$

$$R'_2 \left(\frac{1-s}{s} \right) = a^2 R_2 \left(\frac{1-s}{s} \right) \tag{6.58}$$

Due to the presence of the air gap in the induction machine, the magnetizing impedance is low and therefore the exciting current I_e is high (about 30 to 50 percent of full-load current). The leakage reactance X_1 is also high. The equivalent circuit of the induction machine can be simplified by omitting resistance R_c and lumping the corresponding core loss with the friction and windage losses. The error involved is negligible. Figure 6.16 shows the resultant equivalent circuit.[17] Therefore, if core loss is assumed to be constant, then such an equivalent circuit should be used. Note that all stator-referred rotor quantities are shown without prime notation as is customary. However, from now on *it should be understood that they are stator-referred.*

6.7.4 THE APPROXIMATE EQUIVALENT CIRCUIT

The computation can be simplified with very little loss of accuracy, by moving the magnetizing (shunt) branch (i.e., R_c and X_m) to the machine terminals as shown in Figure 6.17. This modification is based mainly on the assumption

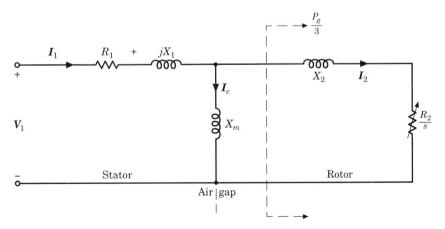

Figure 6.16   Exact equivalent circuit with the core-loss resistor omitted.

---

[17] It is recommended by IEEE and known as the **Steinmetz model of one phase of a three-phase induction machine**.

that $V_1 = E_1 = E_2$. A further simplification can be achieved by also omitting the resistance R_c.

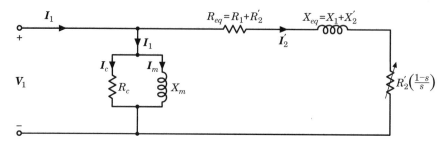

Figure 6.17   Approximate equivalent circuit.

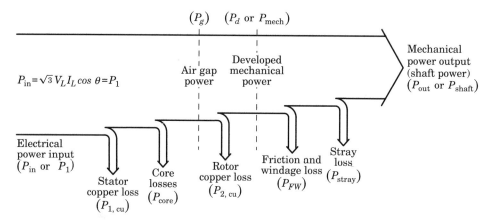

Figure 6.18   Power-flow diagram of an induction motor.

6.8 PERFORMANCE CALCULATIONS

Figure 6.18 shows the power-flow diagram of an induction motor. It is based on the equivalent circuit shown in Figure 6.15(c). The input power is the electrical power input to the stator of the motor. Therefore,

$$P_{in} = P_1 = \sqrt{3}V_L I_L \cos\theta \tag{6.59}$$

The total stator copper losses are

$$P_{1,cu} = 3I_1^2 R_1 \tag{6.60}$$

The total core losses can be found from

$$P_{core} = 3E_1^2 G_c \tag{6.61a}$$

or

$$P_{core} = \frac{3E_1^2}{R_c} \tag{6.61b}$$

Therefore, the total air-gap power can be given as

$$P_g = P_{in} - P_{1,cu} - P_{core} \tag{6.62a}$$

or

$$P_g = 3I_2^2 \frac{R_2}{s} \tag{6.62b}$$

The total rotor copper losses are

$$P_{2,cu} = 3I_2^2 R_2 \tag{6.63a}$$

or

$$P_{2,cu} = sP_g \tag{6.63b}$$

Thus the total mechanical power developed can be found by

$$P_d = P_{mech} = P_g - P_{2,cu} \tag{6.64a}$$

or

$$P_d = P_{mech} = P_g(1-s) \tag{6.64b}$$

or

$$P_d = P_{mech} = \frac{1-s}{s} P_{2,cu} \tag{6.64c}$$

If the friction and windage losses and the stray losses[18] are known, the output power (or shaft power) can be determined from

$$P_{out} = P_{shaft} = P_d - P_{FW} - P_{stray} \tag{6.65}$$

If the core losses are assumed to be constant, they can be lumped with the friction and windage losses, and the stray losses. Their sum is called the **rotational losses**. Thus, the rotational loss is given as

[18] **Stray losses** consist of all losses not otherwise included above, e.g., the losses due to nonuniform current distribution in the copper, and additional core losses developed in the iron core as a result of distortion in the magnetic flux by the load current. They may also include losses due to harmonic fields. The stray losses are also called **miscellaneous losses**.

$$P_{rot} = P_{core} + P_{FW} + P_{stray} \tag{6.66}$$

Therefore, the corresponding output power can be found from

$$P_{out} = P_d - P_{rot} \tag{6.67a}$$

or

$$P_{out} = P_d - (P_{core} + P_{FW} + P_{stray}) \tag{6.67b}$$

The corresponding equivalent circuit is shown in Figure 6.16.

The developed torque is defined as the mechanical torque developed by the electromagnetic energy conversion process. It can be found by dividing the developed power by the shaft speed. Therefore, the developed torque[19] can be expressed as

$$T_d = \frac{P_d}{\omega_m} \tag{6.68a}$$

or

$$T_d = \frac{P_g(1-s)}{\omega_s(1-s)} \tag{6.68b}$$

or

$$T_d = \frac{P_g}{\omega_s} \tag{6.68c}$$

or

$$T_d = \frac{3I_2^2 R_2}{s\omega_s} \tag{6.68d}$$

The output torque (or shaft torque) is

$$T_{out} = \frac{P_{out}}{\omega_m} \tag{6.69}$$

The efficiency of the induction motor can be determined from

---

[19] Because the developed torque can be expressed by equation (6.68c), the air-gap power is also called the **torque in synchronous watts**.

$$\eta = \frac{P_{out}}{P_{in}} \qquad (6.70a)$$

or

$$\eta = \frac{P_{out}}{P_{out} + P_{loss}} \qquad (6.70b)$$

where P_{loss} represents the total losses.

Example 6.2

A three-phase 480-V 50-hp induction motor is supplied 70 A at 0.8 lagging power factor. Its stator and rotor copper losses are 4257.53 W and 1000 W, respectively. Its core losses are 3000 W, the friction and windage losses are 800 W, and the stray losses are 200 W. Determine the following:
(a) The air-gap power.
(b) The mechanical power developed.
(c) The shaft output power.
(d) The efficiency of the motor.

SOLUTION:
(a) Since

$$P_{in} = P_1 = \sqrt{3} V_L I_L \cos \theta$$

$$= \sqrt{3}(480\ \text{V})(70\ \text{A})0.8 = 46{,}557.53\ \text{W}$$

therefore, the air-gap power is

$$P_g = P_{in} - P_{1,\,cu} - P_{core}$$

$$= 46{,}557.53 - 4{,}257.53 - 3{,}000 = 39{,}300\ \text{W}$$

(b) The developed mechanical power is the same as the developed power. Thus,

$$P_d = P_{mech} = P_g - P_{2,\,cu}$$

$$= 39{,}300 - 1{,}000 = 38{,}300\ \text{W}$$

(c) The shaft output power can be found as

$$P_{out} = P_d - P_{FW} - P_{stray}$$

$$= 38{,}300 - 800 - 200 = 37{,}300\ \text{W}$$

or in horsepower,

$$P_{out} = (37,300\text{ W})\left(\frac{1\text{ hp}}{746\text{ W}}\right) = 50\text{ hp}$$

(d) The efficiency of the motor is

$$\eta = \frac{37,300\text{ W}}{46,557.53\text{ W}} \times 100 = 80.1\text{ \%}$$

Example 6.3

A three-phase two-pole 50-hp 480-V 60-Hz wye-connected induction motor has the following constants in ohms per phase referred to the stator:

$$R_1 = 0.322\text{ }\Omega \qquad\qquad R_2 = 0.196\text{ }\Omega$$
$$X_1 = 0.675\text{ }\Omega \qquad\qquad X_2 = 0.510\text{ }\Omega$$
$$X_m = 12.5\text{ }\Omega$$

The total rotational losses are 1850 W and are assumed to be constant. The core loss is lumped in with the rotational losses. For a rotor slip of 3 percent at the rated voltage and rated frequency, determine:
(a) The speed in rpm and in rad/s.
(b) The stator current.
(c) The power factor.
(d) The developed power and output power.
(e) The developed torque and output torque.
(f) The efficiency.

SOLUTION:
(a) The synchronous speed is

$$n_s = \frac{120f_1}{p}$$

$$= \frac{120(60\text{ Hz})}{2} = 3600\text{ rev/min}$$

or

$$\omega_s = (3600\text{ rev/min})\left(\frac{2\pi\text{ rad}}{1\text{ rew}}\right)\left(\frac{1\text{ min}}{60\text{ s}}\right) = 376.99\text{ rad/s}$$

Thus, the rotor's mechanical shaft speed is

$$n_m = (1-s)n_s$$

$$= (1-0.03)3600 = 3492\text{ rpm}$$

or

$$\omega_m = (1-s)\omega_s$$

$$= (1-0.03)376.99 = 365.68 \text{ rad/s}$$

(b) Since the core loss is assumed to be constant, the appropriate equivalent circuit for the motor is the one shown in Figure 6.16. To determine the stator current, the equivalent impedance of the circuit has to be found. Therefore, the referred rotor impedance is found from

$$\mathbf{Z}_2 = \frac{R_2}{s} + jX_2$$

$$= \frac{0.196}{0.03} + j0.510 = 6.55\angle 4.46° \; \Omega$$

Since this rotor impedance is in parallel with the magnetization branch, the corresponding impedance is

$$\mathbf{Z}_{eq} = \frac{1}{\dfrac{1}{jX_m} + \dfrac{1}{\mathbf{Z}_2}}$$

$$= \frac{1}{\dfrac{1}{j12.5} + \dfrac{1}{6.55\angle 4.46°}} = 5.63\angle 31.13° \; \Omega$$

Therefore, the total impedance is

$$\mathbf{Z}_{tot} = (R_1 + jX_1) + \mathbf{Z}_{eq}$$

$$= (0.322 + j0.675) + 5.63\angle 31.13° = 6.265\angle 34.89° \; \Omega$$

Thus, the stator current is

$$\mathbf{I}_1 = \frac{\mathbf{V}_1}{\mathbf{Z}_{tot}}$$

$$= \frac{\left(\dfrac{480}{\sqrt{3}}\right)\angle 0°}{6.265\angle 34.89°} = 44.24\angle -34.89° \text{ A}$$

(c) The power factor of the motor is

$$PF = \cos 34.89° = 0.82 \text{ lagging}$$

(d) The input power to the motor is

$$P_{in} = P_1 = \sqrt{3}V_1 I_1 \cos\theta$$

$$= \sqrt{3}(480 \text{ V})(44.24\,\text{A})(0.82) = 30{,}166.38 \text{ W}$$

The stator copper losses are

$$P_{1,\,cu} = 3I_1^2 R_1$$

$$= 3(44.24 \text{ A})^2(0.322 \text{ }\Omega) = 1890.27 \text{ W}$$

The air-gap power is

$$P_g = P_{in} - P_{1,\,cu}$$

$$= 30{,}186.33 - 1890.27 = 28{,}276.2 \text{ W}$$

Thus, the developed power is

$$P_d = (1 - s)P_g$$

$$= (1 - 0.03)28{,}276.1 = 27{,}427.82 \text{ W}$$

Therefore, the output power is

$$P_{out} = P_d - P_{rot}$$

$$= 27{,}427.82 - 1850 = 25{,}577.82 \text{ W}$$

(e) The developed torque is

$$T_d = \frac{P_g}{\omega_s}$$

$$= \frac{28{,}276.1}{376.99} = 75 \text{ N} \cdot \text{m}$$

The output torque is

$$T_{out} = \frac{P_{out}}{\omega_m}$$

$$= \frac{25,577.82}{365.58} = 69.96 \, \text{N} \cdot \text{m}$$

(f) The motor's efficiency at this operating condition is

$$\eta = \frac{P_{out}}{P_{in}} \times 100$$

$$= \frac{25,577.82}{365.68} \times 100 = 84.79 \, \%$$

30, 166.38

6.9 THE EQUIVALENT CIRCUIT AT START-UP

At start-up the rotor is at standstill and therefore the slip of the motor is 1.0. The corresponding equivalent circuit is the same as the one shown in Figure 6.16, except that all the values of slip are set to a value of 1.0. All powers and torques can be found as shown before, except for the output quantities. Since the motor is at standstill, there are no windage and friction losses. Furthermore, since ω_m is zero, T_{out} is undefined. Similarly, if equation (6.68a) is used, T_d is also undefined. However, T_d can be found by using equations (6.68c) and (6.68d) and setting s equal to 1.0. Therefore, the starting torque can be determined from

$$T_{d, \, start} = \frac{P_g}{\omega_m} \tag{6.71}$$

or

$$T_{d, \, start} = \frac{3I_2^2 R_2}{\omega_s} \tag{6.72}$$

Example 6.4

Consider the induction motor given in Example 6.3 and assume that the rotor is at standstill. Determine the following:
(a) The speed at start-up.
(b) The stator current at start-up.
(c) The power factor at start-up.
(d) The developed power and output power at start-up.
(e) The developed torque and output torque at start-up.

SOLUTION:
(a) The synchronous speed is

$$n_S = 3600 \text{ rev/min}$$

or

$$\omega_S = 376.99 \text{ rad/s}$$

However, the rotor's mechanical shaft speed is

$$n_m = (1-s)n_S = (1-1)3600 = 0$$

or

$$\omega_m = (1-s)\omega_S = (1-1)376.99 = 0$$

(b) The referred rotor impedance is

$$\boldsymbol{Z}_2 = \frac{R_2}{s} + jX_2$$

$$= \frac{0.196}{1.0} + j0.510 = 0.5464\angle 68.98° \ \Omega$$

Also

$$\boldsymbol{Z}_{eq} = \frac{1}{\dfrac{1}{jX_m} + \dfrac{1}{\boldsymbol{Z}_2}}$$

$$= \frac{1}{\dfrac{1}{j12.5} + \dfrac{1}{0.5464\angle 68.98°}} = 0.558\angle 83.75° \ \Omega$$

Thus, the total impedance is

$$\boldsymbol{Z}_{tot} = (R_1 + jX_1) + \boldsymbol{Z}_{eq}$$

$$= (0.322 + j0.675) + 0.558\angle 83.75° = 1.29\angle 72.71° \ \Omega$$

Therefore, the stator current is

$$\boldsymbol{I}_1 = \frac{\boldsymbol{V}_1}{\boldsymbol{Z}_{tot}}$$

$$= \frac{\dfrac{480}{\sqrt{3}}\angle 0°}{1.29\angle 72.71°} = 214.83\angle -72.71° \text{ A}$$

Notice that the starting current is almost five times the load current found in Example 6.3. (*Such a starting current would blow the fuses!*)

(c) The power factor of the motor is

$$PF = \cos 72.71° = 0.3 \text{ lagging}$$

(d) The input power to the motor is

$$P_{in} = P_1 = \sqrt{3}\, V_1 I_1 \cos \theta$$

$$= \sqrt{3}\,(480 \text{ V})(214.83 \text{ A})0.3 = 53{,}083.27 \text{ W}$$

The stator copper losses are

$$P_{1,\,cu} = 3I_1^2 R_1$$

$$= 3(214.83)^2 0.322 = 44{,}582.76 \text{ W}$$

The air-gap power is

$$P_g = P_{in} - P_{1,\,cu}$$

$$= 53{,}083.27 - 44{,}582.76 = 8{,}500.5 \text{ W}$$

Therefore, the developed power is

$$P_d = (1 - s)P_g$$

$$= (1 - 1)8{,}500.5 = 0$$

Thus, the output power is

$$P_{out} = P_d - P_{rot} = 0$$

(e) The developed torque is

$$T_{d,\,start} = \frac{P_g}{\omega_s}$$

$$= \frac{8{,}500.5}{376.99} = 22.55 \text{ N} \cdot \text{m}$$

Example 6.5

A three-phase two-pole 60-Hz induction motor provides 25 hp to a load at a speed of 3420 rpm. If the mechanical losses are zero, determine the following:
(a) The slip of the motor in percent.

(b) The developed torque.

(c) The shaft speed of the motor, if its torque is doubled.

(d) The output power of the motor, if its torque is doubled.

SOLUTION:

(a) Since

$$n_s = \frac{120 f_1}{p}$$

$$= \frac{120 \times 60}{2} = 3600 \text{ rpm}$$

the slip is

$$s = \frac{n_s - n_m}{n_s} \times 100$$

$$= \frac{3600 - 3420}{3600} \times 100 = 5 \text{ \%}$$

(b) Since the mechanical losses are zero,

$$T_d = T_{load} = T_{out}$$

and

$$P_d = P_{load} = P_{out}$$

the developed torque is

$$T_d = \frac{P_d}{\omega_m}$$

$$= \frac{(25 \text{ hp})(746 \text{ W/hp})}{(3420 \text{ rpm})(2\pi \text{ rad/rev})(1 \text{ min/60 s})} = 52.07 \text{ N} \cdot \text{m}$$

or in English units,

$$T_d = \frac{5252 P_d}{n_m}$$

$$= \frac{5252(25 \text{ hp})}{4320 \text{ rpm}} = 38.4 \text{ lb} \cdot \text{ft}$$

Alternatively, the torque in lb·ft can be found directly from

$$T_d = \frac{550}{746}(T_d \text{ N·m})$$

$$= \frac{550(52.07 \text{ N·m})}{746} = 38.4 \text{ lb·ft}$$

(c) The developed torque is proportional to the slip. If the developed torque is doubled, then the slip also doubles and the new slip is

$$s = 2 \times 0.05 = 0.10$$

Hence, the shaft speed becomes

$$n_m = (1-s)n_s$$

$$= (1-0.10)3600 = 3240 \text{ rpm}$$

(d) Since

$$P_d = T_d \omega_m$$

the power supplied by the motor is

$$P_d = (2 \times 52.07)[(3240 \text{ rpm})(2\pi \text{ rad/rew})(1 \text{ min/60 s})]$$

$$= 35,333.9 \text{ W}$$

or in English units,

$$P_d = \frac{T_d n_m}{5252}$$

$$= \frac{(2 \times 38.4)(3240 \text{ rpm})}{5252} = 47.4 \text{ hp}$$

6.10 DETERMINATION OF POWER AND TORQUE BY USE OF THÉVENIN'S EQUIVALENT CIRCUIT

According to Thévenin's theorem, a network of linear impedances and voltage sources can be represented by a single voltage source and a single impedance as viewed from two terminals. The equivalent voltage source is the voltage that appears across these terminals when the terminals are open-circuited. The equivalent impedance is the impedance that can be found by looking into the network from the terminals with all voltage sources short-circuited.

Therefore, to find the current I_2 in Figure 6.19(a), Thévenin's theorem can be applied to the induction-motor equivalent circuit. The Thévenin voltage can be found by separating the stator circuit from the rotor circuit, as indicated in the figure. Thus, by voltage division,

$$V_{th} = V_1 \frac{jX_m}{R_1 + jX_1 + jX_m} \tag{6.73}$$

The magnitude of the Thévenin voltage is

$$V_{th} = V_1 \frac{X_m}{[R_1^2 + (X_1 + X_m)^2]^{1/2}} \tag{6.74}$$

However, since $R_1^2 \ll (X_1 + X_m)^2$, the voltage is approximately

$$V_{th} \cong V_1 \frac{X_m}{X_1 + X_m} \tag{6.75}$$

To find the Thévenin impedance, the V_1 voltage source has to be replaced by a short circuit, as shown in Figure 6.19(b). Therefore, the Thévenin impedance is

$$Z_2 = R_{th} + jX_{th} = \frac{jX_m(R_1 + jX_1)}{R_1 + j(X_1 + X_m)} \tag{6.76}$$

Since $X_1 \ll X_m$ and $R_1^2 \ll (X_1 + X_m)^2$, the Thévenin resistance and reactance are approximately

$$R_{th} \cong R_1 \left(\frac{X_m}{X_1 + X_m} \right)^{1/2} \tag{6.77}$$

and

$$X_{th} \cong X_1 \tag{6.78}$$

Figure 6.19(c) shows the resultant equivalent circuit of the induction motor. Here, the rotor current can be found from

$$I_2 = \frac{V_{th}}{Z_{th} + Z_{th}} \tag{6.79a}$$

or

$$I_2 = \frac{V_{th}}{R_{th} + R_2/s + j(X_{th} + X_2)} \tag{6.79b}$$

The magnitude of the rotor current is

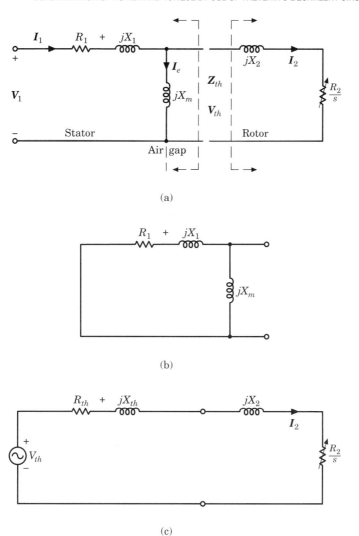

(a)

(b)

(c)

Figure 6.19   (a) Application of Thevénin's theorem to the induction motor circuit model; (b) the stator circuit used to determine the Thevénin-equivalent impedance of the stator circuit; (c) the resultant induction-motor equivalent circuit simplified by Thevénin's theorem.

$$I_2 = \frac{V_{th}}{[(R_{th} + R_2/s)^2 + (X_{th} + X_2)^2]^{1/2}} \tag{6.80}$$

Thus, the corresponding air-gap power is

$$P_g = 3I_2^2 \frac{R_2}{s} \tag{6.81a}$$

or

$$P_g = \frac{3V_{th}^2 \dfrac{R_2}{s}}{\omega_s[(R_{th} + R_2/s)^2 + (X_{th} + X_2)^2]} \qquad (6.81\text{b})$$

Therefore, the developed torque is

$$T_d = \frac{P_g}{\omega_s} = \frac{3I_2^2 \dfrac{R_2}{s}}{\omega_s} \qquad (6.82\text{a})$$

or

$$T_d = \frac{3V_{th}^2 \dfrac{R_2}{s}}{\omega_s[(R_{th} + R_2/s)^2 + (X_{th} + X_2)^2]} \qquad (6.82\text{b})$$

Since at start-up the slip is unity, the developed starting torque is

$$T_{start} = \frac{3V_{th}^2 R_2}{\omega_s[(R_{th} + R_2)^2 + (X_{th} + X_2)^2]} \qquad (6.83)$$

6.11 PERFORMANCE CHARACTERISTICS

The performance characteristics of the induction machine include starting torque, maximum (or pull-out) torque,[20] maximum power, current, power factor, and efficiency. The maximum torque can be determined by using Thévenin's equivalent circuit. Since

$$T_d = \frac{P_g}{\omega_s} \qquad (6.84)$$

the developed torque will be maximum when the air-gap power is a maximum. The air-gap power can be found from equation (6.81). To find at what value of the variable R_2/s the maximum P_g takes place, the derivative of the right side of equation (6.81) with respect to R_2/s must be determined and set equal to zero. Thus,

$$\frac{3V_{th}^2[R_{th}^2 - (R_2/s)^2 + (X_{th} + X_2)^2]}{[(R_{th} + R_2/s)^2 + (X_{th} + X_2)^2]^2} = 0 \qquad (6.85)$$

by setting the numerator of this equation equal to zero,

---

[20] It is also known as the **maximum internal** or **breakdown torque**.

$$R_{th}^2 - (R_2/s)^2 + (X_{th} + X_2)^2 = 0 \qquad (6.86)$$

from which

$$\frac{R_2}{s} = [R_{th}^2 + (X_{th} - X_2)^2]^{1/2} \qquad (6.87)$$

That is, the maximum power is transferred to the air-gap power resistor $R_2/2$ when this resistor is equal to the impedance looking back into the source. Therefore, the slip $s_{\max T}$ at which the maximum (or pull-out) torque is developed is

$$s_{\max T} = \frac{R_2}{[R_{th}^2 + (X_{th} + X_2)^2]^{1/2}} \qquad (6.88)$$

The slip at which maximum torque takes place is directly proportional to the rotor resistance and may be increased by using a larger rotor resistance. Thus, the **maximum** or **pull-out torque** can be found, by inserting equation (6.88) into equation (6.82), as

$$T_{\max} = \frac{3V_{th}^2}{2\omega_s\{R_{th} + [R_{th}^2 + (X_{th} + X_2)^2]^{1/2}\}} \qquad (6.89)$$

The maximum torque is proportional to the square of the supply voltage and is also inversely related to the size of the stator resistance and reactance, and the rotor reactance.

Figure 6.20 shows the torque-speed or torque-slip characteristic curve of an induction motor. As shown in the figure, the full-load torque of an induction motor is less than its starting torque. If the value of the supply voltage is halved, both the starting torque and the maximum torque become one-fourth of their respective full-voltage values. Figure 6.21 shows the torque versus speed (or slip) characteristic curves of an induction motor at full and half of the supply voltage.

According to equation (6.89), the maximum torque developed by the induction motor is independent of the rotor winding resistance. Note that the value of the rotor winding resistance R_2 determines the speed at which the maximum torque will take place, as suggested by equation (6.88). In other words, increasing the rotor winding resistance by inserting external resistance increases the slip at which the maximum (or pull-out) torque occurs, but leaves its magnitude unchanged. Figure 6.22 shows the effect of increasing the rotor resistance on the torque-speed characteristics of a wound rotor induction motor. Notice that as the rotor resistance increases, the curve becomes flatter. Also notice that the starting torque and the maximum torque are the same at the given rotor resistance value.

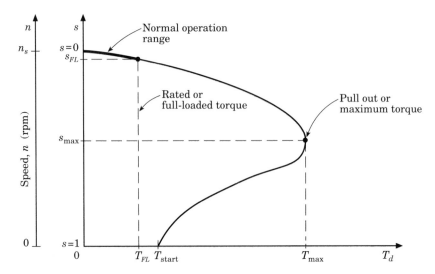

Figure 6.20 Torque-speed or torque-slip characteristic curve of an induction motor.

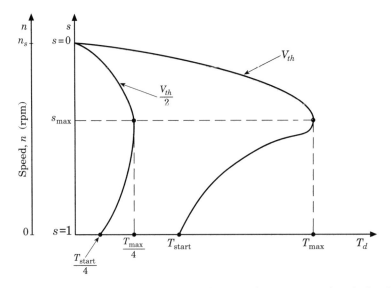

Figure 6.21 Torque versus speed (or slip) characteristic curves of an induction motor at full and halved supply voltage.

As can be observed in Figure 6.16, the input impedance of an induction motor is

$$\boldsymbol{Z}_1 = R_1 + jX_1 + \frac{jX_m(R_2/s + jX_2)}{R_2/s + j(X_m + X_2)} \qquad (6.90a)$$

or

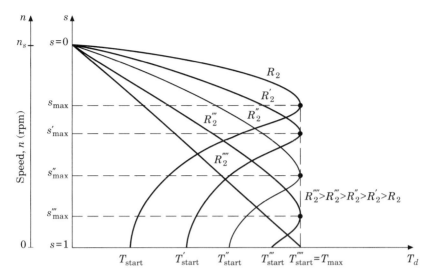

Figure 6.22   The effect of increased rotor resistance on the torque-speed characteristics of a wound-rotor induction motor.

$$\boldsymbol{Z}_1 = |\boldsymbol{Z}_1| \angle \theta_1 \qquad\qquad (6.90\text{b})$$

Therefore, the stator current is

$$\boldsymbol{I}_1 = \frac{\boldsymbol{V}_1}{\boldsymbol{Z}_1} = \boldsymbol{I}_e + \boldsymbol{I}_2 \qquad\qquad (6.91)$$

At synchronous speed (i.e., at $s = 0$), the resistance R_2/s becomes infinite and therefore $\boldsymbol{I}_2$ is zero. Thus, the stator current $\boldsymbol{I}_1$ becomes equal to the excitation current $\boldsymbol{I}_e$. At greater values of slip $R_2/s + jX_2$ is low and hence the resultant $\boldsymbol{I}_2$ and $\boldsymbol{I}_1$ are larger. For example, as illustrated in Figure 6.23, the typical starting current (i.e., at $s = 1$) is 500 to 700 percent of the rated (full-load) current. At synchronous speed, the typical stator current is 25 to 50 percent of the full-load current.

The power factor of an induction motor is $\cos \theta_1$ where θ_1 is the phase angle of the stator current $\boldsymbol{I}_1$. This phase angle θ_1 is the same as the input impedance angle of the equivalent circuit shown in Figure 6.16 and given by equation (6.90b). Figure 6.24 shows the typical power factor variation as a function of output power and slip. Note that the figure is not drawn to scale.

Figure 5.25 shows the efficiency of a typical induction motor as a function of output power and slip. The full-load efficiency of a large induction motor may be as high as 95 percent. As an induction motor is loaded beyond its rated output power, its efficiency decreases considerably.

Example 6.6

The induction motor given in Example 6.3 has a wound rotor. Determine the following:

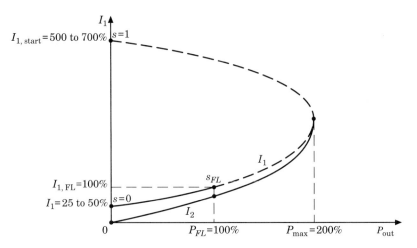

Figure 6.23   Stator and rotor currents as a function of output power and slip.

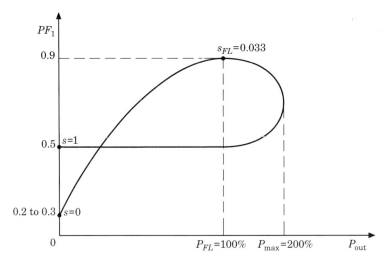

Figure 6.24   Power factor of a typical induction motor as a function of output power and slip.

(a) The slip at which the maximum torque is developed.
(b) The speed at which the maximum torque is developed.
(c) The maximum torque developed.
(d) The starting torque developed.
(e) The speed at which the maximum torque is developed, if the rotor resistance is doubled.
(f) The maximum torque developed, if the rotor resistance is doubled.
(g) The starting torque developed, if the rotor resistance is doubled.

SOLUTION:
The Thévenin voltage of the motor is

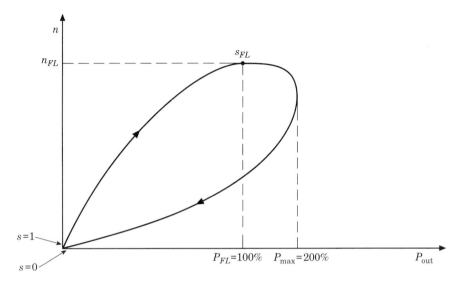

Figure 6.25   The efficiency of a typical induction motor as a function of output power and slip.

$$V_{th} = V_1 \frac{X_m}{[R_1^2 + (X_1 + X_m)^2]^{1/2}}$$

$$= \frac{(480/\sqrt{3})12.5}{[0.322^2 + (0.675 + 12.5)^2]^{1/2}} = 262.85 \text{ V}$$

The Thévenin resistance is

$$R_{th} \cong R_1 \left(\frac{X_m}{X_1 + X_m} \right)^2$$

$$= 0.322 \left(\frac{12.5}{0.675 + 12.5} \right)^2 = 0.29 \ \Omega$$

The Thévenin reactance is

$$X_{th} \cong X_1 = 0.675 \ \Omega$$

(a) The slip at which the maximum torque is developed is

$$s_{\max T} = \frac{R_2}{[R_{th}^2 + (X_{th} + X_2)^2]^{1/2}}$$

$$= \frac{0.196}{0.29^2 + [(0.675 + 0.51)^2]^{1/2}} = 0.1607$$

(b) The speed at which the maximum torque is developed is

$$n_m = (1 - s)n_s$$

$$= (1 - 0.1607)3600 = 3021.6 \text{ rpm}$$

(c) The maximum torque developed is

$$T_{max} = \frac{3V_{th}^2}{2\omega_s\{R_{th} + [R_{th}^2 + (X_{th} + X_2)^2]^{1/2}\}}$$

$$= \frac{3(262.85)^2}{2(377\text{rad/s})\{0.29 + [0.29^2 + (0.675 + 0.51)^2]^{1/2}\}} = 182\text{N} \quad \sim \! m$$

(d) The starting torque developed is

$$T_{start} = \frac{3V_{th}^2 R_2}{\omega_s[(R_{th} + R_2)^2 + (X_{th} + X_2)^2]}$$

$$= \frac{3(262.85)^2 0.196}{377[(0.29 + 0.196)^2 + (0.675 + 0.51)^2]} = 107.3 \text{ N} \cdot \text{m}$$

(e) Since the rotor resistance is doubled, the slip at which the maximum torque occurs also doubles. Thus,

$$s_{max} = 2(0.1607) = 0.3124$$

and the speed at maximum torque is

$$n_m = (1 - s)n_s$$

$$= (1 - 0.3214)3600 = 2443 \text{ rpm}$$

(f) The maximum torque still remains at

$$T_{max} = 182 \text{ N} \cdot \text{m}$$

(g) If the rotor resistance is doubled, the developed starting torque becomes

$$T_{start} = \frac{3(262.85)^2 0.392}{377[(0.29 + 0.196)^2 + (0.675 + 0.51)^2]} = 131.4 \text{ N} \cdot \text{m}$$

6.12 CONTROL OF MOTOR CHARACTERISTICS BY SQUIRREL-CAGE ROTOR DESIGN

In principle, an increase in rotor resistance decreases the speed at which a given torque is found, increases the starting torque, and lowers the motor efficiency. Also, it decreases the starting current and increases the power factor. If the increase in the power factor is greater than the decrease in the starting current, it results in a better starting torque.

In general, in a squirrel-cage rotor design, the rotor resistance value is determined by a compromise between conflicting requirements of good speed regulation and good starting torque. Usually, the resistance of a squirrel-cage motor, referred to the stator, is less than that of a wound-rotor machine of the same size. The rotor resistance of a squirrel-cage motor can be increased by decreasing the cross-sectional area of the end-rings. The resistance of the stator windings should be minimal to reduce both the stator's copper losses and its internal voltage drop. Also, an increase in the stator resistance causes the maximum torque to decrease.

The leakage reactances are affected by changes in the air gap and slot openings. As the air gap is increased, these reactances decrease, causing a greater excitation current to flow at a lesser power factor. Open slots cause the same thing. Also, an increase in reactance decreases the pull out torque. Therefore, induction motors are designed with as small an air gap as possible to reduce the excitation current. The rotor frequency changes with speed and at standstill is the same as the stator frequency. As the motor speeds up, the rotor frequency decreases in value to 1 or 3 Hz at full load in a typical 60-Hz motor. By using suitable shapes and arrangements for rotor bars, it is possible to design squirrel-cage rotors so that their effective resistance at 60 Hz is several times their resistance at 1 or 3 Hz. This results from the inductive effect of the slot-leakage flux on the current distribution in the rotor bars. The change in the resistance of the rotor bars is due to what is commonly known as the **skin effect**.

Various slot shapes for squirrel-cage induction motor rotors which produce the National Electrical Manufacturers Association (NEMA) design characteristics[21] are shown in Figure 6.26. The rotor bars of the NEMA design class A motor shown in Figure 6.26(a) are quite large and located near the surface of the rotor. They have low resistance (because of their large cross sections) and a low leakage reactance X_2 (because the bars are located near the stator). Such motors are also called **class A** motors. They have low slip at full load, high running efficiency, and high pull-out torque (due to the low rotor resistance). However, because R_2 is small, the starting torque of the motor is small, but its starting current is large. These motors are usually used in constant-speed drives to drive pumps, fans, lathes, blowers, and other devices.

Figure 6.26(b) shows the deep-bar rotor slots of the **NEMA design class B** motor. Rotor bars embedded in deep slots provide a high effective resistance

---

[21] These standard designs are commonly called **design classes**. Each of them provides different torque-speed curves. Recently, the International Electrotechnical Commission (IEC) in Europe adopted similar design classes for motors.

and a large torque at startup. Due to the skin effect, the current has a tendency to concentrate at the top of the bars at startup, when the frequency of the rotor currents is high. Under normal operating conditions with low slips (since the frequency of the rotor currents is much smaller) the skin effect is negligible and the current tends to distribute almost uniformly throughout the entire rotor-bar cross section. Thus, the rotor resistance decreases to a small value causing a higher efficiency. Such motors are used in applications that are comparable to those for design class A. Due to their lower starting current requirements, they are usually preferred to design class A motors.

Figure 6.26(c) shows a **NEMA design class C** rotor slot shape which is an alternative design for the deep-bar rotor. Such a double-cage arrangement is used to attain greater starting torque and better running efficiency. The squirrel-cage winding is made up of two layers of bars short-circuited by end rings. The inner cage, consisting of low-resistance bottom bars, is deeply embedded in the rotor's iron core. However, the outer cage has relatively high-resistance bars located close to the inner stator surface. At startup, the frequency of the rotor currents is relatively high (almost equal to stator frequency) and the leakage reactance of the cage made up of the larger (inner) rotor bars is also high, suppressing the current in that cage. Therefore, the outer cage with the small-

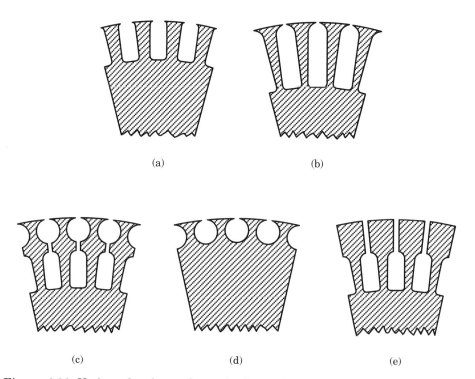

(a)   (b)

(c)   (d)   (e)

Figure 6.26 Various slot shapes for squirrel-cage induction motor rotors which produce NEMA design-class characteristics: (a) NEMA Design Class A; (b) NEMA Design Class B; (c) NEMA Design Class C; (d) NEMA Design Class D; (e) old NEMA Design Class F.

er bars, because of its higher resistance and lower leakage inductance (because of skin effect), predominates during startup, producing high startup torque. At the steady-state (i.e., normal) operation of the motor, its speed is normal. Thus, the rotor frequency is so low that the leakage reactance of the low-resistance cage is substantially lower than its resistance, and the current densities in the two cages are practically equal. Therefore, during the normal running period, because of the negligible skin effect, the current penetrates the full depth of the lower cage causing an efficient steady-state operation. In summary, such a design results in high rotor resistance on startup and low resistance at normal speed. These motors are more expensive than the others. They are used in applications involving high-starting-torque loads, such as compressors and conveyors that are fully loaded when started, and loaded pumps.

Figure 6.26(d) shows **NEMA design class D** rotor slot shapes. Class D motors are characterized by high starting torque, low starting current, and high operating slip. The rotor cage bars are made of higher-resistance material such as brass. The maximum torque takes place at a slip of 0.5 or higher. Because of high rotor resistance, the full-load slip for theses motors is high (about 7 to 17 percent or more). Thus, the running efficiency is low. The high losses in the rotor circuit dictate that the machine be large and is therefore expensive for a given power. They are ideal for loads with rapid acceleration or high impact such as punch presses and shears.

In addition to the four design classes reviewed so far, NEMA used to also have design classes E and F. They were called **soft-start** induction motors, with very low starting currents and torques, but are no longer in use. Figure 6.26(e) shows old NEMA design class F. It was once used in large motors designed for very easily started loads such as industrial fans. Figure 6.27 shows torque-speed characteristics for the four design classes.

6.13 STARTING OF INDUCTION MOTORS

A wound-rotor induction motor can be connected directly to the line with its rotor open-circuited, or with relatively high resistances connected to the slip-ring brushes. At full operating speed, the brushes are shorted together so that there is zero external resistance in each phase. However, induction motors with squirrel-cage rotors can be started by a number of different methods, depending on the size and type of the motor involved. In general, the basic methods of starting can be classified as *direct-on-line starting*, *reduced voltage starting*, and *current limiting by series resistance or impedance*. Other methods of starting include *part-winding starting* and *multicircuit starting*. In these methods, the motor is connected asymmetrically during the starting period.

6.13.1 DIRECT-ON-LINE STARTING

In general, most induction motors are rugged enough so that they can be started across the line without any resultant damage to the motor windings, even though about five to seven times the rated current flows through the sta-

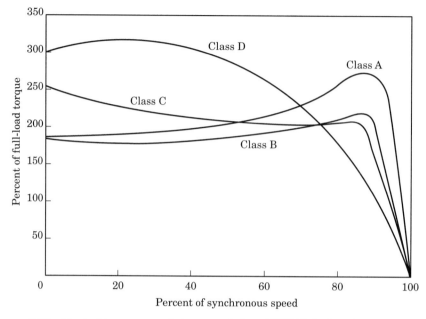

Figure 6.27 Typical torque-speed characteristics of NEMA design class squirrel-cage motors.

tor at rated voltage at standstill. However, such across-the-line starting of large motors is not advisable for two reasons: (1) the lines supplying the induction motor may not have enough capacity, and (2) the large starting current may cause a large *voltage dip*,[22] resulting in reduced voltage across the motor. Since the torque changes approximately with the square of the voltage, the starting torque can become so small at the reduced line voltage that the motor may not even start on load.

To estimate the starting current, all squirrel-cage motors now have a starting *code letter* (not to be mixed up with their *design class letter*) on their nameplates. The code letter gives the maximum limit of current for the motor at starting conditions. Table 6.1 gives the starting kVA per hp at starting conditions for the machine. To find the starting current for an induction motor, its rated voltage, horsepower, and code letter are read from its nameplate. Therefore, its starting apparent power is

$$S_{start} = \text{(rated horsepower)(code letter factor)}$$

and its starting current is

$$I_L = \frac{S_{start}}{\sqrt{3}V_t}$$

---

[22] For further information, see Gönen, T., *Electric Power Distribution System Engineering*, McGraw-Hill Book Company, New York, 1986.

TABLE 6.1 The NEMA Code Letters

Code Letter	Locked-Rotor kVA/hp	Code Letter	Locked-Rotor kVA/hp
A	0-3.14	L	9.00-9.99
B	3.15-3.54	M	10.00-11.19
C	3.55-3.99	N	11.20-12.49
D	4.00-4.49	P	12.50-13.99
E	4.50-4.99	R	14.00-15.99
F	5.00-5.59	S	16.00-17.99
G	5.60-6.29	T	18.00-19.99
H	6.30-7.09	U	20.00-22.39
J	7.10-7.99	V	22.40 and up
K	8.00-8.99		

Source: Reproduced by permission from *Motors and Generators*, NEMA Publications MG-1, Copyright 1982 by NEMA.

6.13.2 REDUCED-VOLTAGE STARTING

At the time of starting, a reduced voltage is supplied to the stator and slowly increased to the rated value when the motor is within approximately 25 percent of its rated speed. Such reduced-voltage starting can be achieved by means of *wye-delta starting, autotransformer starting,* and *solid-state voltage controller starting.*

1. Wye-delta starting. In this starting, both ends of each phase of the stator winding must be brought out to the terminals of a wye-delta switch so that at start the stator windings are connected in wye. As the motor approaches full speed, the switch is operated and the stator windings are connected in delta, and the motor runs at full speed. Here, the motor used has to be designed for delta operation and is connected in wye only during the starting period. Since the impedance between the line terminals for the wye connection is three times that of the delta connection for the same line voltage, the line current is reduced to one-third of its value for the delta connection. That is, when the motor is connected in wye, it takes one-third as much starting current and develops one-third as much torque. A wye-delta starter is equivalent to an autotransformer with a ratio of $n = 1/\sqrt{3}$.

2. Autotransformer starting. Here, the setting of the autotransformer can be predetermined to limit the starting current to any given value. Therefore, an autotransformer, which reduces the voltage applied to the motor to x times the normal voltage, will reduce the starting current in the supply system as well as the starting torque of the motor to x^2 times the normal values. Here, the x is known as the *compensator turns ratio*. When the motor reaches about 80 percent of normal speed, the connections are changed so that the autotransformers are deenergized and the motor is connected to full line voltage. Most compensators are provided with three sets of standard taps in order to apply 80, 65, or 50 percent of the line voltage to the motor. To achieve a satisfactory starting the lowest tap is normally used. Figure 6.28 shows an autotransformer starter for squirrel-cage motors. Such starters are also called *starting compensators*.

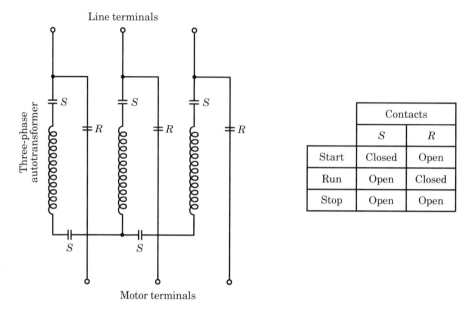

Figure 6.28 An autotransformer starter for an induction motor.

3. Solid-state voltage controller starting. In this starting a solid-state voltage controller is used as a reduced-voltage starter, as shown in Figure 6.29. The advantage of this method is that the solid-state voltage controller provides a smooth starting and also controls the speed of the induction motor during running.

6.13.3 CURRENT LIMITING BY SERIES RESISTANCE OR IMPEDANCE

Such a technique may be used if the starting torque requirement is not too great. In this method, series resistances, or impedances, are inserted in the three lines to limit the starting current. These resistors, or impedances, are

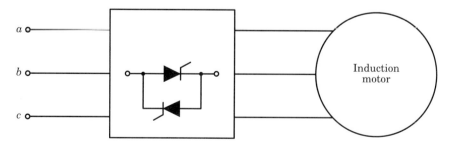

Figure 6.29 Solid-state voltage controller starting.

shorted out when the motor gains speed. Because of the extra power losses in the external resistances during startup, it is an inefficient method.

Example 6.7

A 20-hp, 480-V three-phase induction motor has a code letter H on its nameplate. Determine its starting current.

SOLUTION:
From Table 6.2, the maximum kVA per hp is 7.09. Thus, the maximum starting kVA of this motor is

$$S_{start} = (20 \text{ hp})(7.09) = 141.9 \text{ kVA}$$

Therefore, the starting current is

$$I_L = \frac{S_{start}}{\sqrt{3}V_t}$$

$$= \frac{141,800 \text{ VA}}{\sqrt{3}(480 \text{ V})} = 170.6 \text{ A}$$

Example 6.8

A three-phase eight-pole 100-hp 440-V 60-Hz wye-connected induction motor has the following constant in ohms per phase referred to the stator: $R_1 = 0.085 \text{ }\Omega$, $X_1 = 0.196 \text{ }\Omega$, $R_2 = 0.067 \text{ }\Omega$, $X_2 = 0.161 \text{ }\Omega$ and $X_m = 6.65 \text{ }\Omega$. The motor has a wound rotor with a turns ratio of 2.0. To produce maximum torque at starting (i.e., $T_{start} = T_{max}$), the wound-rotor windings are connected to external resistors. The Thévenin resistance and reactance of the motor are $0.0802 \text{ }\Omega$ and $0.191 \text{ }\Omega$, respectively. Determine the value of the external rotor resistance under the following conditions:
(a) If they are referred to the stator and connected in wye.
(b) If they are referred to the rotor and connected in wye.
(c) If they are referred to the stator and connected in delta.
(d) If they are referred to the rotor and connected in delta.

SOLUTION:
(a) The external rotor-circuit resistance is adjusted until the motor produces its maximum torque at start so that $T_{start} = T_{max}$. Since T_{max} occurs at starting $s_{maxT} = 1.0$, from equation (6.88)

$$\frac{R_{2, tot}}{s_{maxT}} = [R_{th}^2 + (X_{th} + X_2)^2]^{1/2}$$

or

$$\frac{R_{2,\,tot}}{1.0} = [0.0802^2 + (0.191 + 0.161)^2]^{1/2} = 0.361$$

Therefore, $R_{2,\,tot} = 0.361\ \Omega$ referred to the stator is found. Since this value represents the total resistance in the rotor circuit, the value of the external rotor resistance referred to the stator and connected in wye is

$$R'_{2,\,ext} = R_{2,\,tot} - R_2$$

$$= 0.361 - 0.067 = 0.294\ \Omega\ per\ phase$$

(b) The value of the external resistance referred to the rotor (i.e., in terms of physical resistance) and connected in wye is

$$R_{2,\,ext} = \frac{R'_{2,\,ext}}{2}$$

$$= \frac{0.294}{2^2} = 0.0735\ \Omega\ per\ phase$$

(c) The value of the external rotor resistance referred to the stator and connected in delta is

$$R'_{2,\,ext} = 3(0.294\ \Omega) = 0.882\ \Omega$$

(d) The value of the external rotor resistance referred to the rotor and connected in delta is

$$R_{2,\,ext} = 3(0.0735\ \Omega) = 0.2295\ \Omega$$

or

$$R_{2,\,ext} = \frac{R'_{2,\,ext}}{a^2}$$

$$= \frac{0.882}{2^2} = 0.2205\ \Omega$$

Example 6.9

Consider the induction motor given in Example 6.8 and assume that either its wound rotor slip rings have been short-circuited or its wound rotor has been replaced by an equivalent squirrel-cage rotor. Investigate the line-voltage start-

ing of the motor. Neglect jX_m as a reasonable simplifying assumption because of the large slip (i.e., $s_{start} = 1.0$) at starting. Determine the following:
(a) The stator current at starting.
(b) The rotor current at starting.
(c) The air-gap power at starting.
(d) The developed power at starting.
(e) The developed torque at starting.

SOLUTION:
(a) The stator current at starting is

$$I_{1,\ start} = \frac{V_1}{\left(R_1 + \dfrac{R_2}{s}\right) + j(X_1 + X_2)}$$

$$= \frac{254 \angle 0°}{\left(0.085 + \dfrac{0.067}{1.0}\right) + j(0.196 + 0.161)} = 655 \angle -66.9° \text{ A}$$

(b) Since

$$I_{2,\ start} \cong I_{1,\ start} = 655 \angle -66.9° \text{ A}$$

(c) The air-gap power at starting is

$$P_{g,\ start} = 3 I_{2,\ start}^2 \frac{R_2}{s}$$

$$= 3(655)^2 0.067 \cong 86 \text{ kW}$$

(d) The developed power at starting is

$$P_{d,\ start} = 3 I_{2,\ start}^2 R_2 \left(\frac{1-s}{s}\right)$$

$$= 3(655)^2 0.067 \left(\frac{1-1}{1}\right) = 0$$

(e) The developed torque at starting is

$$T_{d,\ start} = \frac{P_g}{\omega_s}$$

$$= \frac{86,139 \text{ W}}{94.24 \text{ rad/s}} = 914 \text{ N} \cdot \text{m}$$

where

$$\omega_s = \frac{2\pi}{60}\left(\frac{120f_1}{p}\right)$$

$$= \frac{2\pi}{60}\left(\frac{120\times 60}{8}\right) = 94.24 \ \text{rad/s}$$

Since $P_{d, start} = 0$ and $\omega_m = 0$, the torque equation (6.68a) cannot be used to determine the developed torque at starting.

Example 6.10

Consider autotransformer-type reduced voltage starting with either the squirrel-cage rotor or the equivalent wound rotor given in Example 6.9, ignoring the shunt branch X_m. Apply ideal autotransformer theory, find and tabulate the motor's stator current $I_{1, start}$, the line current into the autotransformer starter I_L, and the developed starting torque. Determine the above values as a function of the autotransformer taps, which are 50, 65, and 80 percent of the line voltage.

SOLUTION:
Figure 6.30 shows the application of the autotransformer-type reduced voltage starting. The results are given in Table 6.2 and are based on the results of Example 6.9. Note that such a starting technique is the most suitable for centrifugal type loads. However, it fails in applications such as elevators where constant load torque is required.

TABLE 6.2 Results of Example 6.10

V_1 (%)	N_2/N_1	$(N_2/N_1)^2$	$I_{1, start} = 655(N_2/N_1)$	$I_{L, start} = 655(N_2/N_1)^2$	$T_{d, start} = 914(N_2/N_1)^2$
100*	1.0	1.0	655 A	655 A	914 N·m
80	0.80	0.64	524	419	585
65	0.65	0.4225	426	277	386
50	0.50	0.50	0.25	164	229

*From Example 6.8.

Example 6.11

Consider the induction motor given in Example 6.10 and replace the autotransformer-type reduced voltage starting either by primary resistor starting or by primary reactor starting as shown in Figure 6.31. The existence of the resistor and the reactor are mutually exclusive, that is, there is either R or X in the box shown but both cannot be present. It is required that the starting torque

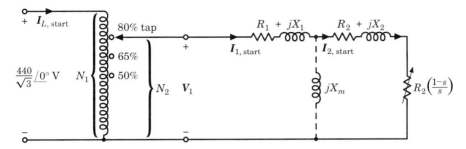

Figure 6.30   The autotransformer starting used for Example 6.9.

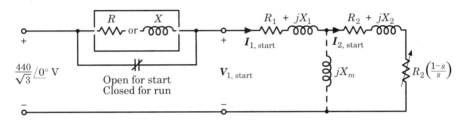

Figure 6.31   The primary resistor or primary reactor starting used for Example 6.10.

at such reduced voltage be 25 percent of the starting torque at full voltage. Determine the following:

(a) The starting torque at such a reduced voltage.
(b) The primary voltage at starting.
(c) The stator current at starting.
(d) The total input impedance at starting.
(e) The value of the resistor R in ohms.
(f) The value of the reactor X in ohms.

SOLUTION:

(a) In Example 6.10, the starting torque at full voltage was 914 N•m. Therefore, the new starting torque will be

$$T_{start} = \frac{914 \text{ N} \cdot \text{m}}{4} = 228.5 \text{ N} \cdot \text{m}$$

(b) If the value of the supply voltage is halved, the starting torque becomes one-forth of its full-voltage value. Therefore, the primary voltage at starting is

$$V_{1,\,start} = \frac{1}{2}\left(\frac{440 \text{ V}}{\sqrt{3}}\right) = 127 \text{ V}$$

(c) Therefore, the stator current at starting is

$$I_{1,\,start} = \frac{1}{2}(655\text{ A}) = 327.5\text{ A}$$

(d) Thus, the total input impedance at starting is

$$Z_{input} = \frac{V_{1,\,start}}{I_{1,\,start}} = \frac{254\text{ V}}{327.5\text{ A}} = 0.776\,\Omega$$

(e) Since the shunt branch jX_m is neglected, the input impedance, for the primary resistor starting, can be expressed as

$$Z_{input} = (R_1 + R_2 + R) + j(X_1 + X_2)$$

Since the magnitude of this input impedance is 0.776 Ω, the value of R is 0.536 Ω.

(f) Since the input impedance for the primary reactor starting can be expressed as

$$Z_{input} = (R_1 + R_2) + j(X_1 + X_2 + X)$$

where the magnitude of this input impedance is 0.776 Ω, the value of X from the above equation can be found as 0.404 Ω. This technique is often used to reduce the acceleration torque T_a or the developed torque T_d. However it should not be used to reduce the stator current I_1.

6.14 SPEED CONTROL

An induction motor is basically a constant-speed motor when it is connected to a constant-voltage and constant-frequency power supply. Even though a large number of industrial drives run at constant speed, there are many applications in which variable speed is a requirement. Examples include elevators, conveyors, and hoists. Traditionally, dc motors have been used in such adjustable-speed drive systems. However, dc motors are expensive, require frequent maintenance of commutators and brushes and should not be used in hazardous environments. The synchronous speed of an induction motor can be changed by changing the number of poles or varying the line frequency. The operating slip can be changed by varying the line voltage, varying the rotor resistance, or applying voltages with appropriate frequency to the rotor circuits.

1. Pole-Changing Method. In this method, the stator winding of the motor can be designed so that by simple changes in coil connections the number of poles can be changed by the ratio of 2 to 1. In this way, two synchronous speeds can be obtained. This method is not suitable for wound-rotor motors, since the rotor windings would also have to be reconnected to have the same number of poles as the stator. However, a squirrel-cage rotor automatically develops a number of magnetic poles equal to those of the air-gap field. With two independent sets of stator windings, each arranged for pole changing, as many as four synchronous speeds can be achieved in a squirrel-cage motor. For example, 600, 1200, 1800, and 3600 rev/min can be attained for a 60-Hz operation. In addition, the motor phases can be connected either in wye or delta, resulting in eight possible combinations.

2. Variable-Frequency Method. The synchronous speed of an induction motor can be controlled by changing the line frequency. The change in speed is continuous or discrete depending upon whether or not supply frequency is continuous or discrete. To maintain approximately constant flux density, the line voltage must also be changed with the frequency. Therefore, the maximum torque remains nearly constant. This type of control is known as **constant volts per hertz** and is possible only if a variable-frequency supply is available. A wound-rotor induction machine can be used as a frequency changer. The arrival of solid-state devices with relatively large power ratings has made it possible to use solid-state frequency converters.[23]

3. Variable Line-Voltage Method. The torque developed by an induction motor is proportional to the square of the applied voltage. Therefore, the speed of the motor can be controlled over a limited range by changing the line voltage. If the voltage can be varied continuously from V_1 to V_2, the speed of the motor can also be varied continuously between speeds n_1 to n_2 for a given load. This method is used for small squirrel-cage motors driving fans and pumps.

4. Variable Rotor-Resistance Method. This method can only be used with wound-rotor motors. By varying the external resistance connected to the rotor through the slip rings, the torque-speed characteristics of a wound-rotor induction motor can be controlled. The high available torque permits reduced starting voltage to be used while maintaining a sufficiently high starting torque. In addition, the maximum torque and the starting torque may be made the same. By continuous variation of rotor-circuit resistance, continuous variation of speed can also be achieved. The disadvantages of this method include low efficiency at reduced speeds and poor speed regulation with respect to changes in the load.

5. Variable-Slip Method. Without sacrificing efficiency at low-speed operation or affecting the speed with load variation, the induction motor speed can be controlled by using semiconductor converters.

6. Speed Control by Solid-State Switching. With the exception of the cycloconverter or inverter-driven motor, the speed of a wound-rotor motor is con-

---

[23] One such arrangement is a silicon-controlled rectifier (SCR) supplying dc voltage to a static inverter, with solid-state components which in turn supplies the variable frequency to the motor. This arrangement is called a **cycloconverter**.

trolled by the inverter in the rotor circuit or by regulating the stator voltage with solid-state switching devices such as power transistors or silicon-controlled rectifiers (SCRs or thyristors). In general, SCR-based control provides a wider range of operation and is more efficient than other slip-control methods.

6.15 TESTS TO DETERMINE EQUIVALENT CIRCUIT PARAMETERS

The parameters of the equivalent circuit of the induction motor can be found from the **no-load** and **blocked-rotor tests**.[24] These tests correspond to the no-load and short-circuit tests done on the transformer. The stator resistance can be determined from the **dc test**.

6.15.1 THE NO-LOAD TEST

Rated balanced voltage at rated frequency is applied to the stator and the motor is permitted to run without a load. The voltage, current, and power input to the stator are measured. At no load, R_2 is very small with respect to $R_2(1 - s)/s$; therefore, the no-load rotor copper loss is negligible. The no-load input power is the sum of the stator copper loss and the rotational losses. Thus,

$$P_{n\ell} = P_{1,cu} + P_{rot} \tag{6.92a}$$

or

$$P_{n\ell} = 3I_{1,n\ell}^2 R_1 + P_{rot} \tag{6.92b}$$

where

$$P_{n\ell} = P_{core} + P_{FW} \tag{6.93}$$

assuming that the stray losses are negligible. Hence, the rotational losses can be found from

$$P_{rot} = P_{n\ell} - 3I_{1,n\ell}^2 R_1 \tag{6.94}$$

where $P_{n\ell}$ is the total three-phase power input to the machine at rated voltage and frequency and $I_{1,n\ell}$ is the average of the three line currents.

Under no load conditions, R_1 is small with respect to X_m, and the overall input power factor is very small, about 0.1. The equivalent input impedance is

$$|Z_{eq}| = |Z_{n\ell}| = \frac{V_{n\ell}}{\sqrt{3}I_{n\ell}} \cong X_1 + X_m \tag{6.95}$$

---

[24] For further information, see *IEEE Standard Test Procedures for Polyphase Induction Motors and Generators*, IEEE Standard 112, 1984, Institute of Electrical and Electronics Engineers, Inc.

where $V_{n\ell}$ is the line-to-line terminal voltage. Thus,

$$X_m = \frac{V_{n\ell}}{\sqrt{3}I_{n\ell}} - X_1 \qquad (6.96)$$

6.15.2 THE DC TEST

The stator resistance R_1 can be considered equal to its *dc* value. Thus, it can be measured independently of the rotor impedance. In such a test, a *dc* power supply is connected to two of the three terminals of a wye-connected induction motor. The current in the stator windings is adjusted to the rated value, and the voltage between the terminals is measured. Since the current flows through two of the wye-connected windings, the total resistance in the path is $2R_1$. Hence,

$$2R_1 = \frac{V_{dc}}{I_{dc}} \qquad (6.97)$$

or

$$R_1 = \frac{V_{dc}}{2I_{dc}} \qquad (6.98)$$

This value of R_1 may now be used in equation (6.94) to determine the stator copper loss as well as the rotational losses. Usually, the calculated R_1 has to be corrected for the skin effect and the temperature of the windings during the short-circuit test.[25] Therefore, the *ac* resistance is found by multiplying the *dc* resistance by a factor which varies from 1.2 to about 1.8, depending on the frequency and other factors.

6.15.3 THE BLOCKED-ROTOR TEST

This test corresponds to the short-circuit test of a transformer and is also called the **locked-rotor test**. Here, the rotor of the machine is blocked ($s = 1.0$) to prevent it from moving. A reduced voltage[26] is applied to the machine so that the rated current flows through the stator windings. This input power, voltage, and current suggest a blocked-rotor test frequency of 25 percent of the rated frequency. The input power to the motor is

$$P_{br} = \sqrt{3}V_{br}I_{br}\cos\theta \qquad (6.99)$$

---

[25] For further information see reference [42].

[26] If full voltage at the rated frequency were applied, the current would be five to eight or more times the rated value. Because of this, blocked-rotor tests are **not done at full voltage** except for small motors. Even then, such tests are made as rapidly as possible to prevent overheating of the windings.

so that the blocked-rotor power factor can be expressed as

$$\cos \theta = \frac{P_{br}}{\sqrt{3} V_{br} I_{br}} \tag{6.100}$$

The magnitude of the total impedance in the motor can be expressed as

$$|Z_{br}| = \frac{V_{\phi}}{I_1} = \frac{V_{br}}{\sqrt{3} I_{br}} \tag{6.101}$$

Since the impedance angle is θ,

$$Z_{br} = R_{br} + jX_{br} \tag{6.102a}$$

or

$$Z_{br} = Z_{br} \cos \theta + j Z_{br} \sin \theta \tag{6.102b}$$

Since R_1 is found by the *dc* test, the blocked-rotor resistance is

$$R_{br} = R_1 + R_2 \tag{6.103}$$

and the blocked-rotor reactance is

$$X'_{br} = X'_1 + X'_2 \tag{6.104}$$

where X'_1 and X'_2 represent the stator and rotor reactances *at the test frequency*, respectively. Alternatively, this blocked-rotor reactance can be expressed as

$$X'_{br} = Z_{br} \sin \theta \tag{6.105a}$$

or

$$X'_{br} = (Z_{br}^2 - R_{br}^2)^{1/2} \tag{6.105b}$$

or

$$X'_{br} = \left[\left(\frac{V_{br}}{\sqrt{3} I_{br}} \right)^2 - \left(\frac{P_{br}}{3 I_{br}^2} \right)^2 \right]^{1/2} \tag{6.105c}$$

The rotor resistance R_2 can be found from

$$R_2 = R_{br} - R_1 \tag{6.106}$$

where R_1 is found from the *dc* test. Therefore,

$$R_2 = \frac{P_{br}}{3I_{br}^2} - \frac{V_{dc}}{2I_{dc}} \tag{6.107}$$

Since the reactance is directly proportional to the frequency, the total equivalent reactance at the normal operating frequency can be expressed as

$$X_{br} = \frac{f_{rated}}{f_{test}} X_{br}' = X_1 + X_2 \tag{6.108}$$

Unfortunately, there is no simple way to determine the stator and rotor reactances. This information is known only by the designer of the machine. Table 6.3 gives the approximate values of X_1 and X_2 as fractions of X_{br}. In general how X_{br} is divided between X_1 and X_2 is not that important, but what is important is the amount of X_{br} since it affects the breakdown torque. Also note that Design B and C rotors are designed so that their rotor resistances change with frequency.

TABLE 6.3 Determination of X_1 and X_2 from Blocked-Rotor Reactance at the Rated Frequency

Rotor Design	X_1 and X_2 as fractions of X_{br}	
	X_1	X_2
Wound rotor	$0.5X_{br}$	$0.5X_{br}$
Design A	$0.5X_{br}$	$0.5X_{br}$
Design B	$0.4X_{br}$	$0.6X_{br}$
Design C	$0.3X_{br}$	$0.7X_{br}$
Design D	$0.5X_{br}$	$0.5X_{br}$

Example 6.12

The following test data were taken on a three-phase four-pole 150-hp 480-V 60-Hz Design B wye-connected induction motor with a rated current of 101.3 A.
Dc test:

$\qquad V_{dc} = 20.26 \qquad\qquad I_d = 101.3\text{ A}$

No-load test:

$\qquad V_{n\ell} = 480\text{ V} \qquad\qquad f = 60\text{ Hz}$
$\qquad I_a = 34.8\text{ A} \qquad\qquad P_{n\ell} = 3617.5\text{ W}$
$\qquad I_b = 35\text{ A}$
$\qquad I_c = 35.2\text{ A}$

Blocked-rotor test:

$\qquad V_{br} = 51.3\text{ V} \qquad\qquad f_{test} = 15\text{ Hz}$
$\qquad I_a = 101.3\text{ A} \qquad\qquad P_{br} = 5200\text{ W}$
$\qquad I_b = 100.4\text{ A}$
$\qquad I_c = 102.5\text{ A}$

Determine the following:

(a) The R_1 and R_2 of the motor. Use a factor of 1.5 in computing the effective ac armature resistance per phase.

(b) The X_1, X_2, and X_m of the motor.

SOLUTION:

(a) From the dc test,

$$R_1 = \frac{V_{dc}}{2I_{dc}} = \frac{20.26 \text{ V}}{2(101.3 \text{ A})} = 0.1 \ \Omega$$

Thus, this resistance in ac is $(0.1 \ \Omega)1.5 = 0.15 \ \Omega$. From the no-load test,

$$I_{1, n\ell} = \frac{I_a + I_b + I_c}{3} = \frac{34.8 + 35 + 35.2}{3} = 35 \text{ A}$$

and since

$$|\boldsymbol{Z}_{n\ell}| = \frac{V_{n\ell}}{\sqrt{3}I_{n\ell}} \cong X_1 + X_m$$

then

$$|\boldsymbol{Z}_{n\ell}| = \frac{480 \text{ V}}{\sqrt{3}(35 \text{ A})} = 7.918 \ \Omega = X_1 + X_m$$

so that when X_1 is known, X_m can be determined. The stator copper losses are

$$P_{1, cu} = 3I_{1, n\ell}^2 R_1 = 3(35 \text{ A})^2 (0.15 \ \Omega) = 551.25 \text{ W}$$

Hence, the no-load rotational losses are

$$P_{rot} = P_{n\ell} - P_{1, cu}$$

$$= 3617.5 - 551.25 = 3066.25 \text{ W}$$

From the blocked-rotor test,

$$I_L = \frac{101.3 + 100.4 + 102.5}{3} = 101.4 \text{ A}$$

The blocked-rotor impedance can be found from

$$|\boldsymbol{Z}_{br}| = \frac{V_\phi}{I_1} = \frac{V_{br}}{\sqrt{3}I_{br}}$$

as

$$|\mathbf{Z}_{br}| = \frac{51.3 \text{ V}}{\sqrt{3}(101.4 \text{ A})} = 0.292 \ \Omega$$

and the impedance angle θ is

$$\theta = \cos^{-1}\left(\frac{P_{br}}{\sqrt{3}V_{br}I_{br}}\right)$$

$$= \cos^{-1}\left(\frac{5200 \text{ W}}{\sqrt{3}(51.3 \text{ V})(101.4 \text{ A})}\right) = 54.75°$$

Thus,

$$R_{br} = Z_{br}\cos\theta$$

$$= 0.292\cos 54.75° = 0.169 \ \Omega$$

Since

$$R_{br} = R_1 + R_2$$

then

$$R_2 = R_{br} - R_1 = 0.169 - 0.15 = 0.019 \ \Omega$$

(b) The reactance at 15 Hz is

$$X'_{br} = Z_{br}\sin\theta$$

$$= 0.292\sin 54.75° = 0.239 \ \Omega$$

The equivalent reactance at 60 Hz is

$$X_{br} = \frac{f_{rated}}{f_{test}}X'_{br}$$

$$= \left(\frac{60 \text{ Hz}}{15 \text{ Hz}}\right)(0.239 \ \Omega) = 0.956 \ \Omega$$

Since it is a design class B induction motor,

$$X_1 = 0.4X_{br} = 0.4(0.956) = 0.382 \ \Omega$$

and

$$X_2 = 0.6X_{br} = 0.6(0.956) = 0.574 \ \Omega$$

Also,

$$X_m = 7.92 - X_1 = 7.92 - 0.382 = 7.538 \ \Omega$$

PROBLEMS

6.1   A 50-hp three-phase 60-Hz wye-connected induction motor operates at a shaft speed of almost 900 rpm at no load and 873 rpm at full load. Determine the following:
(a)   The number of poles of the motor.
(b)   The per-unit and percent slip at full load.
(c)   The slip frequency of the motor.
(d)   The speed of the rotor field with respect to the rotor itself.
(e)   The speed of the rotor field with respect to the stator.
(f)   The speed of the rotor field with respect to the stator field.
(g)   The shaft torque of the motor at full load.

6.2   A three-phase 60-Hz wye-connected induction motor operates at a shaft speed of almost 3600 rpm at no load and 3420 rpm at full load. Determine the following:
(a)   The number of poles of the motor.
(b)   The per-unit and percent slip at full load.
(c)   The slip frequency of the motor.
(d)   The speed of the rotor field with respect to the rotor itself.
(e)   The speed of the rotor field with respect to the stator.
(f)   The speed of the rotor field with respect to the stator field.

6.3   Solve Problem 6.1 but assume that the 10-hp motor runs at a shaft speed of almost 1800 rpm at no load and 1761 rpm at full load.

6.4   Solve Problem 6.2 but assume that the motor runs at a shaft speed of almost 120 rpm at no load and 114 rpm at full load.

6.5   A three-phase 480-V 25-hp two-pole 60-Hz induction motor has a full load slip of 5 percent. Determine the following:
(a)   The synchronous speed of the motor.
(b)   The rotor speed at full load.
(c)   The slip frequency at full load.
(d)   The shaft torque at full load.

6.6   A three-phase 480-V 50-hp four-pole 60-Hz induction motor has a full load slip of 3.5 percent. Determine the following:

(a) The synchronous speed of the motor.
(b) The rotor speed at full load.
(c) The slip frequency at full load.
(d) The shaft torque at full load.

6.7   A three-phase 208-V 25-hp induction motor is supplied with 75 A at 0.85 PF lagging. Its stator and rotor copper losses are 1,867 W and 650 W, respectively. Its core losses are 1,500 W, the friction and windage losses are 300 W, and the stray losses are negligible. Determine the following:
(a) The air-gap power.
(b) The mechanical power developed.
(c) The shaft output power.
(d) The efficiency of the motor.

6.8   An induction motor draws 50 A from a 480-V three-phase line at a lagging power factor of 0.85. Its stator and rotor copper losses are 1,000 W and 500 W, respectively. Its core losses are 500 W, the friction and windage losses are 250 W, and the stray losses are 250 W.
Determine the following:
(a) The air-gap power.
(b) The mechanical power developed.
(c) The shaft output power in horsepower.
(d) The efficiency of the motor.

6.9   Consider the data given in Problem 6.8. If the frequency of the power source is 60 Hz and the induction motor has two poles, determine the following:
(a) The slip in percent.
(b) The operating speed in rad/s and in rpm.
(c) The developed torque.
(d) The output torque.

6.10   Show (i.e., prove by derivation) that if rotor copper loss were the only loss in an induction motor, the efficiency of the machine would be $\eta = 1 - s$, where s = per-unit slip (i.e., slip given as a fraction).

6.11   An induction motor draws 50 A from a 380-V three-phase line at a lagging power factor of 0.90. Its stator and rotor copper losses are 1,000 W and 500 W, respectively. Its core losses are 650 W, the friction and windage losses are 200 W, and the stray losses are 250 W.
Determine the following:
(a) The air-gap power.
(b) The mechanical power developed.
(c) The shaft output power in W and hp.
(d) The efficiency of the motor.

6.12   Consider the data given in Problem 6.11. If the frequency of the power
source is 50 Hz, and the machine has four poles, determine the following:
(a) The slip in percent.
(b) The operating speed in rad/s and rpm.
(c) The developed torque.
(d) The output torque.

6.13   A two-pole 60-Hz induction motor has its full-load torque at a speed of
3,492 rpm. Determine the following:
(a) Its speed at half rated torque.
(b) Its speed at half rated torque and half rated voltage, if rotor resistance per phase is doubled.

6.14   Consider a 10-hp 60-Hz induction motor. If the motor is operated at
120-Hz, determine:
(a) The amount of voltage that should be applied to the motor to maintain the normal degree of iron saturation.
(b) The approximate value of the rated horsepower at such a frequency.

6.15   The input to the rotor of a 208-V three-phase 60-Hz 24-pole induction
motor is 20 kW. Determine the following:
(a) The developed (i.e., electromagnetic) torque in N·m and lb·ft.
(b) The speed in rpm and rad/s, and the hp output of the motor, if the
rotor current is 63.25 A per phase and the rotor resistance is 0.05 Ω
per phase. Ignore the rotational losses.

6.16   A three-phase four-pole 100-hp 480-V 60-Hz wye-connected induction motor
has the following constants in ohms per phase referred to the stator:

$$R_1 = 0.1 \ \Omega \qquad\qquad R_2 = 0.079 \ \Omega$$
$$X_1 = 0.205 \ \Omega \qquad\qquad X_2 = 0.186 \ \Omega$$
$$X_m = 7.15 \ \Omega$$

The total rotational losses are 2,950 W and are assumed to be constant.
The core loss is lumped in with the rotational losses. For a rotor slip of
3.33 percent at the rated voltage and rated frequency, determine:
(a) The speed in rpm and in rad/s.
(b) The stator current.
(c) The power factor.
(d) The developed power and output power.
(e) The developed torque and output torque.
(f) The efficiency of the motor.

6.17   A three-phase two-pole 25-hp 380-V 60-Hz wye-connected induction motor has the following constants in ohms per phase referred to the stator:

$$R_1 = 0.525 \ \Omega \qquad\qquad R_2 = 0.295 \ \Omega$$
$$X_1 = 1.75 \ \Omega \qquad\qquad X_2 = 0.8 \ \Omega$$
$$X_m = 20.5 \ \Omega$$

The total rotational losses are 1,850 W and are assumed to be constant. The core loss is lumped in with the rotational losses. For a rotor slip of 3.33 percent at the rated voltage and rated frequency, determine:
(a) The speed in rpm and in rad/s.
(b) The stator current.
(c) The power factor.
(e) The developed power and output power.
(d) The developed torque and output torque.
(f) The efficiency of the motor.

6.18 A three-phase four-pole 150-hp 480-V 60-Hz wye-connected induction motor has the following constants in ohms per phase referred to the stator:

$$R_1 = 0.1 \ \Omega \qquad\qquad R_2 = 0.085 \ \Omega$$
$$X_1 = 0.25 \ \Omega \qquad\qquad X_2 = 0.175 \ \Omega$$
$$X_m = 6.25 \ \Omega$$

The total rotational losses are 3,250 W and are assumed to be constant. The core loss is lumped in with the rotational losses. For a rotor slip of 3 percent at the rated voltage and rated frequency, determine:
(a) The speed in rpm and in rad/s.
(b) The stator current.
(c) The power factor.
(d) The developed power and output power.
(e) The developed torque and output torque.
(f) The efficiency of the motor.

6.19 A three-phase four-pole 60-Hz induction motor supplies 50-hp to a load at a speed of 1,701 rpm. Assume that the mechanical losses are zero and determine the following:
(a) The slip of the motor in percent.
(b) The developed torque in $N \cdot m$ and $lb \cdot ft$.
(c) The shaft speed of the motor, if its torque is doubled.
(d) The output power of the motor in W and hp, if its torque is doubled.

6.20 A three-phase sixty-pole 60-Hz induction motor supplies 30-hp to a load at a speed of 114 rpm. Assume that the mechanical losses are zero and determine the following:
(a) The slip of the motor in percent.
(b) The developed torque in $N \cdot m$ and $lb \cdot ft$.
(c) The shaft speed of the motor, if its torque is doubled.
(d) The output power of the motor in W and hp, if its torque is doubled.

6.21 A three-phase four-pole 60-Hz induction motor supplies 100-hp to a load at a speed of 1,701 rpm. Assume that the mechanical losses are zero and determine the following:
(a) The slip of the motor in percent.
(b) The developed torque in $N \cdot m$ and $lb \cdot ft$.

(c) The shaft speed of the motor, if its torque is doubled.
(d) The output power of the motor in W and hp, if its torque is doubled.

6.22 A three-phase thirty-pole 60-Hz induction motor supplies 50-hp to a load at a speed of 233 rpm. Assume that the mechanical losses are zero and determine the following:
(a) The slip of the motor in percent.
(b) The developed torque in N•m and lb•ft.
(c) The shaft speed of the motor, if its torque is doubled.
(d) The output power of the motor in W and hp, if its torque is doubled.

6.23 Consider the induction motor given in Problem 6.16 and assume that it has a wound rotor. Determine the following:
(a) The slip at which the maximum torque is developed.
(b) The speed at which the maximum torque developed.
(c) The maximum torque developed.
(d) The starting torque developed.
(e) The speed at which the maximum torque is doubled, if the rotor resistance is doubled.
(f) The maximum torque developed, if the rotor resistance is doubled.
(g) The starting torque developed, if the rotor resistance is doubled.

6.24 A three-phase six-pole 75-hp 480-V 60-Hz wye-connected induction motor has the following constants in ohms per phase referred to the stator:
$$R_1 = 0.245 \ \Omega \qquad\qquad R_2 = 0.198 \ \Omega$$
$$X_1 = 0.975 \ \Omega \qquad\qquad X_2 = 0.72 \ \Omega$$
$$X_m = 14.5 \ \Omega$$
Determine the following:
(a) The slip at which the maximum torque is developed.
(b) The speed at which the maximum torque developed.
(c) The maximum torque developed.
(d) The starting torque developed.
(e) The speed at which the maximum torque is doubled, if the rotor resistance is doubled.
(f) The maximum torque developed, if the rotor resistance is doubled.
(g) The starting torque developed, if the rotor resistance is doubled.

6.25 A three-phase six-pole 150-hp 380-V 50-Hz wye-connected induction motor has the following constants in ohms per phase referred to the stator:
$$R_1 = 0.09 \ \Omega \qquad\qquad R_2 = 0.07 \ \Omega$$
$$X_1 = 0.195 \ \Omega \qquad\qquad X_2 = 0.172 \ \Omega$$
$$X_m = 6.83 \ \Omega$$
Determine the following:
(a) The slip at which the maximum torque is developed.
(b) The speed at which the maximum torque is developed.

(c) The maximum torque developed.
(d) The starting torque developed.
(e) The speed at which the maximum torque is doubled, if the rotor resistance is doubled.
(f) The maximum torque developed, if the rotor resistance is doubled.
(g) The starting torque developed, if the rotor resistance is doubled.

6.26 A three-phase eight-pole 100-hp 440-V 60-Hz wye-connected induction motor has the following constants in ohms per phase referred to the stator: $R_1 = 0.085\ \Omega$, $X_1 = 0.196\ \Omega$, $R_2 = 0.067\ \Omega$, $X_2 = 0.161\ \Omega$, and $X_m = 6.65\ \Omega$. The total rotational losses are 3,200 W and are assumed to be constant. The core losses are lumped in with the rotational losses. For a rotor slip of 3.33 percent at the rated voltage and rated frequency, determine the following:
(a) The equivalent (input) impedance of the motor.
(b) The stator (or starting) current.
(c) The power factor of the motor.
(d) The induced voltage in the stator winding. *6.46*
(e) The rotor current. *6.80*
(f) The developed torque in N·m and lb·ft. *6.826*
(g) The total losses of the motor.
(h) The output power in W and hp.
(i) Efficiency.

6.27 Consider the data given in Problem 6.26 and determine the following:
(a) The maximum torque and slip at which the maximum torque is developed.
(b) Neglect the jX_2 and determine the maximum torque and the slip at which the maximum torque is developed. (Note that ignoring X_m at the maximum slip is reasonable because at the maximum slip I_2 is much greater than I_e.)

6.28 Consider the induction motor given in Example 6.8 and assume that the motor is now supplied from a power line that has an impedance of $Z_L = R_L + jX_L = 0.02 + j0.05\ \Omega$ per phase and that the constant voltage source connected at the sending-end of the line has a voltage of $V_L = 440/\sqrt{3}$ V. Determine the following:
(a) The slip at which the maximum torque is developed, if the jX_m is neglected.
(b) The developed torque at starting.
(c) The developed torque at starting, if the jX_m is not neglected.

6.29 A wound-rotor induction motor is operating from a constant line voltage V_1 and with variable external rotor circuit resistance $R_{2,\ ext}$. The rotor winding resistance per phase is R_2 so that $R_{2,\ tot} = R_2 + R_{2,\ ext}$. If the motor is mechanically loaded with a constant torque load, that is T_d = constant, then explain analytically the reasons why the following are true:

(a) Rotor current I_2 is constant as $R_{2,\ ext}$ is varied.

(b) $\dfrac{R_{2,\ tot}}{s}$ is constant.

6.30   Consider the induction motor given in Example 6.8 and assume that
the motor is uncoupled from its mechanical load and operated at the
rated stator terminal voltage. Determine the following:
(a) The slip at which the motor is operating.
(b) The stator phasor current I_1.
(c) The three-phase power input to the stator, that is, P_1.

6.31   Figure P6.31 shows a system that can be used to convert balanced 60-Hz
voltages to other frequencies. The synchronous motor has four poles
and drives the interconnecting shaft in a clockwise direction. The induction motor has eight poles, and its stator windings are connected to
the lines to produce a counterclockwise rotating field (i.e., in the opposite direction to the synchronous motor). As shown in the figure, the
induction machine has a wound rotor with terminals brought out through
slip rings. Determine the following:
(a) The speed at which the motor runs.
(b) The frequency of the rotor voltages in the induction machine.

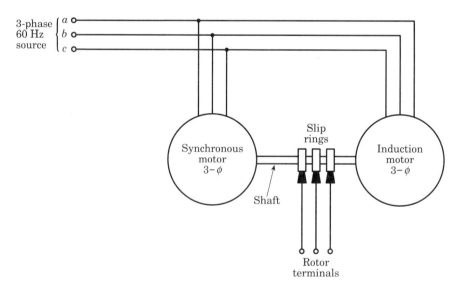

Figure P6.31

6.32   The following test data were taken on a three-phase four-pole 100-hp
480-V 60-Hz design A wye-connected induction motor with a rated current of 116.3 A:

Dc test:

$$V_{dc} = 23.26 \text{ V} \qquad\qquad I_{dc} = 116.3 \text{ A}$$

No-load test:

$$V_{nl} = 480 \text{ V} \qquad\qquad I_a = 24.5 \text{ A}$$
$$f = 60 \text{ Hz} \qquad\qquad I_b = 24.1 \text{ A}$$
$$P_{nl} = 4{,}900 \text{ W} \qquad\qquad I_c = 24.7 \text{ A}$$

Blocked-rotor test:

$$V_{br} = 42.2 \text{ V} \qquad\qquad I_a = 116.4 \text{ A}$$
$$f_{test} = 15 \text{ Hz} \qquad\qquad I_b = 116.4 \text{ A}$$
$$P_{br} = 6{,}100 \text{ W} \qquad\qquad I_c = 116.2 \text{ A}$$

Determine the following:

(a) The resistances R_1 and R_2 of the motor. Use a factor of 1.2 in computing the effective (ac) armature resistance per phase.

(b) The reactances X_1, X_2, and X_m of the motor.

CHAPTER 7

SYNCHRONOUS MACHINES

Who neglects learning in his youth, loses the past and is dead for the future.
Euripides, 438 B.C.

You cannot teach a crab to walk straight.
Aristophanes, 421 B.C.

But you can teach a man to walk like a crab.
Turan Gönen

7.1 INTRODUCTION

Almost all three-phase power is generated by three-phase synchronous machines operated as generators. Synchronous generators are also called **alternators** and are normally large machines producing electrical power at hydro, nuclear, or thermal power plants. Efficiency and economy-of-scale dictate the use of very large generators. Because of this, synchronous generators rated in excess of 1000 MVA (mega-volt-amperes) are quite commonly used in generating stations. Large synchronous generators have a high efficiency which at ratings greater than 50 MVA usually exceeds 98 percent. The term **synchronous** refers to the fact that these machines operate at constant speeds and frequencies under steady-state operations.

A given synchronous machine can operate as a generator or as a motor. Such machines are used as motors in constant-speed drives in industrial applications and also for pumped-storage stations. In small sizes with only fractional horsepower, they are used in electric clocks, timers, record players, and in other applications which require constant speed. Synchronous motors with frequency changers such as inverters or cycloconverters can also be used in variable-speed drive applications. An overexcited synchronous motor with no load can be used as a **synchronous capacitor** or

synchronous condenser[1] to correct power factors. A linear version of a synchronous motor can develop linear or translational motion. Presently, such a *linear synchronous motor* (LSM) is being developed for future high-speed public transportation systems in Japan. However, in general the LSM is not being used as much as the *linear induction motor* (LIM).

7.2 CONSTRUCTION OF SYNCHRONOUS MACHINES

In a synchronous machine, the **armature**[2] winding is on the stator and the **field** winding is on the rotor. In normal operation the three-phase stator currents (in the three-phase distributed stator winding) set up a rotating magnetic field. The synchronous machine **rotors** are simply rotating electromagnets which have the same number of poles as the stator winding. The rotor winding is supplied from an external dc source through slip rings and brushes; therefore, it produces a rotor magnetic field. Since the rotor rotates in synchronism with the stator magnetic field, the total magnetic field is the result of these two fields.

A synchronous machine is a constant-speed (i.e., synchronous speed) machine. Its rotor structure therefore depends on its speed rating. For this reason, high-speed machines have cylindrical (or non-salient pole) rotors, whereas low-speed machines have salient[3] pole rotors. With a cylindrical rotor the reluctance of the magnetic circuit of the field is independent of its actual direction and relative to the direct axis. However, with salient poles the reluctance is lowest when the field is along the direct axis where the air-gap is the minimum. It is highest when the field is directly halfway between the poles, that is, along the quadrature axis.

Since the rotor field structure depends upon the speed rating of the synchronous machine, turbogenerators (also known as turbo-alternators or turbine-generators), which are high-speed machines, have cylindrical rotors with two or four poles. Figure 7.1 shows a four-pole cylindrical rotor. Hydroelectric and diesel-electric generators are low-speed machines that have salient-pole rotors with four or more poles. Figure 7.2 shows a salient-pole rotor of a hydroelectric generator. This type of rotor structure typically has a relatively short axial length and a relatively large diameter.

The stator of a synchronous machine is basically similar to that of a three-phase induction machine. The stator winding is the source of voltage and electric power when the machine is operating as a generator, and the input winding when it is operating as a motor. It is usually made of preformed stator coils in a double-layer winding. The winding itself is distributed and chorded to reduce the harmonic content of the output voltages and currents. Figure 7.3 shows the stator of a 650 MW synchronous generator. Figure 7.4 shows the stator winding of a synchronous generator that is being wound.

---

[1] **Condenser** is an old name for capacitor.
[2] In rotating machinery, the term **armature** refers to the machine part in which an alternating voltage is generated due to relative motion with respect to a magnetic flux field.
[3] The word **salient** means "protruding" or "sticking out." Thus, in a salient pole rotor, a magnetic pole protrudes from the surface of the rotor; whereas a non-salient pole is built flush with the surface of the rotor, and its winding laid in slots in the rotor periphery.

Figure 7.1   Insertion of a four-pole cylindrical rotor into a 1200 MW nuclear generator unit. (Courtesy of ABB Corporation.)

Figure 7.2   A salient-pole type hydroelectric generator rotor being lowered into a turbine pit at an electric utility site. (Courtesy of General Electric Canada, Inc.)

Figure 7.3   Internal measurement of the stator of a 650 MW synchronous generator. (Courtesy of ABB Corporation.)

In salient-pole synchronous machines with laminated rotor construction, where induced currents are not allowed to flow in the rotor body, heavy copper bars are installed in slots in the pole faces. These bars are all shorted together at both ends of the rotor similar to the squirrel-cage rotor of an induction motor. Such a winding is known as the **amortisseur** or **damper winding**. Damper windings are installed in almost all synchronous machines that have **salient** poles. When the load on a synchronous machine changes, the load angle also changes. As a result, oscillations in the load angle and corresponding mechanical oscillations in the synchronous rotation of the shaft take place. These rotor oscillations are known as the **hunting**. The damper windings produce damping torques to eliminate these rotor oscillations caused by such transients and starting torques in synchronous motors. Cylindrical-rotor machines are formed from solid-steel forgings. Because transient rotor currents can be induced in the solid-rotor body itself, there is no need for a damper winding in such a machine.

7.3 FIELD EXCITATION OF SYNCHRONOUS MACHINES

In a synchronous machine, the rotor poles have constant polarity and must be supplied with direct current. The current can be supplied by an external dc generator or by a rectifier. Figures 7.5(a) and 7.5(b) show physical arrangements for a shaft-mounted exciter, and for a shaft-mounted exciter and a pilot exciter, respectively. The arrangement shown in Figure 7.5(b) is usually used in slow-speed machines with large ratings, such as hydrogenerators. Here the exciter may not be self-excited; instead, a self-excited or permanent-magnet type **pilot exciter** may be used to activate the exciter. Figure 7.6(a) shows a conventional shaft-mounted exciter that is a self-excited dc generator mounted on the same shaft as the rotor of the synchronous machine. In such an arrangement, the generator is the **exciter**. Stationary contacts called **brushes** ride on the slip rings to provide current from the dc source to the rotating field windings.

Figure 7.4 Winding the stator of a synchronous generator. (Courtesy of MagneTex.)

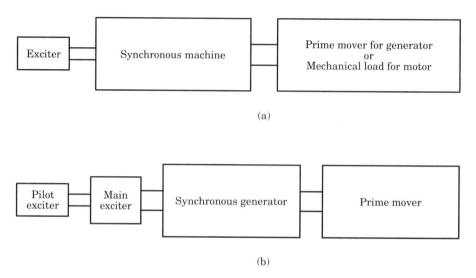

(a)

(b)

Figure 7.5 Conventional excitation systems for synchronous machines: (a) physical arrangement for a shaft-mounted exciter; (b) physical arrangement for a shaft-mounted exciter and pilot exciter.

The slips are metal rings which completely encircle the shaft of a machine, but are insulated from it. The brushes are made of a carbon compound which provides good contact with low mechanical friction.

An alternative form of excitation is to mount the armature of a relatively small exciter alternator on the shaft of a synchronous machine with a stationary field mounted on the stator. The three-phase output of the exciter generator is rectified to direct current by a three-phase rectifier circuit mounted on the shaft of the generator. It is then supplied to the main dc field circuit, as shown in Figure 7.6(b). This means of supplying field current to the rotor coils is called **brushless excitation** and is now used in most large synchronous machines.

7.4 STATOR WINDINGS

The stator (i.e., armature) winding of an ac generator is the source of the induced voltage. The coils of these windings are distributed in space, as shown in Figure 7.7, so that the voltages induced in the coils are not in phase, but are displaced from each other by the **slot angle** α.

The coils may have **full pitch** or **fractional pitch**. A full-pitch[4] coil spans 180 electrical degrees (i.e., the peripheral distance from the center line of a north pole to the center line of an adjacent south pole); whereas a fractional-pitch coil spans less than 180°, but seldom less than 120°. Polyphase stator

---

[4] It is also called **pole pitch** and is $2\pi/p$ radians, where p is the number of poles. Thus, the pole pitch is always π electrical radians or 180 electrical degrees, regardless of the number of poles on the machine.

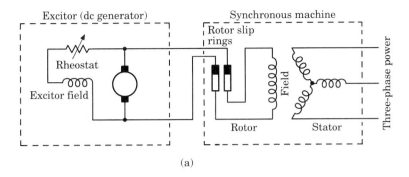

(a)

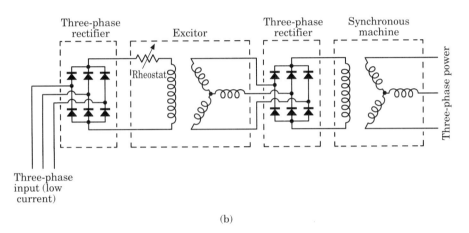

(b)

Figure 7.6   Circuit diagram for: (a) conventional shaft-mounted exciter; (b) brushless exciter.

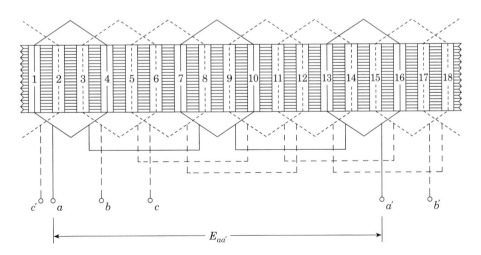

Figure 7.7   The coil interconnection of a three-phase, six-pole single-layer armature (stator) winding.

windings are usually **double-layer**, with the two layers arranged one above the other in the slot. In such an arrangement, the number of coils equals the number of slots. Thus if there are n slots per pole per phase, each pole has n coils. The **single-layer** polyphase winding is employed in small induction motors. The main advantage of the two-layer winding is that of accommodating fractional-pitch coils, which have lower resistance turns or end connections than full-pitch coils. They also have a lower resistance without a corresponding decrease in their flux linkage. The windings can be of the **lap** or **wave** type. Wave windings are usually used in induction motors. Figure 7.7 shows the coil interconnection of a three-phase, six-pole single-layer stator winding. Here, the armature has three slots per pole, which corresponds to one slot per phase per pole for a three-phase winding. The three-phase winding results from the addition of another two sets of armature coils displaced by 120° and 240° (in electrical degrees) from the first phase to produce a system of three voltages equal in magnitude and displaced from each other by 120°.

As previously stated, the armature winding coils are distributed over several slots in space.[5] Thus, the induced voltages in the coils are not in phase but are displaced from each other by the slot angle α. Hence, the winding voltage is the phasor sum of the coil voltages, as shown in Figure 7.8. The **distribution factor**[6] k_d is defined as

$$k_d = \frac{phasor\ sum\ of\ coil\ voltages}{arithmetic\ sum\ of\ coil\ voltages} \tag{7.1}$$

or

$$k_d = \frac{E_g}{nE_c} \tag{7.2}$$

where E_g = resultant group emf given by the phasor sum of coil voltages
$\quad\ \ E_c$ = individual coil voltage
$\quad\ \ n$ = number of slots per pole per phase

Therefore, from the geometry shown in Figure 7.8, the distribution factor can be reexpressed as

$$k_d = \frac{\sin(n\alpha/2)}{n\sin(\alpha/2)} \tag{7.3}$$

Note that the slot angle α is also called the **slot pitch** and is given in electrical degrees. Therefore, if q is the number of stator phases and s is the number of slots that exist in the stator, then the slot angle can be determined from

---

[5] Such a winding is called a **distributed winding**.
[6] It is also called the **breadth factor**.

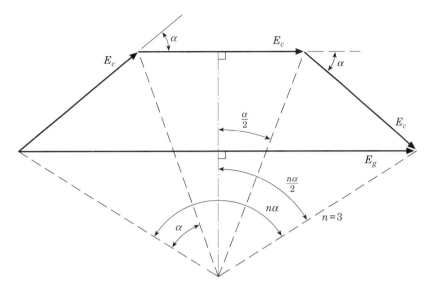

Figure 7.8 Phasor addition of individual coil voltages to determine the distribution factor.

$$\alpha = \frac{180°}{nq} \qquad (7.4)$$

Here,

$$n = \frac{s}{pq} \qquad (7.5)$$

Since there are p coil groups, the phase voltages are

$$E_a = pE_g \qquad (7.6)$$

Therefore, the groups are interconnected in such a way that all of the group voltages add in phase. If there are n slots per pole per phase, then each pole group has n coils.

If the individual coils of a winding are short pitched (i.e., if the coil span is less than a pole pitch), the induced voltage is less than the voltage that would be induced if the coil span were a full pole pitch. The **pitch factor** k_p of the coil is defined as

$$k_p = \frac{voltage\ induced\ in\ short\text{-}pitch\ coil}{voltage\ induced\ in\ full\text{-}pitch\ coil} \qquad (7.7)$$

or

$$k_p = \sin \frac{\rho}{2} \tag{7.8a}$$

$$= \sin \left\{ \frac{coil\ span\ in\ electrical\ degrees}{2} \right\} \tag{7.8b}$$

where ρ is the **coil pitch** or **coil span** in electrical degrees.

If the coils of a winding are distributed in several slots and the coils are short-pitched, the voltage induced in the winding will be affected by both factors k_p and k_d. They can be combined into a single winding factor k_w. The winding factor of a stator is given by

$$k_w = k_p k_d \tag{7.9}$$

Example 7.1

A three-phase, six-pole, synchronous machine has a stator with 36 slots. Each coil in this double-layer winding has a span of five slots. Determine the following:
(a) The number of slots per pole per phase.
(b) The slot pitch (or slot angle) in electrical degrees.
(c) The coil pitch.
(d) The pitch factor.
(e) The number of coils in a phase group.
(f) The distribution factor.

SOLUTION:
(a) The number of slots per pole per phase is

$$n = \frac{s}{pq} = \frac{36}{(6)(3)} = 2$$

(b) The slot pitch or slot angle is

$$\alpha = \frac{180°}{nq} = \frac{180°}{(2)(3)} = 30°$$

(c) The coil pitch (i.e., coil span) is

$$\rho = (5)(30°) = 150°$$

(d) The pitch factor is

$$k_p = \sin \frac{\rho}{2} = \sin \frac{150°}{2} = 0.9659$$

(e) The number of coils in a phase group in such a double-layer winding is equal to the number of slots per pole per phase, which is 2.

(f) The distribution factor is

$$k_d = \frac{\sin(n\alpha/2)}{n\sin(\alpha/2)} = \frac{\sin[2(30°)/2]}{2\sin(30°/2)} = 0.9659$$

(g) The winding factor is

$$k_w = k_p k_d = (0.9659)(0.9659) = 0.933$$

7.5 SYNCHRONOUS SPEED

A synchronous machine operates only at **synchronous speed**, a constant speed that can be determined by the number of poles and the frequency of alternation of the armature-winding voltage. Synchronous machines are called **synchronous** because their speed is directly related to the stator electrical frequency. Therefore, the synchronous speed can be expressed as

$$\omega_s = \frac{\omega}{p/2} = \frac{2\pi f}{p/2} = \frac{4\pi f}{p} \text{ rad/s} \tag{7.10}$$

or

$$n_s = \frac{120f}{p} \text{ rpm} \tag{7.11}$$

where ω_s = angular speed of the magnetic field (which is equal to the *angular rotor speed of the synchronous machine*)

ω = angular frequency of the electrical system

f = electrical frequency, Hz

p = number of poles

Since the rotor rotates at the same speed as the magnetic field, the stator electrical frequency can be expressed as

$$f = \frac{pn_s}{120} \tag{7.12}$$

Note that the frequency in hertz for a two-pole machine is the same as the speed of the rotor in revolutions per second; that is, the electrical frequency is synchronized with the mechanical speed of rotation. Therefore, a two-pole synchronous machine **must** rotate at 60 rps or 3600 rpm to produce a 60 Hz voltage. Alternatively, the radian frequency ω of the voltage wave in terms of ω_m, the mechanical speed in radians per second, is given as

$$\omega = \frac{p}{2}\omega_m \tag{7.13}$$

or

$$\theta = \frac{p}{2}\theta_m \tag{7.14}$$

where θ is in electrical measure and θ_m is in mechanical measure.

7.6 SYNCHRONOUS GENERATOR OPERATION

Consider the elementary synchronous generator shown in Figure 2.1(a). It has three identical stator coils (aa', bb', cc'), of one or more turns, displaced by 120° in space with respect to each other. When the field current I_f flows through the rotor field winding, it establishes a sinusoidally distributed flux in the air gap. If the rotor is now driven counterclockwise at a constant speed by the prime mover, a revolving magnetic field is developed in the air gap. This magnetic field is called the excitation field due to the fact that it is produced by the excitation current I_f. The rotating flux will vary the flux linkage of the armature windings aa', bb', cc' and will induce voltages in these stator windings. As shown in Figure 2.1(b), these induced voltages have the same magnitudes but are phase-shifted by 120 electrical degrees. Therefore, the resultant voltages in each of the three coils can be expressed as

$$e_{aa'}(t) = E_{max}sin\ \omega t \tag{7.15a}$$

$$e_{bb'}(t) = E_{max}sin\ (\omega t - 120°) \tag{7.15b}$$

$$e_{cc'}(t) = E_{max}sin\ (\omega t - 240°) \tag{7.15c}$$

The peak voltage in any phase of a three-phase stator is

$$E_{max} = \omega N\Phi \tag{7.16}$$

However, if the winding is distributed over several slots, the induced voltage is less and is given as

$$E_{max} = \omega N\Phi k_w \tag{7.17}$$

since $\omega = 2\pi f$, then

$$E_{max} = 2\pi f N\Phi k_w \tag{7.18}$$

where N = number of turns in each phase winding
  Φ = flux per pole due to the excitation current I_f
  k_w = winding factor[7]

---

[7] Its value is less than unity and depends on the winding arrangement.

Thus, the rms voltage of any phase of this three-phase stator is

$$E_a = \frac{2\pi}{\sqrt{2}} fN\Phi k_w \tag{7.19}$$

or

$$E_a = 4.44\, fN\Phi k_w \tag{7.20}$$

This voltage is a function of the frequency or speed of rotation, the flux that exists in the machine, and of course the construction of the machine itself. Therefore, it is possible to rewrite equation (7.19) as

$$E_a = K\Phi\omega \tag{7.21}$$

where K is a constant representing the construction of the machine. Thus,

$$K = \frac{Nk_w}{\sqrt{2}} \tag{7.22}$$

if ω is given in **electrical** radians per second. Alternatively,

$$K = \frac{Npk_w}{2\sqrt{2}} \tag{7.23}$$

if ω is given in **mechanical** radians per second.

Note that E_a is the **internal generated voltage**,[8] or simply the generated voltage. Its value depends upon the flux and the speed of the machine. However, the flux itself depends upon the current I_f flowing in the rotor field circuit. Therefore, for a synchronous generator operating at a constant synchronous speed, E_a is a function of the field current, as shown in Figure 7.9. Such a curve is known as the **open-circuit characteristic** (OCC) or **magnetization curve** of the synchronous machine. Contrary to the plot shown in Figure 7.9, at $I_f = 0$ the internal generated voltage (i.e., the induced voltage) is not zero due to the residual magnetism. At the beginning the voltage rises linearly with the field current. As the field current is increased further, the flux Φ does not increase linearly with I_f (as suggested by the air gap line) due to saturation of the magnetic circuit and E_a levels off. If the machine terminals are kept open, the internal generated voltage E_a is the same as the terminal voltage V_t and can be determined using a voltmeter.

---

[8] However, the **internal generated voltage** E_a is also known as the **excitation voltage** E_f. Since the **excitation voltage** (sometimes called the **field voltage**) can be confused with the dc voltage across the field winding, it is preferable to use the former.

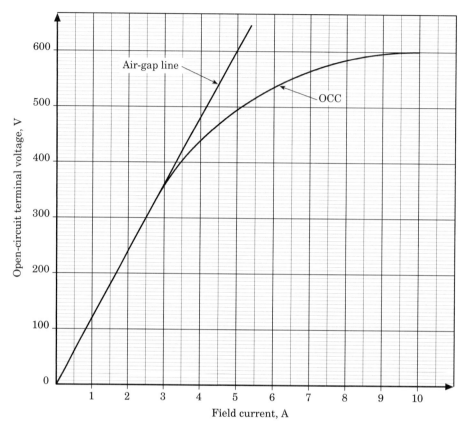

Figure 7.9 Open-circuit characteristic (OCC) or magnetization curve of a synchronous machine.

Example 7.2

An elementary two-pole three-phase 50-Hz alternator has a rotating flux of 0.0516 Wb. The number of turns in each phase coil is 20. Its shaft speed is 3000 rev/min and its stator winding factor is 0.96. Determine the following:

(a) The angular speed of the rotor.
(b) The three phase voltages as a function of time.
(c) The rms phase voltage of this generator if the stator windings are connected in delta.
(d) The rms terminal voltage if the stator windings are connected in wye.

SOLUTION:
(a) The angular speed of the rotor is

$$\omega = (3000 \text{ rev/min})(2\pi \text{ rad/rev})(1 \text{ min/60 s}) = 314.1593 \text{ rad/s}$$

(b) The magnitudes of the peak phase voltages are

$E_{max} = \omega N \Phi k_w$

$$= (314.1593 \text{ rad/s})(20)(0.0516 \text{ Wb})(0.96) = 311.127 \text{ V}$$

Thus, the three phase voltages are

$$e_{aa'}(t) = 311.127\sin 314.1593t \text{ V}$$

$$e_{bb'}(t) = 311.127\sin (314.1593t - 120°) \text{ V}$$

$$e_{cc'}(t) = 311.127\sin (314.1593t - 240°) \text{ V}$$

(c) If the stator windings are delta-connected, the rms phase voltage of the generator is

$$E_a = \frac{E_{max}}{\sqrt{2}} = \frac{311.127 \text{ V}}{\sqrt{2}} = 220 \text{ V}$$

(d) If the stator windings are wye-connected,

$$V_t = \sqrt{3}\,E_a = \sqrt{3}(220 \text{ V}) = 380 \text{ V}$$

Example 7.3

Determine the value of the K constant of the generator given in Example 7.2.
(a) If ω is in *electrical* radians per second.
(b) If ω is in *mechanical* radians per second.
(c) Determine the value of the internal generated voltage E_a.

SOLUTION:
(a) If ω is in *electrical* radians per second, then

$$K = \frac{Nk_w}{\sqrt{2}} = \frac{(20)(0.96)}{\sqrt{2}} = 13.5765$$

(b) If ω is in *mechanical* radians per second, then

$$K = \frac{Npk_w}{2\sqrt{2}} = \frac{(20)(2)(0.96)}{2\sqrt{2}} = 13.5765$$

(c) The value of the internal generated voltage is

$$E_a = K\Phi\omega = (13.5765)(0.0516 \text{ Wb})(314.1593) = 220 \text{ V}$$

7.7 EQUIVALENT CIRCUITS

Figure 7.10 shows the complete equivalent circuit representation of a three-phase synchronous generator. A dc power source supplies the rotor field circuit. The field current I_f is controlled by a rheostat connected in series with the field winding. Each phase has an internal generated voltage with series resistance R_a and series reactance X_s. Assuming balanced operation of the machine, the rms phase currents are equal to each other and 120° apart in phase. The same thing is also true for the voltages. In other respects the three phases are identical to each other. Therefore, the armature (stator) winding can be analyzed on a *per-phase* basis.

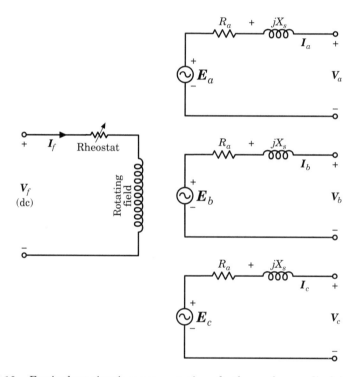

Figure 7.10   Equivalent circuit representation of a three-phase cylindrical rotor synchronous generator.

Figures 7.11(a), 7.11(b), and 7.11(c) show the per-phase equivalent circuits of a synchronous generator. Even though the internal generated voltage E_a is induced in the armature (stator) winding, the voltage that exists at the terminal of the winding is V_ϕ. The reasons for the difference are: (1) the resistance of the armature winding, (2) the leakage reactance of the armature winding, (3) the distortion of the air-gap magnetic field caused by the load current flowing in the armature winding, and (4) the effect of salient-pole rotor shapes if the machine has a salient-pole rotor.

The resistance R_a is the **effective resistance**[9] of the armature winding and is about 1.6 times the dc resistance of the stator winding. It includes the effects of the operating temperature and the **skin effect** caused by the alternating current flowing through the armature winding.

The leakage reactance X_a of the armature winding is caused by the leakage fluxes linking the armature windings due to the currents in the windings. These fluxes do not link with the field winding. For easy calculations, the leakage reactance can be divided into: (1) end-connection leakage reactance, (2) slot-leakage reactance, (3) tooth-top and zigzag (or differential) leakage reactance, and (4) belt-leakage reactance. However, in most large machines the last two reactances are a small portion of the total leakage reactance.

The air-gap magnetic field (caused by the rotor magnetic field) is distorted by the armature (stator) magnetic field because of the load current flowing in the stator. This effect is known as the **armature reaction** and the resultant reactance X_{ar} is called the **armature reactance**.[10]

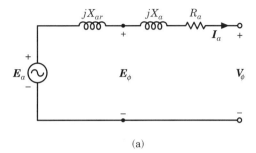

(a)

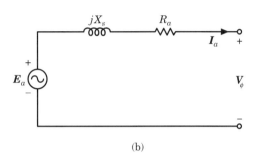

(b)

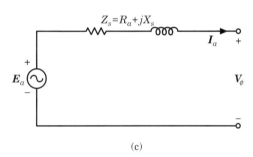

(c)

Figure 7.11   The per-phase equivalent circuits of a cylindrical-rotor synchronous generator.

The two reactances X_{ar} and X_a are combined into one reactance and called the **synchronous reactance**[11] X_s, which can be expressed as

$$X_s = X_a + X_{ar} \tag{7.24}$$

Therefore, as shown in Figure 7.11(b), the **synchronous impedance** becomes

---

[9] It is also called the **ac resistance**.
[10] It is also called as the **magnetizing reactance**.
[11] It is also called the **direct-axis synchronous reactance** and denoted by X_d.

$$\boldsymbol{Z}_s = R_a + jX_s \tag{7.25}$$

In general, as the machine size increases the per-unit resistance decreases but the per-unit synchronous reactance increases. Thus, the magnitude of the synchronous impedance becomes

$$\boldsymbol{Z}_s = (R_a^2 + X_s^2)^{1/2} \cong X_s \tag{7.26}$$

Because of this, R_a is omitted[12] from many analyses of synchronous machine operations.

Figures 7.12(a), 7.12(b), and 7.13 show the phasor diagrams[13] of a cylindrical-rotor synchronous generator operating at a lagging, leading, and unity power factor, respectively. Note that the dc current I_f in the field winding produces the mmf $\mathscr{F}_a$ in the air gap and that the ac load current flowing in the stator produces the mmf $\mathscr{F}_{ar}$ due to the armature reaction. The vector sum of the two mmfs gives the resultant mmf of $\mathscr{F}_r$. The flux produced by an mmf is in phase with the mmf and the voltage induced[14] by a certain flux is behind the corresponding mmf by 90°. Thus, as shown in Figure 7.12(a), the mmf $\mathscr{F}_f$ is ahead of E_a by 90° and the mmf $\mathscr{F}_{ar}$ is in phase with $\boldsymbol{I}_a$. The resultant mmf $\mathscr{F}_r$ is ahead of E_ϕ by 90°. The **armature reaction voltage** E_{ar} can be determined from

$$\boldsymbol{E}_{ar} = -jX_{ar}\,\boldsymbol{I}_a \tag{7.27}$$

From Figure 7.14(a),

$$\boldsymbol{E}_\phi = \boldsymbol{E}_a + \boldsymbol{E}_{ar} \tag{7.28}$$

or

$$\boldsymbol{E}_a = \boldsymbol{E}_\phi - \boldsymbol{E}_{ar} \tag{7.29}$$

Thus

$$\boldsymbol{E}_a = \boldsymbol{E}_\phi + jX_{ar}\boldsymbol{I}_a \tag{7.30}$$

The phase voltage V_ϕ is found from

$$\boldsymbol{V}_\phi = \boldsymbol{E}_a - R_a\boldsymbol{I}_a - j(X_a + X_{ar})\boldsymbol{I}_a \tag{7.31a}$$

---

[12] The magnitude of R_a is about 0.5% to 2% of X_s and therefore can be ignored except in efficiency computations.

[13] In the phasor diagrams, the length of the armature resistance voltage drop phasor has been shown larger than it should be in order to make it noticeable.

[14] According to Lenz's Law, $e = -N\dfrac{d\phi}{dt}$.

$$= \boldsymbol{E}_a - R_a\boldsymbol{I}_a - jX_s\boldsymbol{I}_a \qquad (7.31\text{b})$$

$$= \boldsymbol{E}_a - (R_a + jX_s)\boldsymbol{I}_a \qquad (7.31\text{c})$$

$$= \boldsymbol{E}_a - \boldsymbol{Z}_s\boldsymbol{I}_a \qquad (7.31\text{d})$$

Alternatively, the air-gap voltage and the internal generated voltage[15] can be expressed, respectively, as

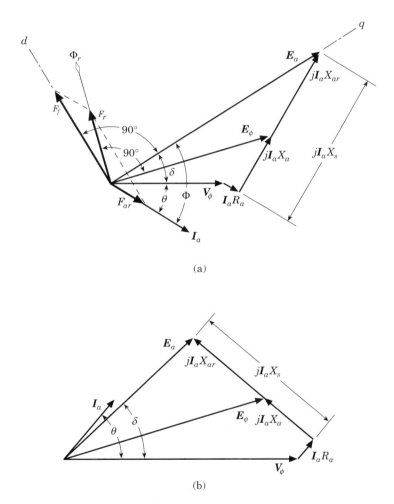

(a)

(b)

Figure 7.12 Phasor diagrams of a cylindrical-rotor synchronous generator operating at: (a) lagging (overexcited) and (b) leading (underexcited) power factor. (The diagrams shown are not drawn to scale.)

---

[15] It is interesting to note that even though synchronous motors, almost without exception, have salient poles, they are often treated as cylindrical-rotor machines. Therefore, equation (7.33) can also be used for motors as long as the sign of current I_a is made negative to yield
$$\boldsymbol{V}_\phi = \boldsymbol{E}_a + (R_a + jX_s)\,\boldsymbol{I}_a$$

$$\boldsymbol{E}_{\phi} = \boldsymbol{V}_{\phi} + (R_a + jX_a)\boldsymbol{I}_a \tag{7.32}$$

and

$$\boldsymbol{E}_a = \boldsymbol{V}_{\phi} + (R_a + jX_s)\boldsymbol{I}_a \tag{7.33}$$

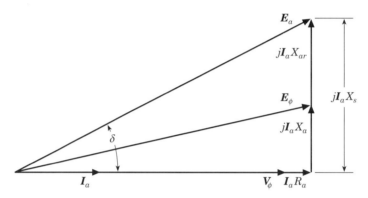

Figure 7.13   Phasor diagram of a cylindrical-rotor synchronous generator operating at unity power factor. (The diagram shown is not drawn to scale.)

In a synchronous machine the internal generated voltage $\boldsymbol{E}_a$ takes into account the flux produced by the field current, whereas the synchronous reactance takes into account all the flux produced by the balanced three-phase armature currents. For a given unsaturated cylindrical-rotor machine operating at a constant frequency, the synchronous reactance is a constant and the internal generated voltage E_a is proportional to the field current I_f. As can be observed in Figure 7.12, for a given phase voltage V_f and armature current I_a, a greater E_a is required for lagging loads than for leading loads. Since

$$E_a = K\Phi\omega \tag{7.21}$$

then a greater field current **(overexcitation)** is required with lagging loads to keep the V_{ϕ} constant. (Here, ω has to be constant to hold the frequency constant.) However for leading loads, a smaller E_a is needed and thus a smaller field current **(underexcitation)** is required. Succinctly put, the synchronous machine is said to be **overexcited** when voltage E_a exceeds voltage V_{ϕ}; otherwise, it is said to be **underexcited**. The angle δ between voltage E_a and voltage V_{ϕ} is called the **torque angle** or the **power angle** of the synchronous machine.

The **voltage regulation** of a synchronous generator at full-load, power factor, and rated speed is defined as

$$V\,Reg = \frac{E_a - V_{\phi}}{V_{\phi}} \tag{7.34}$$

It is often expressed as percent voltage regulation. Thus,

$$\% \ V \ Reg = \frac{E_a - V_\phi}{V_\phi} \times 100 \qquad (7.35)$$

where V_ϕ is the voltage at full-load and E_a is the internal generated voltage (i.e., the V_ϕ voltage) at no load. The voltage regulation is a useful measure in comparing the voltage behavior of generators. It is positive for an inductive load since the voltage rises when the load is removed, and is negative for a capacitive load if the load angle is large enough for the voltage to drop.

Example 7.4

A three-phase, 13.2-kV, 60-Hz, 50-MVA, wye-connected cylindrical-rotor synchronous generator has an armature reactance of 2.19 Ω per phase. The leakage reactance is 0.137 times the armature reactance. The armature resistance is small enough to be negligible. Also ignore the saturation. Assume that the generator delivers full-load current at the rated voltage and 0.8 lagging power factor. Determine the following:
(a) The synchronous reactance in ohms per phase.
(b) The rated load current.
(c) The air gap voltage.
(d) The internal generated voltage.
(e) The power angle.
(f) The voltage regulation.

SOLUTION:
(a) The leakage reactance per phase is

$$X_a = 0.137 X_{ar} = 0.137(2.19 \ \Omega) \cong 0.3 \ \Omega$$

Therefore the synchronous reactance per phase is

$$X_s = X_a + X_{ar} = 0.3 + 2.19 = 2.49 \ \Omega$$

(b) The rated load (or full load) current is

$$I_a = \frac{S}{\sqrt{3} V_t} = \frac{50 \times 10^6}{\sqrt{3}(13,200)} = 2,186.93 \ \text{A}$$

and when expressed as a phasor,

$$\boldsymbol{I}_a = I_a(\cos \theta - j \sin \theta) = 2,186.93(0.8 - j0.6)$$

$$= 1,749.55 - j1,312.16 = 2,186.93 \angle - 36.87^\circ \ \text{A}$$

(c) The air-gap voltage (also known as the *voltage behind the leakage reactance*) is

$$E_\phi = V_\phi + jX_a I_a$$

$$= \frac{13,200}{\sqrt{3}} \angle 0° + j0.3(2,186.93\angle - 36.87°) = 8,031.84\angle 3.75° \text{ V}$$

(d) The internal generated voltage (also known as the *voltage behind the synchronous reactance*) is

$$E_a = V_\phi + jX_s I_a$$

$$= \frac{13,200}{\sqrt{3}} \angle 0° + j2.49(2,186.93\angle - 36.87°) = 9,102.80\angle 11.08° \text{ V}$$

(e) The power angle (also called the *torque angle*) is

$$\delta = 11.08°$$

(f) The voltage regulation at full load is

$$V\,Reg = \frac{E_a - V_\phi}{V_\phi} = \frac{9,102.80 - 7,621.02}{7,621.02} = 0.194 \text{ or } 19.4 \text{ percent}$$

7.8 SYNCHRONOUS MOTOR OPERATION

A given synchronous machine can also operate as a motor. However, when the synchronous machine makes the transition from generator to motor action, reversal of power flow takes place. Instead of current flowing *out of* the armature (stator) terminals, the current flows *into* the armature terminals. The speed of the synchronous motor is constant as long as the source frequency is constant. Thus, the equivalent circuit of a synchronous motor is exactly the same as the equivalent circuit of a synchronous generator (as shown in Figure 7.11), with one exception: the direction of the current I_a is reversed. The corresponding KVL equations for the motor are

$$V_\phi = E_a + (R_a + jX_s)I_a \tag{7.36}$$

$$V_\phi = E_\phi + (R_a + jX_a)I_a \tag{7.37}$$

or

$$E_a = V_\phi - (R_a + jX_s)I_a \tag{7.38}$$

$$E_\phi = V_\phi - (R_a + jX_a)I_a \tag{7.39}$$

The power output of a synchronous motor depends totally on the mechanical load on the shaft. As previously stated, its speed depends on the source frequency. Since the speed does not vary as the field current I_f is changed, varying I_f has no effect on the output power. However, changing I_f affects the E_a (i.e., E_a increases when I_f increases) and the power factor of the current I_a drawn from the three-phase source. Figures 7.14(a), 7.14(b), and 7.14(c) show phasor diagrams of a synchronous motor operating at unity, leading, and lagging power factor, respectively. When the magnitude of E_a is equal to V_ϕ this condition is referred to as **100% excitation** (i.e., operating at *unity* power factor). When $E_a > V_\phi$, it is called **overexcitation** (i.e., operating at *leading* power factor). Finally, when $E_a < V_\phi$, it is called **underexcitation** (i.e., operating at *lagging* power factor).

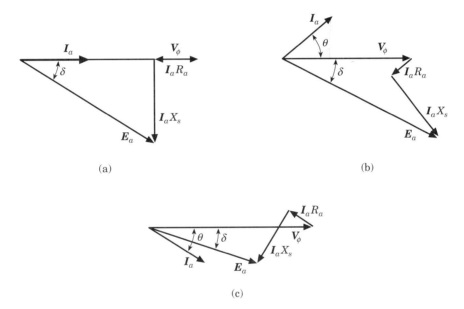

Figure 7.14   Phasor diagram of a synchronous motor operating at: (a) unity; (b) leading (overexcited); (c) lagging (underexcited) power factor.

7.9 POWER AND TORQUE CHARACTERISTICS

In a synchronous generator the input power is provided by a prime mover in terms of shaft power. The input mechanical power of the generator can be expressed as

$$P_{in} = P_{shaft} = T_{in}\omega_m \qquad (7.40)$$

where T_{in} represents the applied torque to the shaft by the prime mover and ω_m denotes the mechanical speed of the shaft rotation. In contrast, the power internally developed from mechanical to electrical form can be expressed as

$$P_d = T_d \omega_m \qquad (7.41)$$

or

$$P_d = 3E_a I_a \cos \Phi \qquad (7.42)$$

where Φ is the angle between E_a and I_a as shown in Figure 7.12(a). Therefore

$$\Phi = \theta + \delta \qquad (7.43)$$

The difference between output power and input power gives the losses of the machine; whereas the difference between the input power and the developed power gives the mechanical and core losses of the generator. The electrical output power of the generator can be found in terms of line quantities as

$$P_{out} = \sqrt{3}\, V_t I_L \cos \theta \qquad (7.44)$$

or in phase quantities as

$$P_{out} = 3V_\phi I_a \cos \theta \qquad (7.45)$$

Similarly, the reactive power can be found in terms of line quantities as

$$Q_{out} = \sqrt{3}\, V_t I_L \sin \theta \qquad (7.46)$$

or in phase quantities as

$$Q_{out} = 3V_\phi I_a \sin \theta \qquad (7.47)$$

The real and reactive power output of a synchronous generator can also be expressed as a function of the terminal voltage, the internal generated voltage, the synchronous impedance, and the **power angle** or **torque angle** δ. This is also true for the real and reactive power received by a synchronous motor. Since X_s is much greater than R_a, then it can be proven easily that

$$E_a \sin \delta = X_s I_a \cos \theta \qquad (7.48)$$

Thus,

$$I_a \cos \theta = \frac{E_a \sin \delta}{X_s} \qquad (7.49)$$

and substituting this equation into Equation (7.45),

$$P = \frac{3E_a V_\phi}{X_s} \sin \delta \qquad (7.50)$$

Because the stator losses are ignored (i.e., R_a is assumed to be zero), P represents both the developed power P_d (or the air-gap power) and the output power P_{out}.

The power output of the synchronous generator depends on the angle between E_a and V_ϕ. If the angle δ is increased gradually, the real power output increases, reaching a maximum when δ is 90°. Therefore the *maximum power* becomes

$$P_{max} = \frac{3E_a V_\phi}{X_s} \qquad (7.51)$$

This is also known as the **steady-state power limit** or the **static stability limit**. From equation (7.42) the developed torque of the synchronous machine can be found as

$$T_d = \frac{P}{\omega_{sn}} \qquad (7.52a)$$

$$= \frac{3E_a V_\phi}{\omega_m X_s} \sin \delta \qquad (7.52b)$$

$$= T_{max} \sin \delta \qquad (7.52c)$$

where the maximum torque is

$$T_{max} = \frac{P_{max}}{\omega_m} \qquad (7.53a)$$

$$= \frac{3E_a V_\phi}{\omega_m X_s} \qquad (7.53b)$$

Therefore, any increase in the mechanical power to the synchronous generator or in the mechanical output of the synchronous motor[16] after δ has reached 90° produces a decrease in real electrical power. The generator accelerates while the motor decelerates and either way the result is a **loss of synchronism**.[17] The maximum torque T_{max} is also known as the **pull-out torque**.[18] Figure 7.15 shows the steady-state power-angle or torque-angle characteristic of a synchronous machine with negligible armature resistance. Note that when δ becomes negative, the power flow reverses. In other words, when power flows into the

---

[16] Even though the rotors of three-phase synchronous motors are salient-pole rather than cylindrical type, applying cylindrical-rotor theory yields a good degree of approximation.

[17] It is also known as **pulling out of step**.

[18] More precisely, the pull-out torque is the maximum sustained torque that the motor will develop at synchronous speed for one minute, with rated voltage applied at rated frequency and with normal excitation.

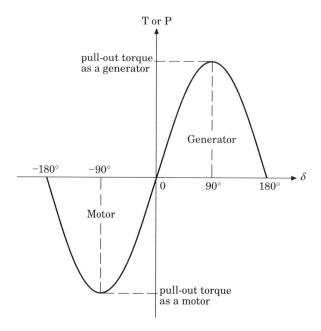

Figure 7.15   Synchronous machine power or torque as a function of power angle, δ.

electrical terminals, the machine starts acting as a motor with a negative δ. In the generator mode the power flows out of the electrical terminals and the angle δ becomes positive. This behavior can be explained by equation (7.51). Similarly, the torque reverses direction (sign) when the machine goes from generator operation to motor operation according to equation (7.53b). In generator mode the torque is positive, that is, a **counter torque** and therefore the T_d is opposite to ω_m. In motor mode the torque is negative which means that it is in the same direction as ω_m. Figure 7.16 shows the superimposed power and torque-angle characteristics of a synchronous machine operating in generator mode. Note that both the maximum power and the maximum torque take place when δ is 90°. If the prime mover tends to drive the generator to supersynchronous speed by excessive driving torque, the field current can be increased to develop more counter torque to overcome such a tendency. Similarly, if a synchronous motor is apt to pull out of synchronism due to excessive load torque, the field current can be increased to produce greater torque and prevent a loss of synchronism.

The reactive power of a synchronous machine can be expressed as

$$Q = \frac{3(E_a V_t \cos \delta - V_\phi)^2}{X_s} \tag{7.54}$$

Here, positive Q means supplying inductive vars in the generator mode or receiving inductive vars in the motor mode; and negative Q means supplying capacitive vars in the generator mode or receiving capacitive vars in the motor mode.

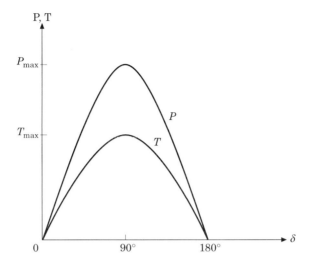

Figure 7.16 The superimposed power and torque-angle characteristics of a synchronous generator.

Example 7.5

A three-phase, 100-hp, 60-Hz, 480-V, four-pole, wye-connected, cylindrical-rotor synchronous motor has an armature resistance of 0.15 Ω and a synchronous reactance of 2 Ω per phase, respectively. At the rated load and a leading power factor of 0.8, the motor efficiency is 0.95. Determine the following:
(a) The internal generated voltage.
(b) The torque angle δ.
(c) The maximum torque.

SOLUTION:
(a) The motor input power is

$$P_{in} = \frac{(100 \text{ hp})(746 \text{ W/hp})}{0.95} = 78,526.32 \text{ W}$$

The rated load current is

$$I_a = \frac{P_{in}}{\sqrt{3}V_L \cos\theta} = \frac{78,526.32 \text{ W}}{\sqrt{3}(480 \text{ V})0.8} = 118.07 \text{ A}$$

The voltage per phase is

$$V_\phi = \frac{480 \text{ V}}{\sqrt{3}} = 277.13 \text{ V}$$

and $\theta = \cos^{-1} 0.8 = 36.87°$ leading

$$E_a = V_\phi - (R_a + jX_s)I_a$$

$$= 277.13\angle 0° - (0.15 + j2)(118.07\angle 36.87°) = 451.17\angle - 26.25° \text{ V}$$

(b) The negative sign indicates that E_a lags V_ϕ (which is used as the reference phasor). The $26.25°$ also represents the torque angle δ. Alternatively, the torque angle can be found from

$$\tan(\theta + \delta) = \frac{V_\phi \sin\theta + I_a X_s}{V_\phi \cos\theta - I_a R_a}$$

$$= \frac{277.13 \sin 36.87° + (118.07)2}{277.13 \cos 36.87° - (118.07)0.15} = 1.97$$

Thus,

$$\theta + \delta = \tan^{-1}(1.97) = 63.12°$$

and

$$\delta = 63.12° - 36.87° = 26.25°$$

(c) Since the machine has four poles, its speed is 1800 rpm or 188.495 rad/s. Thus the maximum torque is

$$T_{max} = \frac{P_{max}}{\omega_m} = \frac{3E_a V_\phi}{\omega_m X_s}$$

$$= \frac{3(451.17)277.13}{(188.495)2} = 994.98 \text{ N} \cdot \text{m}$$

7.10 STIFFNESS OF SYNCHRONOUS MACHINES

The ability of a synchronous machine to endure the forces that tend to pull it out of synchronism is called **stiffness**. Stiffness, which represents the slope of the power-angle curve at a given operating point, can be determined by taking the partial derivative of the power delivered with respect to the torque angle. The unit of such a rate of power is W per radian. Since

$$P = \frac{3E_a V_\phi}{X_s} \sin\delta \tag{7.50}$$

For small displacements of $\Delta\delta$ the change in power is ΔP. Also,

$$K_s = \frac{\Delta P}{\Delta \delta} \cong \frac{dP}{d\delta} \tag{7.55}$$

Thus, the stiffness can be expressed as

$$k_s \quad P_s = \frac{3E_a V_\phi}{X_s} \cos \delta \tag{7.56}$$

The maximum stiffness is referred to as **synchronizing power**. Of course, at pull-out the stiffness of the machine is zero.

Example 7.6

Consider the synchronous generator given in Example 7.4 and assume that the machine has eight poles. Determine the following:
(a) The synchronizing power in MW per electrical radian and in MW per mechanical degree.
(b) The synchronizing power in MW per mechanical degree.
(c) The synchronizing torque in MW per mechanical degree.

SOLUTION:
(a) The synchronizing power is

$$P_s = \frac{3E_a V_\phi}{X_s} \cos \delta$$

$$= \left(\frac{3(9,102.80)(7621.02)}{2.49} \right) \cos 11.08°$$

$$= 82.024 \text{ MW per electrical radian}$$

$$= 1.4316 \text{ MW per electrical degree}$$

(b) Since the machine has four pole pairs, there are four electrical cycles for each mechanical revolution. Thus,

$$P_s = 4 \times 1.4316 = 5.7264 \text{ MW per mechanical degree}$$

(c) Since the synchronous speed is

$$n_s = \frac{60}{4} = 15 \text{ rev/s}$$

the synchronizing torque is

$$T_s = \frac{P_s}{\omega_m}$$

$$= \frac{5,726,318.4 \text{ W}}{2\pi(15 \text{ rev/s})} = 60,758.1 \text{ N} \cdot \text{m per mechanical degree.}$$

7.11 EFFECT OF CHANGES IN EXCITATION

One of the important characteristics of the synchronous machine is that its power factor can be controlled by the field current. In other words, the power factor of the stator (or line) current can be controlled by changing the field excitation. However, the behavior of a synchronous generator (alternator) connected to an infinite bus (large system) is quite different from that of one operating alone.

7.11.1 A SYNCHRONOUS MACHINE CONNECTED TO AN INFINITE BUS

Assume a constant-power operation of a synchronous machine connected to an infinite bus so that it operates at constant frequency and terminal voltage. Under such circumstances the power factor is determined by the field current. In the generator mode, the amount of power generated and the frequency (or speed) are determined by the prime mover. In the motor mode, the speed is determined by the line frequency and the output depends on the mechanical load on the shaft.

Figures 7.17(a), 7.17(b), and 7.17(c) show the phasor diagrams of a synchronous machine operating as an overexcited generator, a normal excited generator, and an underexcited motor, respectively. The figures show that the locus of the current I_a for constant real power is a vertical line while the locus of the internal generated voltage E_a is a horizontal line. Notice that the variation in the power-factor angle θ is very significant, but the variation in the torque angle δ is almost insignificant. As shown in Figure 7.17(a), when the machine is operating as an overexcited generator, it has a lagging power factor due to a high field current. The maximum power P_{max} is large and therefore the machine operation is stable. As shown in Figure 7.17(b), when the machine is operating as a normal excited generator it has a unity power factor as a result of normal field current. Finally, as shown in Figure 7.17(c), when the machine is operating as an underexcited generator, it has leading power factor due to a low field current. Therefore the maximum power P_{max} is small and hence the machine operation is less stable.

Figures 7.17(d), 7.17(e), and 7.17(f) show the phasor diagrams of a synchronous machine operating as an overexcited motor, normal excited motor, and underexcited motor, respectively. The figures show that the locus of the current I_a for constant real power is a vertical line while the locus of the internal generated voltage E_a is a horizontal line. As shown in Figure 7.17(d), when the machine is operating as an overexcited motor it has leading power factor due to a high field current. Therefore, the maximum power P_{max} is large and the machine operation is stable. As shown in Figure 7.17(e) when the machine is operating as a normal excited motor it has a unity power factor due to a normal field current. Under such conditions the motor draws the minimum stator current I_a. Finally, as shown in Figure 7.17(f), when the machine is operating as an underexcited motor, it has lagging power factor due to a low field current. Here the maximum power P_{max} is small and hence the machine operation is less stable.

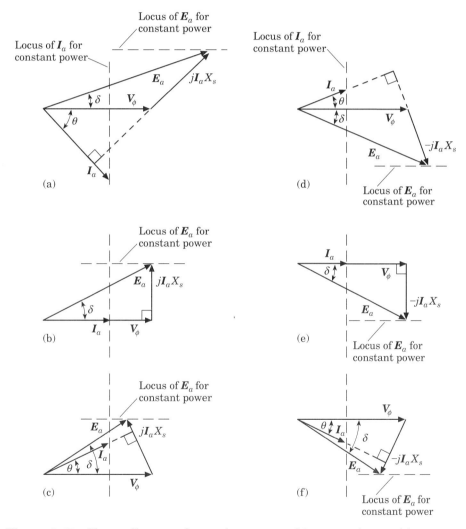

Figure 7.17   Phasor diagram of a synchronous machine operating as: (a) an overexcited generator; (b) a normally excited generator; (c) an underexcited generator; (d) an overexcited motor; (e) a normally excited motor; (f) an underexcited motor.

From the phasor diagrams shown in Figure 7.17, one can observe that the voltage E_a leads the voltage V_ϕ when the synchronous machine operates as a generator, and *lags* when it operates as a motor. Also, note that the torque angle or the power angle δ is positive when generating and negative when motoring.

Succinctly put, the power factor at which a synchronous machine operates and its stator (armature) current can be controlled by changing its field excitation.[19] The curve showing the relationship between the stator current and the field current at a constant terminal voltage with a constant real power is called a **synchronous machine V curve** because of its shape. The V curves can be developed for synchronous generators as well as for synchronous motors and will be almost identical.[20] Figure 7.18(a) shows a family of V curves for a synchronous motor. Note that there are three V curves in the figure corresponding to full load, half load, and no load. The dashed lines are loci for constant power and are called **compounding curves**. Notice that minimum armature (i.e., stator or line) current is always associated with a unity power factor. The corresponding field current is indicated as **normal** excitation. Also notice that the region to the right of the unity-power-factor compounding curve is associated with overexcitation and a leading power factor and that the region to the left is associated with underexcitation and a lagging power factor. Figure 7.18(b) shows the correlation between the power factor and the field current. These curves show that a synchronous motor can be overexcited and carry a substantial leading power factor. Also notice that both curves, in Figures 7.18(a) and 7.18(b), show that a slightly increased field current is needed to produce normal excitation as the load increases.

7.11.2 A SYNCHRONOUS GENERATOR OPERATING ALONE

In general, synchronous machines operating in generator mode or in motor mode are connected to an infinite bus. However there are many applications in which synchronous generators may be used to supply an isolated (independent) power system. For such applications, the infinite bus theory (i.e., having constant voltage and constant frequency) cannot be used, since there are no other generators connected in parallel to compensate for changes in field excitation and prime-mover output in order to keep the terminal voltage and frequency constant. Here, the prime mover is most likely a diesel or gasoline engine. The frequency depends totally on the speed of the prime mover. Thus, a **governor** is needed to maintain the constant frequency. The power factor is the load power factor and changes as the load changes. Hence, the power factor and armature current cannot be controlled at the generator site. In fact the only control that can be used at the generator site is that of the field current. Thus, at con-

---

[19] In the event that the synchronous machine is not transferring any power but is simply **floating** on the infinite bus, the machine power factor is zero. In other words, the armature current either lags or leads the terminal voltage by 90°.

[20] If it were not for the small effects of armature resistance.

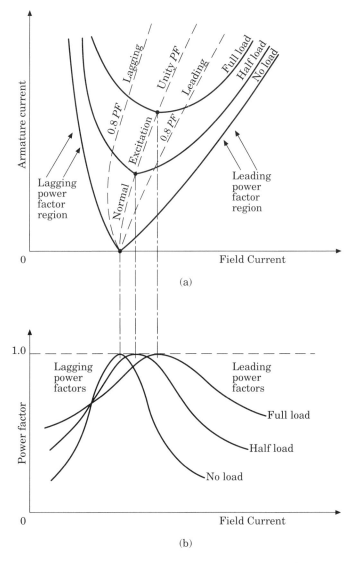

Figure 7.18   Synchronous-motor V curves: (a) stator current versus field current; (b) power factor versus field current.

stant speed, if the field current is increased, the terminal voltage will increase, as shown in Figure 7.19(a). Since the terminal voltage changes drastically as the load changes, an automatic voltage regulator is required to control I_f so that the terminal voltage can be kept constant with a changing load. Otherwise, as shown in Figure 7.19(b), as the load current I_a is increased, the terminal voltage drops sharply with a drop in the load power factor. From the study of characteristics of a synchronous generator, given in Figure 7.19, one can conclude: (1) the addition of inductive loads causes the terminal voltage to

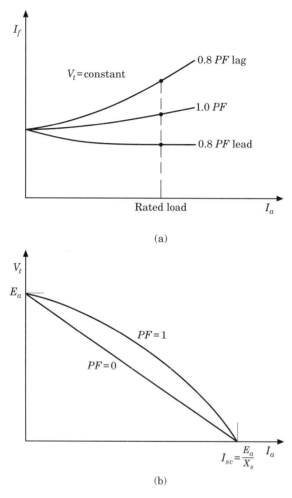

Figure 7.19   Characteristics of a generator operating alone: (a) compounding curves; (b) terminal voltage versus load current at constant field current.

drop drastically, (2) the addition of purely resistive loads causes the terminal voltage to drop very little (almost insignificantly), and (3) the addition of capacitive loads causes the terminal voltage to rise drastically. Figure 7.20 shows the constant-field current volt-ampere characteristic curves of a synchronous generator operating alone. Note that the curves shown are for three different values of constant field current and power factors.

7.12 THE USE OF DAMPER WINDINGS TO OVERCOME MECHANICAL OSCILLATIONS

If the load on a synchronous machine varies, the load angle changes from one steady value to another. During such transient phenomena oscillations in

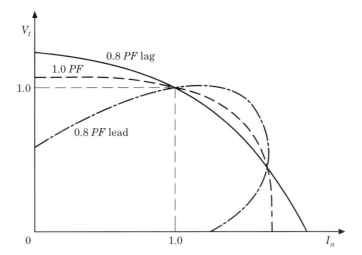

Figure 7.20   The constant-field current volt-ampere characteristic curves of a synchronous generator operating alone.

the load angle and resultant mechanical oscillations[21] of the rotor take place. To dampen out such oscillations, damper windings are used in most salient-pole synchronous machines. The damper windings[22] are made of copper or brass bars located in pole-face slots on the pole shoes of a salient-pole rotor and their ends are connected together. When the rotor speed is different from the synchronous speed, currents are induced in the damper windings. The damper windings behave like the squirrel-cage rotor of an induction motor, developing a torque to eliminate mechanical oscillations and restore the synchronous speed. Note that cylindrical-rotor machines do not have damper windings because the eddy currents that exist in the solid rotor during such transients play the same role as the currents in damper windings in a salient-rotor machine.

7.13 STARTING OF SYNCHRONOUS MOTORS

A synchronous motor is not a self-starter. In other words, if its rotor winding is connected to a dc source and its stator winding is supplied by an ac source, the motor will not start,[23] but simply vibrates. The methods that can be used to start a synchronous motor include: (1) starting the motor as an induction mo-

---

[21] In other words, any variance in load causes an oscillatory motion superimposed on the normal (i.e., synchronous) motion of the machine shaft. This motion is also called **hunting**.

[22] They are also called the **amortisseur (killer) windings**.

[23] To produce the required torque, the rotor must be rotating at the same speed as the armature (stator) field. Therefore, if the rotor is turning (or not turning, in this case) at some other speed, the rotating armature-field poles will be moving past the rotor poles first attracting, then repelling them. Thus, the average torque is zero and the motor cannot start. Such a synchronous motor has no starting torque.

tor, (2) starting it with a variable-frequency supply, and (3) starting it with the help of a dc motor.

The first method is the most practical and therefore the most popular. When the field windings are disconnected from the dc source and the stator windings are connected to its ac source, the motor acts like an induction motor because of its damper windings. Such an induction-motor start brings the machine almost up to synchronous speed and when the dc field windings are excited, the rotor **falls into step**, that is, starts to rotate at the synchronous speed. At synchronous speed there is no current induced in the damper windings and therefore there is no torque produced by them.

The second method involves starting the motor with low-frequency ac voltage by employing a frequency converter. As a result, the armature field rotates slowly to make the rotor poles follow the armature poles. Later, the motor can start operating at its synchronous speed by slowly increasing the supply frequency to its nominal value. The third method involves bringing the motor to its synchronous speed by using a dc motor before connecting the motor to the ac supply.

7.14 OPERATING A SYNCHRONOUS MOTOR AS A SYNCHRONOUS CONDENSER

As previously stated, overexcited[24] synchronous motors can generate reactive power. When synchronous motors are used as synchronous condensers they are manufactured without a shaft extension, since they are operated with no mechanical load. The ac input power supplied to such a motor can only provide for its losses. These losses are very small and the power factor of the motor is almost zero. Therefore, the armature current leads the terminal voltage by close to 90°, as shown in Figure 7.21(a), and the power network perceives the motor as a capacitor bank. As can be seen in Figure 7.21(b), when this motor is overexcited it behaves like a **capacitor** (i.e., **synchronous condenser**), with $E_a > V_\phi$; whereas, when it is underexcited it behaves like an inductor (i.e., a synchronous reactor), with $E_a < V_\phi$. Synchronous condensers are used to correct power factors at load points, or to reduce line voltage drops and thereby improve the voltages at these points, as well as to control reactive power flow. Large synchronous condensers are usually more economical than static capacitors.

7.15 OPERATING A SYNCHRONOUS MOTOR AS A SYNCHRONOUS REACTOR

In general it is not economical to correct the full-load power factor to unity. Therefore, a transmission line usually operates at a lagging power factor. Assume that an overexcited synchronous motor is being used as a synchronous

---

[24] In fact, an overexcited synchronous machine produces reactive power whether or not it is operating as a motor or as a generator.

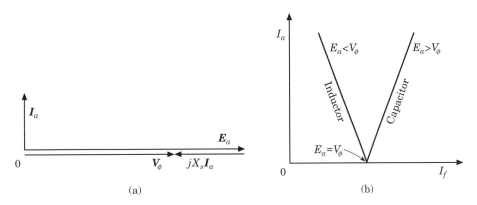

Figure 7.21   Synchronous condenser operation.

capacitor to correct the power factor of a transmission line which is supplying a load with a lagging power factor. Figure 7.22(a) shows the phasor diagram of such a compensated transmission line, with a sending-end voltage of V_S and a receiving-end voltage of V_R, operating under peak-load conditions with $V_S > V_R$. However, under no-load or light-load conditions due to the reactive current I_r, the receiving-end voltage V_R becomes much greater than V_S, that is $V_S < V_R$ as shown in Figure 7.22(b). This condition is known as the **Ferranti Effect**.[25] To prevent this, the dc field excitation of the synchronous motor can be controlled by a voltage regulator and reduced as the load decreases and the V_R increases. As shown in Figure 7.22(d), when the

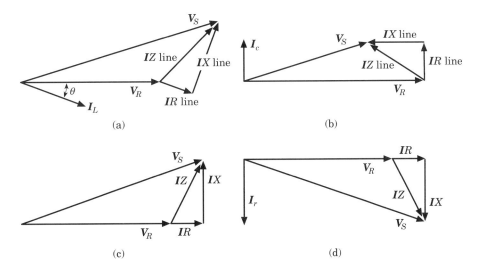

Figure 7.22   The use of the synchronous capacitor as a synchronous reactor.

---

[25] See Gönen [35] for further information.

synchronous motor[26] is underexcited it becomes a synchronous reactor and starts to provide an inductive voltage drop by means of the inductive current I_r to counteract the capacitive line drop. Figures 7.22(c) and 7.22(d) show the corresponding full-load and no-load corrections, respectively. Observe that the relationship between the two receiving-end voltages is about the same. The use of a synchronous condenser provides a constant voltage at the receiving end despite changes in the load current and power factor.

Example 7.7

A three-phase, 750-hp, 4160-V, wye-connected induction motor has a full-load efficiency of 90 percent, a lagging power factor of 0.75, and is connected to a power line. To correct the power factor of such a load to a lagging power factor of 0.85, a synchronous condenser is connected at the load. Determine the reactive power provided by the synchronous capacitor.

SOLUTION:
The input power of the induction motor is

$$P = \frac{(750 \text{ hp})(0.746 \text{ kW/hp})}{0.90} = 621.67 \text{ kW}$$

The reactive power of the motor at the uncorrected power factor is

$$Q_1 = P \tan \theta_1 = 621.67 \tan \left(\cos^{-1} 0.7 \right) = 548.26 \text{ kvar}$$

The reactive power of the motor at the corrected power factor is

$$Q_2 = P \tan \theta_2 = 621.67 \tan \left(\cos^{-1} 0.85 \right) = 385.27 \text{ kvar}$$

Thus the reactive power provided by the synchronous capacitor is

$$Q_c = Q_1 - Q_2 = 548.26 - 385.27 = 162.99 \text{ kvar}$$

7.16 TESTS TO DETERMINE EQUIVALENT-CIRCUIT PARAMETERS

The equivalent-circuit parameters of a synchronous machine can be determined from three tests,[27] namely the open-circuit test, the short-circuit test, and the dc test.

---

[26] The synchronous motors used for this purpose are usually equipped with salient rotors, since a synchronous motor with a cylindrical rotor may step out of synchronism and stop when it is operating with underexcitation. This is due to the fact that decreasing the field current too far can cause the developed torque to be less than the rotational torque required.

[27] It is applicable to both cylindrical-rotor and salient-rotor synchronous machines.

7.16.1 THE OPEN-CIRCUIT TEST

As discussed in Section 7.6, the open-circuit characteristic of a synchronous machine can be developed based on the open-circuit test. As shown in Figure 7.23(a), the machine is driven at synchronous speed with its armature terminals open and its field current set at zero. The open-circuit (line-to-line) terminal voltage V_{oc} is measured[28] as the field current I_f is increased. Since the terminals are open $V_{oc} = E_a = V_t$, assuming that the armature windings are connected in wye. The plot of this voltage with respect to the field excitation current I_f gives the **open-circuit characteristic** (OCC) as shown in Figure 7.23(c). Therefore, the internal generated voltage E_a at any given field current I_f can be found from the OCC characteristic. Observe that as the field current is increased, the OCC starts to separate from the air-gap line due to the saturation of the magnetic core. The no-load rotational losses (i.e., friction, windage, and core losses) can be found by measuring the mechanical power input. While the friction and windage losses remain constant, the core loss is proportional to the open-circuit voltage.

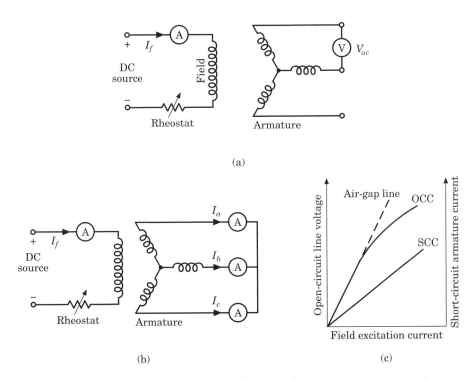

(a)

(b)   (c)

Figure 7.23   Open-circuit and short-circuit tests: (a) connection diagram for open-circuit test; (b) connection diagram for short-circuit test; (c) plots of open-circuit and short-circuit characteristics.

---

[28] If necessary, the voltage may be stepped down by using voltage (potential) transformers for measurement.

7.16.2 THE SHORT-CIRCUIT TEST

As shown in Figure 7.23(b), the armature terminals are short-circuited through suitable ammeters[29] and the field current is set at zero. While the synchronous machine is driven at synchronous speed, its armature current I_a is measured as the field current gradually increases until the armature current is about 150 percent of the rated current. The plot of the average armature current I_a versus the field current I_f gives the **short-circuit characteristic** (SCC) of the machine, as shown in Figure 7.23(c). The SCC is a linear line since the magnetic-circuit iron is unsaturated.[30]

7.16.3 THE DC TEST

If it is necessary, the resistance R_a of the armature winding can be found by applying a dc voltage to two of the three terminals of a wye-connected synchronous machine while it is stationary. Since current flows through two of the wye-connected armature windings, the total resistance of the path is $2R_a$. Thus,

$$2R_a = \frac{V_{dc}}{I_{dc}} \qquad (7.57)$$

or

$$R_a = \frac{V_{dc}}{2I_{dc}} \qquad (7.58)$$

Usually the calculated R_a has to be corrected for the skin effect and the temperature of the winding during the short-circuit test. However, the resistance R_a of synchronous machines with ratings greater than even a few hundred kVA is generally very small and is often ignored except in efficiency computations.

7.16.4 UNSATURATED SYNCHRONOUS REACTANCE

As can be observed in Figure 7.24, if the synchronous machine is unsaturated, the open-circuit line voltage will increase linearly with the field current along the air-gap line. As a result the short-circuit armature current is directly proportional to the field current. Therefore, the unsaturated synchronous impedance for a specific value of the field current can be found from Figure 7.24 as

$$Z_{s,\,un} = \frac{E_{ac}}{\sqrt{3}I_{ab}} = R_a + jX_{s,\,un} \qquad (7.59)$$

---

[29] If necessary, an instrument current transformer can be used with an ammeter in its secondary.
[30] When the short-circuit current is equal to the rated current, the voltage E_a will only be about 20 percent of its rated value. Therefore, the magnetic-circuit iron is unsaturated.

If R_a is small enough to be ignored,

$$X_{s,\,un} \cong \frac{E_{ac}}{\sqrt{3}I_{ab}} \tag{7.60}$$

7.16.5 SATURATED SYNCHRONOUS REACTANCE

Under normal operating conditions, the magnetic circuit is saturated. Therefore, if the field current is changed, the internal generated voltage will vary along the **modified air-gap line** as shown in Figure 7.24. Thus, the saturated synchronous impedance at the rated voltage is given by

$$Z_s = \frac{E_{df}}{\sqrt{3}I_{de}} = R_a + jX_s \tag{7.61}$$

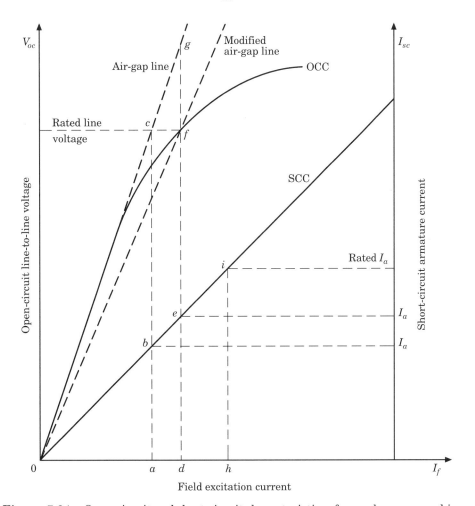

Figure 7.24 Open-circuit and short-circuit characteristics of a synchronous machine.

If R_a is small enough to be ignored,

$$X_s \cong \frac{E_{df}}{\sqrt{3}I_{de}}$$

(7.62)

However, the machine is unsaturated in the short-circuit test. Therefore, the determination of the synchronous reactance based on short-circuit and open-circuit test data is only an approximation[31] at best. Figure 7.25 shows the variation of the synchronous reactance due to saturation.

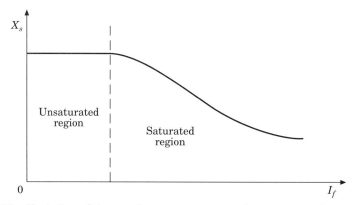

Figure 7.25   Variation of the synchronous reactance due to saturation.

7.16.6 THE SHORT-CIRCUIT RATIO

The **short-circuit ratio** (SCR) of a synchronous machine is the ratio of the field current required to generate rated voltage at the rated speed at open circuit to the field current needed to produce rated armature current at short circuit. Therefore, from Figure 7.24 the short-circuit ratio of the synchronous machine is

$$SCR = \frac{od}{oh}$$

(7.63)

where *od* is the field current that produces rated voltage on the open-circuit characteristic and *oh* is the field current required for the rated short-circuit armature current. The *SCR* is the reciprocal of the per-unit value of the saturated synchronous reactance.

Example 7.8

The following data are taken from the open-circuit and short-circuit characteristics of a 100-kVA, three-phase, wye-connected, 480-V, 60-Hz synchronous machine with negligible armature resistance:

---

[31] See McPherson [61] for a more accurate determination of the saturated synchronous reactance.

From the open-circuit characteristic:
Line-to-line voltage = 480 V
Field current = 3.2 A

From the air-gap line:
Line-to-line voltage = 480 V
Field current = 2.94 A

From the short-circuit characteristic:
Armature current 90.35 A   120.28 A
Field current 3.2 A   4.26 A

Determine the following:
(a) The unsaturated synchronous reactance.
(b) The saturated synchronous reactance at rated voltage.
(c) The short-circuit ratio.

SOLUTION:
(a) The field current of 2.94 A needed for rated line-to-line voltage of 480 V on the air-gap line produces a short-circuit armature current of 2.94(90.35/3.2) = 83.01 A. Therefore, the unsaturated synchronous reactance is

$$X_{s,\,un} = \frac{480\ \text{V}}{\sqrt{3}(83.01\ \text{A})} = 3.34\ \Omega/\text{phase}$$

(b) The field current of 3.2 A produces the rated voltage on the open-circuit characteristic and a short-circuit armature current of 90.35 A. Therefore, the saturated synchronous reactance at the rated voltage is

$$X_s = \frac{480\ \text{V}}{\sqrt{3}(90.35\ \text{A})} = 3.07\ \Omega/\text{phase}$$

(c) The short-circuit ratio is

$$SCR = \frac{3.2}{4.26} = 0.75$$

7.17 THE CAPABILITY CURVE OF A SYNCHRONOUS MACHINE

Figure 7.26(a) shows the capability curve of a synchronous machine. The rated MVA of the synchronous machine is dictated by stator heating in terms of maximum allowable stator current. The upper and lower portions of the area inside the circle, with a radius of maximum S, represent the generator and motor operation, respectively. The maximum allowable field current is limited by the rotor heating. The maximum permissible torque angle is dictated by the steady-state stability limits that exist in the generator and motor modes of operation and further restricts the operation area of the synchronous machine. In the generator mode the power limit is determined by the prime-mover rating.

Figure 7.26(b) shows the capability curve of a synchronous generator. Any point that lies within the area is a safe operating point for the generator from the standpoint of heating and stability. Assume that the operating point S is chosen as shown in Figure 7.26(b), and that the corresponding real and reactive powers are P and Q, respectively. For this operation the power-factor angle can be readily determined from the diagram as θ by drawing a line from the operating point S to the origin. A line drawn from the operating point S to the origin of the I_f axis facilitates finding the power or torque angle δ from that axis.

7.18 PARALLEL OPERATION OF SYNCHRONOUS GENERATORS

The process of connecting a synchronous generator to an infinite bus is called **paralleling with the infinite bus**. The generator to be added to the system is referred to as the one **to be put on line**. As shown in Figure 7.27, to connect the incoming generator to the infinite bus, a definite procedure called the **synchronizing procedure** must be followed before closing the circuit breaker CB to prevent any damage to the generator or generators. Accordingly, the following conditions must be met: (1) the rms voltages of the generator must be the same as the rms voltages of the infinite bus, (2) the phase sequence of the voltages of the generator must be the same as the phase sequence of the infinite bus, (3) the phase voltages of the generator must be in phase with the phase voltages of the infinite bus, and (4) the frequency of the generator must be almost equal to that of the infinite bus.

By using voltmeters, the field current of the incoming generator is increased to a level at which the voltages of the generator are the same as the voltages of the infinite bus. The phase sequence of the incoming generator has to be compared to the phase sequence of the infinite system. This can be done in various ways. One way is to connect three light bulbs as shown in Figure 7.27. If the phase sequence is correct, all three light bulbs will have the same brightness. A second method is to connect a small induction motor first to the terminals of the infinite bus and then to the terminals of the incoming generator. If the motor rotates in the same direction each time, then the phase sequences are the same. A third method is to use an instrument known as a *phase-sequence indicator*. In any case, if the phase sequences are different, then two of the terminals of the incoming generator have to be reversed. The two sets of voltages must be in phase and the three phase voltages have to be almost equal to the voltages of the infinite bus. The frequency of the incoming generator has to be a little higher than the frequency of the infinite bus. The reason for this is that when it is connected it will come on line providing power as a generator rather than being used as a motor. The frequency (or speed) of the incoming generator can also be compared to the frequency of the infinite bus by using an instrument known as a **synchroscope**. When the two frequencies are identical, the pointer of the

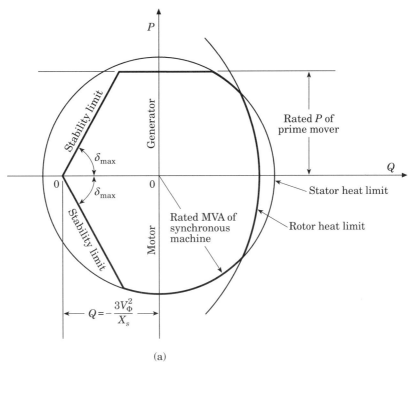

(a)

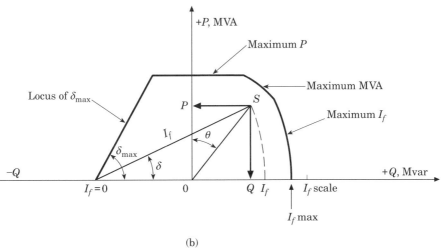

(b)

Figure 7.26 Capability curves of a synchronous machine: (a) construction of the capability curve; (b) the capability curve of a synchronous generator.

synchroscope locks into a vertical position, as shown in Figure 7.28. On the other hand, if the radian frequency of the incoming generator is, for example, 381 rad/s and that of the infinite bus is 377 rad/s, then the pointer rotates at

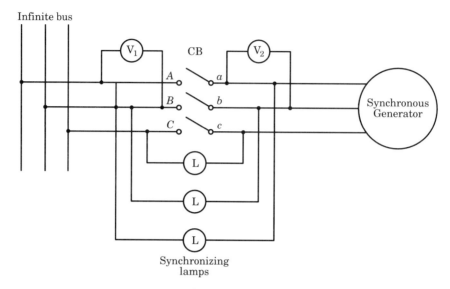

Figure 7.27   Parallel operation of synchronous generators.

Figure 7.28   The face of a synchroscope.

4 rad/s in the direction marked FAST. If the radian frequency of the incoming generator is, for example, 372 rad/s, the pointer rotates at 5 rad/s in the direction marked SLOW. When all of the conditions are met, the incoming generator is connected to the infinite bus by closing the circuit breaker (or switch). Once the breaker is closed, the incoming generator is on line. At this moment it is neither delivering nor receiving power. Having $E_a = V_\phi$, $I_a = 0$, and $\delta = 0$ in each phase, it is simply **floating** as explained in Section 7.11.1. After the generator is connected, the dispatcher determines how much power should be produced by it. The power output of the prime mover is increased until the generator starts to produce the required power. Note that in large generators this operation of putting a new generator on line is done automatically by using computers.

7.19 SALIENT-POLE SYNCHRONOUS MACHINES

Almost always low-speed synchronous machines with four or more poles have salient-pole rotors. Such low-speed alternators are normally driven by water turbines and are often used as hydroelectric alternators, as shown in Figure 7.2.

In cylindrical-rotor synchronous machines there is no saliency. Therefore, the reluctance and thereby the reactances are constants due to the existence of a uniform air gap; but in the salient-pole rotor the area midway between the poles (quadrature or q axis) has a larger air gap than the area between the pole centers (direct or d axis). As a result, the armature mmf produces more flux along the d axis than along the q axis. Because of this saliency, the reactance measured at the terminals of a salient-rotor machine will vary as a function of the rotor position. The reactance associated with the d axis is called the **direct-axis synchronous reactance**, X_d. The one associated with the q axis is called the **quadrature-axis synchronous reactance**, X_q. They can be expressed as

$$X_d = X_a + X_{\phi d} \tag{7.64}$$

$$X_q = X_a + X_{\phi q} \tag{7.65}$$

where X_a is the armature self-reactance and is independent of the rotor angle. The direct- and quadrature-axis magnetizing reactances $X_{\phi d}$ and $X_{\phi q}$ represent the inductive effects of the direct- and quadrature-axis armature-reaction flux waves produced by the d-axis current I_d and the q-axis current I_q, respectively. The reactance X_q is less than the reactance X_d due to the existence of greater reluctance in the air gap of the quadrature axis. The X_q is often between 0.6 and 0.7 of the X_d. Therefore the armature reaction resulting from the armature current I_a can be taken into account by resolving it into two components along the d-axis and the q-axis, namely I_d and I_q. Thus, as shown in Figure 7.29(a), the total voltage in the stator can be expressed as

$$V_\phi = E_a + E_d + E_q - I_a R_a \tag{7.66}$$

where E_d and E_q represent the d-axis component and the q-axis component of armature reaction voltage, respectively. Hence,

$$E_d = -jI_d X_d \tag{7.67}$$

$$E_q = -jI_q X_q \tag{7.68}$$

Therefore,

$$V_\phi = E_a - jI_d X_d - jI_q X_q - I_a R_a \tag{7.69}$$

where the term $I_a R_a$ represents the voltage drop due to the armature resistance. The armature current is

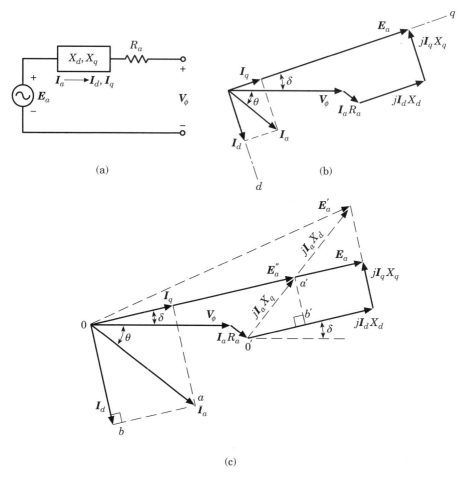

Figure 7.29 Equivalent circuit and phasor diagrams of a salient-pole synchronous generator: (a) equivalent circuit; (b) phasor diagram; (c) drawing the phasor diagram without knowing the power angle δ.

$$I_a = I_d + I_q \qquad (7.70)$$

If the armature resistance is negligibly small, then

$$V_\phi = E_a - jI_dX_d - jI_qX_q \qquad (7.71)$$

or

$$E_a = V_\phi + jI_dX_d + jI_qX_q \qquad (7.72)$$

The equivalent circuit of a salient-pole synchronous generator is shown in Figure 7.29(a). The corresponding phasor diagram is shown in Figure 7.29(b).

Note that this phasor diagram is based on the assumption that the angle between E_a and I_a (i.e., $\theta + \delta$) is known in advance. Normally, however, the internal power factor angle $\theta + \delta$ is not known beforehand, only the power-factor angle θ is known. Figure 7.29(c) shows how to draw the phasor diagram when the power angle δ is not known in advance. Observe that the phasor E_a'' is

$$E_a'' = V_\phi + R_a I_a + j I_a X_q \tag{7.73}$$

The phasor E_a'' and the angle δ can be determined by drawing the phasor $I_a R_a$ parallel to the current I_a and from its tip the phasor $I_a X_q$ perpendicular to the phasor I_a, and finally by drawing a line connecting the phasor origin to the tip of the phasor $I_a X_q$. Once the angle δ is known, then the internal generated voltage can be readily found, as shown in Figure 7.29(c).

Also note that if the armature resistance R_a is ignored, then the angle δ can be found from

$$\tan \delta = \frac{I_a X_q \cos \theta}{V_\phi + I_a X_q \sin \theta} \tag{7.74}$$

However if the armature resistance R_a is not neglected, then the angle δ can be determined from

$$\tan\delta = \frac{I_a X_q \cos \theta - I_a R_a \sin \theta}{V_\phi + I_a X_q \sin \theta + I_a R_a \cos \theta} \tag{7.75}$$

The power output of a three-phase synchronous generator is

$$P = 3V_\phi I_a \cos \theta \tag{7.76a}$$

$$= 3V_\phi I_d \cos (90° - \delta) + 3V_\phi I_q \cos \delta \tag{7.76b}$$

$$= 3V_\phi I_d \sin \delta + 3V_\phi I_q \cos \delta \tag{7.76c}$$

Since

$$V_\phi \sin \delta = I_q X_q$$

or

$$I_q = \frac{V_\phi \sin \delta}{X_q} \tag{7.77}$$

and

$$V_\phi \cos \delta = E_a - I_d X_d$$

or

$$I_d = \frac{E_a - V_\phi \cos \delta}{X_d} \tag{7.78}$$

Substituting for I_d and I_q in equation (7.76c),

$$P = 3V_\phi \frac{E_a - V_\phi \cos \delta}{X_d} \sin \delta + 3V_\phi \frac{V_\phi \sin \delta}{X_q} \cos \delta \tag{7.79}$$

or

$$P = \frac{3V_\phi E_a}{X_d} \sin \delta + 3V_\phi^2 \left(\frac{1}{X_q} - \frac{1}{X_d} \right) \sin \delta \cos \delta \tag{7.80}$$

Since

$$\sin \delta \cos \delta = (\sin 2\delta)/2$$

then

$$P = \frac{3V_\phi E_a}{X_d} \sin \delta + \frac{3V_\phi^2}{2} \left(\frac{1}{X_q} - \frac{1}{X_d} \right) \sin 2\delta \tag{7.81}$$

The first term is the same as the power in a cylindrical rotor synchronous machine, and the second term is the additional power due to the reluctance torque in the machine. Note that the second term depends on the saliency of the machine. Here,

$$\text{Saliency} = \frac{1}{X_q} - \frac{1}{X_d} \tag{7.82}$$

It disappears when X_d is equal to X_q, which is the case for a cylindrical rotor. The developed torque of a synchronous machine can be expressed as

$$T_d = \frac{3V_\phi E_a}{\omega_s X_d} \sin \delta + \frac{3V_\phi^2}{2\omega_s} \left(\frac{1}{X_q} - \frac{1}{X_d} \right) \sin 2\delta \tag{7.83}$$

The torque developed by a salient-pole generator as a function of torque angle δ is plotted in Figure 7.30. Note that the maximum reluctance torque occurs when δ is 45° and is about 20 percent of the available maximum torque. Also note that this type of machine is "stiffer" than a cylindrical rotor machine since its torque increases faster for small values of δ. The maximum power takes place at a value of δ less than 90°.

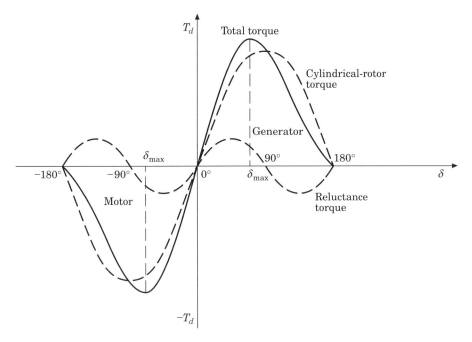

Figure 7.30   Torque developed by a salient-pole synchronous generator as a function of torque angle.

For a salient-pole machine, it is necessary to know the values of both X_d and X_q. These reactances are determined by the slip test. The slip test is performed by applying reduced three-phase voltage to the stator while the rotor is driven slightly above or below synchronous speed with the field winding open-circuited (i.e., unexcited). Oscillograms are taken of the armature terminal voltage, the armature current, and the induced voltage across the open-field winding. Figure 7.31 shows such oscillograms. Approximate values of X_d and X_q can be found from the voltmeter and ammeter readings. For example, when armature resistance is ignored, the X_d and X_q can be determined from

$$X_d = \frac{V_{\max}}{I_{\min}} \tag{7.84}$$

and

$$X_q = \frac{V_{\min}}{I_{\max}} \tag{7.85}$$

Also, the ratio of maximum to minimum armature current gives the ratio of X_d/X_q. For example, assume that from the oscillograms it is found that X_d/X_q is 1.7. If X_d is known from the open- and short-circuit tests for the cylindrical-rotor machine, then X_q can be determined.

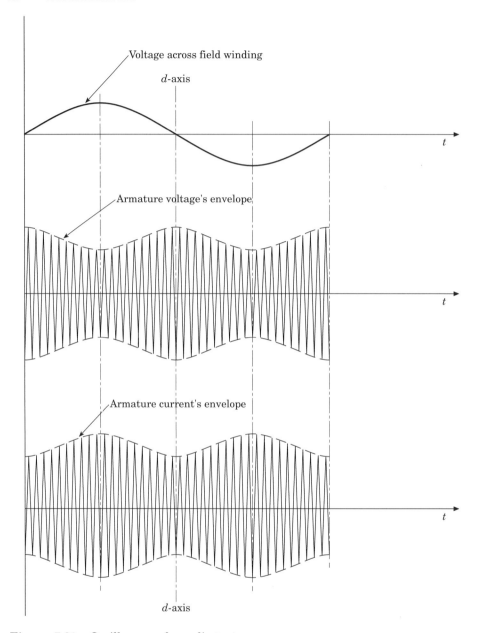

Figure 7.31   Oscillograms from slip test.

Example 7.9

A 50-MVA, 13.2-kV, 60-Hz, wye-connected, three-phase synchronous generator has X_d = 1.52 Ω and X_q = 0.91 Ω. Its resistance is small enough to be ignored. If it operates at full load at a 0.8 lagging power factor, determine the following:

(a) The phase voltage and phase current at full load.
(b) The internal generated voltage E_a if it has a cylindrical rotor.
(c) The internal generated voltage E_a if it has a salient-pole rotor.

SOLUTION:
(a) The phase voltage is

$$V_\phi = \frac{13{,}200}{\sqrt{3}} = 7{,}621 \text{ V or } \boldsymbol{V_\phi} = 7{,}621 \angle 0° \text{ V}$$

The phase current is

$$I_a = \frac{50 \times 10^6}{\sqrt{3}\,(13{,}200)} = 2{,}186.9 \text{ A and } \theta = \cos^{-1}(0.8) = 36.87°$$

Thus

$$\boldsymbol{I_a} = 2{,}186.9 \angle -36.87° \text{ A}$$

(b) If the machine has a cylindrical rotor the internal generated voltage is

$$\boldsymbol{E_a} = \boldsymbol{V_\phi} + j\boldsymbol{I_a}X_s$$

$$= 7{,}621\angle 0° + j1.52\,(2{,}186.9\angle - 36.87°)$$

$$= 7{,}621\angle 0° + (1.52\angle 90°)(2{,}186.9\angle - 36.87°) = 9{,}976.41\angle 15.46° \text{ V}$$

(c) If the machine has a salient-pole rotor,

$$\boldsymbol{E_a} = \boldsymbol{V_\phi} + j\boldsymbol{I_a}X_q$$

$$= 7{,}621\angle 0° + j(2{,}186.9\angle - 36.87°)(0.91) = 8{,}815.14\angle 10.24° \text{ V}$$

Thus, the direction of $\boldsymbol{E_a}$ is $\delta = 10.24°$. Hence the magnitude of the d-axis component of current is

$$I_d = I_a \sin(\theta + \delta)$$

$$= 2{,}186.9 \sin(36.87° + 10.24°) = 1{,}602.26 \text{ A}$$

and the magnitude of the q-axis component of current is

$$I_q = I_a \cos(\theta + \delta)$$

$$= 2{,}186.9 \cos(36.87° + 10.24°) = 1{,}488.39 \text{ A}$$

By combining magnitudes and angles,

$$I_d = 1{,}602.26\angle - 79.76° \text{ A}$$

$$I_q = 1{,}488.39\angle 10.24° \text{ A}$$

and

$$E_a = V_\phi + jI_dX_d + jI_qX_q$$

$$= 7{,}621\angle 0° + j(1{,}602.26\angle - 79.76°)(1.52) + j(1{,}488.39\angle 10.24°)0.91$$

$$= 9{,}935\angle 10.24° \text{ V}$$

Note that the angle δ for the salient-pole rotor is smaller than the one for the cylindrical rotor. Therefore, a salient-pole machine is said to be **stiffer** than one with a cylindrical rotor.

PROBLEMS

7.1 Compute the synchronous speeds in radians per second and revolutions per minute for poles of 2, 4, 8, 10, 12, and 120 and tabulate the results for the following frequencies:
(a) 60 Hz
(b) 50 Hz

7.2 An elementary four-pole, three-phase 60-Hz alternator has a rotating flux of 0.0875 Wb. The number of turns in each phase coil is 25. Its shaft speed is 1800 rev/min. Let its stator winding factor be 0.95 and find:
(a) The angular speed of the rotor.
(b) The three phase voltages as a function of time.
(c) The rms phase voltage of this generator if the stator windings are delta-connected.
(d) The rms terminal voltage if the stator windings are wye-connected.

7.3 Determine by derivation the value of the K constant given in equation (7.21).
(a) If ω is given in *electrical* radians per second.
(b) If ω is given in *mechanical* radians per second.

7.4 A three-phase two-pole synchronous generator has stator windings with a double layer. There are 12 coils per phase with 20 turns in each coil. The windings have an electrical pitch of 150°. The shaft speed is 3600 rev/min. Assume that the flux per pole in the alternator is 0.0203 Wb and determine the following:
(a) The slot pitch of the armature in electrical degrees.
(b) The coil span in terms of slots.

(c)   The magnitude of the phase voltage.
(d)   The terminal voltage of this alternator if the stator windings are connected in wye.

7.5   Consider the synchronous generator given in Example 7.4 and assume that the generator delivers full-load current at the rated voltage and 0.8 leading power factor. Determine the following:
(a)   The rated load current.
(b)   The air gap voltage.
(c)   The internal generated voltage.
(d)   The power angle.
(e)   The voltage regulation.

7.6   A three-phase, 4.17-kV, 60-Hz, wye-connected, cylindrical rotor synchronous generator has a leakage reactance of 2.2 Ω per phase and a winding resistance of 0.15 Ω per phase. If the load connected to the generator is 950 kVA at 0.85 lagging power factor, determine the air gap voltage.

7.7   A three-phase synchronous motor is supplied by 100 A at unity power factor from a bus. Determine the following:
(a)   The current at a leading power factor of 0.85, if the bus voltage is constant.
(b)   The current at a lagging power factor of 0.85 if the bus voltage is constant.
(c)   What usually happens to the bus voltage as the power factor becomes more leading?

7.8   A three-phase wye-connected generator supplies a unity power factor load at 4.16 kV. If the synchronous reactance voltage drop is 262 V per phase and the resistance voltage drop is 25 V per phase, what is the percent regulation?

7.9   A 25-hp synchronous motor has a full-load efficiency of 92 percent and operates at a leading power factor of 0.8. Determine the following:
(a)   The power input to the motor.
(b)   The kVA input to the motor.

7.10   A three-phase, four-pole, 12-kV, 60-Hz, 40-MVA, wye-connected, cylindrical rotor synchronous generator has a synchronous reactance of 0.3 per unit. Ignore the armature resistance and the effects of saturation. Also assume that the generator delivers full-load current at 0.85 lagging power factor. Determine the following:
(a)   The rated (full-load) current.
(b)   The internal generated voltage.
(c)   The torque angle.
(d)   The synchronizing power in W per electrical radian and in W per electrical degree.

(e) The synchronizing power in W per mechanical degree.

(f) The synchronizing torque in W per mechanical degree.

7.11 A three-phase, 100-kVA, 60-Hz, 480-V, four-pole, wye-connected, cylindrical rotor synchronous motor has an armature resistance and a synchronous reactance of 0.09 Ω and 1.5 Ω per phase, respectively. Its combined friction and windage losses are 2.6 kW and its core losses are 1.941 kW. Ignore its dc field losses. If the motor is operating at unity power factor, determine the following:

(a) The internal generated voltage E_a.

(b) The torque angle δ.

(c) The efficiency at full load. $P_{in} - 3I_a^2 R_a$ –

(d) The output torque at full load.

7.12 Solve Problem 7.11 but assume that the machine is operating at 0.85 leading power factor.

7.13 Solve Problem 7.11 but assume that the machine is operating at 0.85 lagging power factor.

7.14 Assume that the following data are obtained from the open-circuit and short-circuit characteristics of a 500-kVA, three-phase, wye-connected, 480-V, 60-Hz synchronous machine with negligible armature resistance:

From the open-circuit characteristic:
Line-to-line voltage = 480 V
    Field current = 16 A

From the air-gap line:
Line-to-line voltage = 480 V
    Field current = 14.7 A

From the short-circuit characteristic:
Armature current 451.76 A   601.41 A
       Field current 16 A   21.3 A
Determine the following:

(a) The unsaturated synchronous reactance.

(b) The saturated synchronous reactance at the rated voltage.

(c) The short-circuit ratio.

7.15 A three-phase, 60-Hz, 480-V, two-pole, delta-connected, cylindrical rotor synchronous generator has a synchronous reactance of 0.12 Ω and an armature resistance of 0.010 Ω. Its open-circuit characteristic is shown in Figure 7.9. Its combined friction and windage losses are 30 kW and its core losses are 20 kW. Neglect its dc field losses and determine:

(a) The amount of field current for the rated voltage of 480 V at no load.

(b) The amount of field current required for the rated terminal voltage of 480 V when the rated load current is 1000 A at a lagging power factor of 0.85.

(c) The efficiency of the generator under the rated load conditions.

(d) The terminal voltage if the load is suddenly disconnected.

(e) The amount of field current required to have the rated terminal voltage of 480 V when the rated load current is 1000 A at a leading power factor of 0.85.

7.16 A three-phase, 60-Hz, 480-V, four-pole, delta-connected, cylindrical rotor synchronous generator has an armature resistance and a synchronous reactance of 0.012 Ω and 0.15 Ω, respectively. Its open-circuit characteristic is shown in Figure 7.9. Its combined friction and windage losses are 35 kW and its core losses are 25 kW. Neglect its dc field losses and determine:

(a) The field current for the rated terminal voltage of 480 V at no load.

(b) The field current for the rated terminal voltage of 480 V when the rated load current is 1100 A at a lagging power factor of 0.90.

(c) The efficiency of the generator under the rated load conditions.

(d) The terminal voltage if the load is suddenly disconnected.

(e) The field current for the rated terminal voltage of 480 V when the rated load current is 1100 A at a leading power factor of 0.90.

7.17 Solve Problem 7.16 but assume that the generator is wye-connected and the rated load current is 500 A.

7.18 A three-phase, 60-Hz, 480-V, two-pole, delta-connected, cylindrical-rotor synchronous generator has an armature resistance and a synchronous reactance of 0.012 Ω and 0.15 Ω, respectively. Its open-circuit characteristic is shown in Figure 7.9. Its combined friction and windage losses are 25 kW and its core losses are 25 kW. Neglect its dc field losses and determine:

(a) The field current for the rated terminal voltage of 480 V at no load.

(b) The field current for the rated terminal voltage of 480 V when the rated load current is 1100 A at a lagging power factor of 0.90.

(c) The efficiency of the generator under the rated load conditions.

(d) The terminal voltage if the load is suddenly disconnected.

(e) The field current for the rated terminal voltage of 480 V when the rated load current is 1100 A at a leading power factor of 0.90.

7.19 A three-phase, 60-Hz, 480-V, four-pole, delta-connected, cylindrical-rotor synchronous generator has an armature resistance and a synchronous reactance of 0.012 Ω and 0.12 Ω, respectively. Its open-circuit characteristic is shown in Figure 7.9. Its combined friction and windage losses are 25 kW and its core losses are 30 kW. Neglect its dc field losses and determine:

(a) The field current for the rated terminal voltage of 480 V at no load.

(b) The field current for the rated terminal voltage of 480 V when the

    rated load current is 1000 A at a lagging power factor of 0.85.
- (c) The efficiency of the generator under the rated load conditions.
- (d) The terminal voltage if the load is suddenly disconnected.
- (e) The field current for the rated terminal voltage of 480 V when the rated load current is 1000 A at a leading power factor of 0.85.

7.20 A three-phase, 60-Hz, 480-V, wye-connected, cylindrical-rotor synchronous generator has a synchronous reactance of 1.2 Ω, per phase. Its armature resistance is small enough to be neglected. Its full-load armature current is 75 A at 0.85 lagging power factor. Its combined friction and windage losses are 1.8 kW and its core losses are 1.1 kW. Neglect its dc field losses and determine:
- (a) Its terminal voltage when it is loaded with the rated current having a power factor of: (1) 0.85 lagging, (2) unity, and (3) 0.85 leading.
- (b) Its efficiency when it is loaded with the rated current at a lagging power factor of 0.85.
- (c) Its input torque and induced counter-torque when it is operating at full load.
- (d) Its voltage regulation when it is operating under full load with a power factor of: (1) 0.85 lagging, (2) unity, and (3) 0.85 leading.

7.21 A three-phase, 60-Hz, 480-V, Y-connected, cylindrical-rotor synchronous generator has a synchronous reactance of 0.95 Ω per phase. Its armature resistance is negligible. Its combined friction and windage losses are 1.3 kW and its core losses are 0.95 kW. Neglect its dc field losses. Its full-load armature current is 55 A at 0.9 PF lagging. Determine:
- (a) Its terminal voltage when it is loaded with the rated current at a power factor of: (1) 0.9 lagging, (2) unity, and (3) 0.9 leading.
- (b) Its efficiency when it is loaded with the rated current at a power factor of 0.9.
- (c) Its input torque and induced counter-torque when it is operating at full load.
- (d) Its voltage regulation when it is operating under full load with a power factor of: (1) 0.9 lagging, (2) unity, and (3) 0.9 leading.

7.22 A three-phase, 60-Hz, 480-V, delta-connected, cylindrical-rotor synchronous generator has a synchronous reactance of 1.5 Ω per phase. Its armature resistance is negligible. Its combined friction and windage losses are 2.1 kW and its core losses are 1.2 kW. Neglect its dc field losses. Its full-load armature current is 100 A at 0.80 PF lagging. Determine the following:
- (a) Its terminal voltage when it is loaded with the rated current at a power factor of: (1) 0.80 lagging, (2) unity, and (3) 0.80 leading.
- (b) Its efficiency when it is loaded with the rated current at a lagging power factor of 0.80.
- (c) Its input torque and induced counter torque when it is operating at full load.

(d) Its voltage regulation when it is operating under full load with a power factor of: (1) 0.80 lagging, (2) unity, and (3) 0.80 leading.

7.23 Solve Problem 7.22 but assume that the generator is wye-connected.

7.24 A three-phase, 60-Hz, 480-V, delta-connected, cylindrical-rotor synchronous generator has a synchronous reactance of 0.95 Ω, per phase. Its armature resistance is negligible. Its combined friction and windage losses are 1.5 kW and its core losses are 1.25 kW. Neglect its dc field losses. Its full-load armature current is 55 A at 0.85 lagging PF. Determine:
(a) Its terminal voltage when it is delivering the rated current at a power factor of: (1) 0.85 lagging, (2) unity, and (3) 0.85 leading.
(b) Its efficiency when it is loaded with the rated current at a lagging power factor of 0.85.
(c) Its input torque and induced counter torque when it is operating at full load.
(d) Its voltage regulation when it is operating under full load with a power factor of: (1) 0.85 lagging, (2) unity, and (3) 0.85 leading.

7.25 Solve Problem 7.24 but assume that the generator is wye-connected.

7.26 A three-phase, 60-Hz, $(480/\sqrt{3})$-V, delta-connected, cylindrical-rotor synchronous motor has a synchronous reactance of 3.5 Ω. Its armature resistance is negligible. Its combined friction and windage losses are 2 kW and its core losses are 1.45 kW. The motor is connected to a 25-hp mechanical load and is operating at a leading power factor of 0.85.
(a) Find the values of I_a, I_L, and E_a of the motor and draw its phasor diagram.
(b) If the mechanical load is increased to 50 hp, draw the new phasor diagram.
(c) Find the values of I_a, I_L, and E_a and the PF of the motor in Part (b).

7.27 A three-phase, 60-Hz, $(480/\sqrt{3})$-V, delta-connected, cylindrical-rotor synchronous motor has a synchronous reactance of 4 Ω. Its armature resistance is negligible. Its combined friction and windage losses are 2.5 kW and its core losses are 2.25 kW. The motor is connected to a 20 hp mechanical load and is operating at a leading power factor of 0.80.
(a) Find the values of I_a, I_L, and E_a of the motor and draw its phasor diagram.
(b) If the mechanical load is increased to 75 hp, draw the new phasor diagram.
(c) Find the values of I_a, I_L, and E_a and the PF of the motor in Part (b)

7.28 Assume that two three-phase induction motors and a three-phase synchronous motor are connected to the same bus. The first induction motor is 150 kW and operating at 0.85 lagging power factor. The second induction motor is 250 kW and operating at 0.70 lagging power factor.

The real power of the synchronous motor is 200 kW. If the bus voltage is 480 V and the synchronous motor is operating at 0.90 lagging power factor, determine the following:
(a) The total real and reactive power at the bus.
(b) The total bus current and its power factor.
(c) If the synchronous motor is operating at 0.90 leading power factor, the new total bus current and its power factor.

7.29   Suppose that two three-phase induction motors and a three-phase synchronous motor are connected to the same bus. The first induction motor is 250 kW and operating at 0.90 lagging power factor. The second induction motor is 350 kW and operating at 0.75 lagging power factor. The real power of the synchronous motor is 300 kW. If the bus voltage is 480 V and the synchronous motor is operating at 0.90 lagging power factor, determine the following:
(a) The total real and reactive power at the bus.
(b) The total bus current and its power factor.
(c) If the synchronous motor is operating at 0.90 leading power factor, the new total bus current and its power factor.

7.30   A manufacturing plant has a load of 500 kW at 0.78 lagging power factor. If a 75-hp synchronous motor is added and operated at 0.85 leading power factor, determine the new total load and new power factor of the plant. Neglect the losses of the synchronous motor.

7.31   A 75-MVA, 20-kV, 60-Hz, wye-connected three-phase synchronous generator has $X_d = 2.1\ \Omega$ and $X_q = 1.0\ \Omega$. Ignore its resistance. Assume that it operates at full load at 0.8 lagging power factor and determine the following:
(a) The phase voltage and phase current at full load.
(b) The internal generated voltage E_a if it has a cylindrical rotor.
(c) The internal generated voltage E_a if it has a salient-pole rotor.

7.32   Suppose that two three-phase induction motors and a three-phase synchronous motor are connected to the same bus. The first induction motor is 275 kW and operating at 0.8 lagging power factor. The second induction motor is 125 kW and operating at 0.8 lagging power factor. If the bus voltage is 480 V and the synchronous motor is operating at 0.95 lagging power factor, determine the following:
(a) The total real and reactive power at the bus.
(b) The total bus current and its power factor.
(c) If the synchronous motor is operating at 0.95 leading power factor, the new total bus current and its power factor.

7.33   A three-phase, 60-Hz, 480-V, wye-connected, cylindrical-rotor synchronous motor has a synchronous reactance of 3 Ω. Its armature resistance is negligible. Its friction and windage losses are 4 kW and its core losses are 3 kW. The motor is connected to a 75 hp mechanical load and is operating at a leading power factor of 0.85.

(a) Find the values of I_a, I_L, and E_a of the motor and draw its phasor diagram.

(b) If the mechanical load is increased to 100 hp, draw the new phasor diagram.

(c) Find the values of I_a, I_L, and E_a and the PF of the motor in Part (b).

7.34 A three-phase, 50-Hz, 380-V, wye-connected, cylindrical-rotor synchronous generator has a synchronous reactance of 0.9 Ω per phase. Its armature resistance is negligible. Its friction and windage losses are 1.0 kW and its core losses are 1.0 kW. Neglect its armature resistance and dc field losses. Its full-load armature current is 50 A at 0.9 PF lagging. Determine:

(a) Its terminal voltage when it is delivering the rated current at a unity power factor.

(b) Its efficiency when it is loaded with the rated current at a lagging power factor of 0.9.

(c) Its input torque and induced counter torque when it is operating at full load.

(d) Its voltage regulation when it is operating under full load at a unity power factor.

CHAPTER 8

DIRECT-CURRENT MACHINES

Nothing is so firmly believed as what we least know.
M.E. De Montaigne, *Essays*, 1580

Talk sense to a fool and he calls you foolish.
Euripides, *The Bacchae*, 407 B.C.

But talk nonsense to a fool and he calls you a genius.
Turan Gönen

8.1 INTRODUCTION

A direct-current (dc) machine is a versatile machine, that is, the same machine can be used as a generator to convert mechanical energy to dc electrical energy or as a motor to convert dc electrical energy into mechanical energy. However, the use of dc machines as dc generators to produce bulk power has rapidly disappeared due to the economic advantages involved in the use of alternating-current generation, transmission, and distribution. This is partly due to the high efficiency and relative simplicity with which transformers convert voltages from one level to another. Today, the need for dc power is often met by the use of solid-state-controlled rectifiers. However, dc motors are used extensively in many industrial applications because they provide constant mechanical power output or constant torque, adjustable motor speed over wide ranges, precise speed or position control, efficient operation over a wide speed range, rapid acceleration and deceleration, and responsiveness to feedback signals. Such machines can vary in size from miniature permanent-magnet motors to machines rated for continuous operation at several thousand horsepower. Examples of small dc motors include those used for small control devices, windshield-wiper motors, fan motors, starter motors, and various servomotors. Application examples for larger dc motors include industrial drive motors in conveyors, pumps, hoists, overhead cranes, forklifts, fans, steel and aluminum rolling mills, paper mills, textile mills, various other rolling mills, golf carts, electrical cars, street cars or trolleys, electric trains, electric elevators, and large earth-moving equipment. Obviously, dc machine applications are very significant, but the advantages of the dc machine must be weighed against its greater initial investment cost and the maintenance problems associated with its brush-commutator system. Figure 8.1 shows the housing of a dc motor being mounted in an industrial mill.

8.2 CONSTRUCTIONAL FEATURES

Figure 8.1   Mounting the housing of a dc mill motor. (Courtesy of Siemens.)

The schematic diagram of the construction of a dc machine is shown in Figure 8.2. The construction has two basic parts, namely the stator (which stands still) and the rotor (which rotates). The stator has salient poles that are excited by one or more field windings. The armature winding of a dc machine is located on the rotor with current flowing through it by carbon brushes making contact with copper commutator segments. Both the main poles and armature core are made up of laminated materials to reduce core losses. With the exception of a few small machines, the dc machines also have **commutating poles**[1] between the main poles of the stator. Each commutating pole has its own winding which is known as the **commutating winding**.

The main (or field) poles are located on the stator and are attached to the stator yoke (or frame). The stator yoke also serves as a return path for the pole flux. Because of this, the yokes are being built with laminations to decrease core losses in solid-state driven motors. The ends of the poles are called the **pole shoes**. The surface of the pole shoe opposite the rotor is called the pole face. The distance between the **pole face** and the rotor surface is called the **air gap**. As shown in the figure, there is a special winding located in the slots of the pole faces called the **compensating winding**.[2]

The field windings are located around the pole cores and are connected in series and/or in shunt (i.e., in parallel) with the armature circuit. The shunt winding is made up of many turns of relatively thin wires, whereas the series winding has only a few turns and is made up of thicker wires. As shown in the

---

[1] They are known as the **interpoles** or **compoles**.

[2] Sometimes it is called the **pole-face winding,** for obvious reasons.

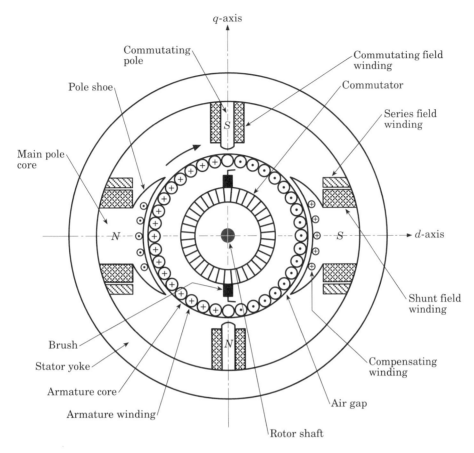

Figure 8.2   Schematic diagram of dc machine construction.

figure, if the field has both windings, the series winding is located on top of the shunt winding. The two windings are separated by extra insulating material, which is usually paper. Figure 8.3 shows the schematic connection diagram for a dc machine with commutating and compensating windings in addition to series and shunt windings. The series and shunt windings are located on the d-axis. This axis is called the **field axis**, or **direct axis**, because the air-gap flux distribution due to the field windings is symmetric at the center line of the field poles. Both the compensating and commutating winding brushes are located on the q-axis. This axis is called the **quadrature axis** because it is 90 electrical degrees from the d-axis and represents the neutral zone. Figures 8.4 and 8.5 show the cutaway views of two different dc machines.

The **commutator** is located on the armature and consists of a number of radial segments assembled into a cylinder which is attached to and insulated from the shaft, as shown in Figures 8.6 and 8.7. These segments are well insulated from each other by mica. The leads of the armature coils are connected to these commutator segments. Current is conducted to the armature coils by car-

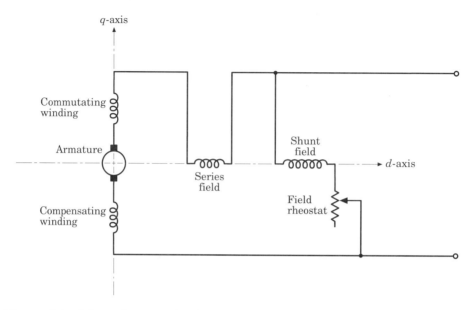

Figure 8.3   Schematic connection diagram for a dc machine having commutating poles and compensating winding in addition to series and shunt windings.

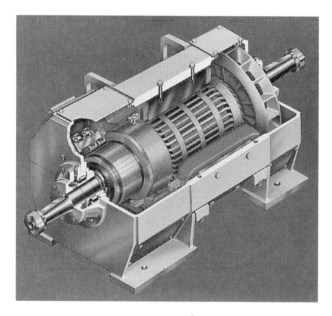

Figure 8.4   Cutaway view of a mill duty dc motor. (Courtesy of General Electric.)

bon brushes that ride on the commutator segments. The brushes are fitted to the surface of the commutator and are held in **brush holders**. These brush holders use springs to push the brushes against the commutator surface to

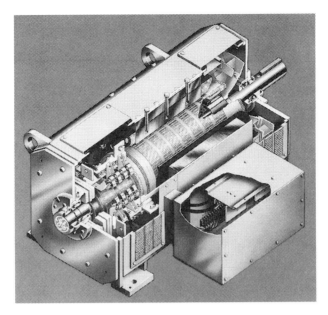

Figure 8.5   Cutaway view of a 400-hp shunt-wound dc motor. (Courtesy of General Electric.)

Figure 8.6   A close-up view of the commutator of a 500 hp dc motor. (Courtesy of MagneTex, Inc.)

maintain constant pressure and problem-free riding. The connection between the brush and brush holder is by a flexible copper cable called a **pigtail**. The rotor itself is mounted on a shaft that rides in the bearings.

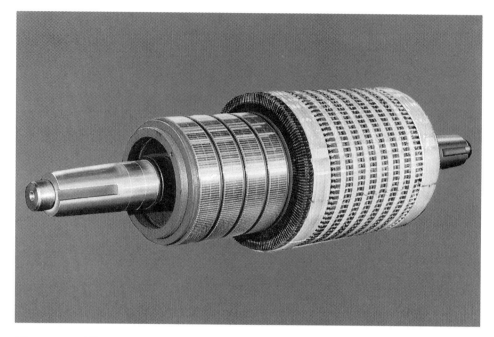

Figure 8.7 The armature of a 500 V, 150 hp, dc motor. (Courtesy of General Electric.)

8.3 A BRIEF REVIEW OF ARMATURE WINDINGS

As previously stated, the armature windings are the windings in which a voltage is induced in a dc machine. The rotor of a dc machine is called an **armature**, because the armature windings are placed in slots of the rotor. Since the armature winding is connected to the commutator, it is also known as the **commutator winding**. This winding is usually built with full-pitch windings.

As shown in Figure 8.8, the armature windings are either the closed continuous type of double-layer **lap windings** or **wave windings**. A winding is formed by connecting several coils in series, and a coil is formed by connecting several turns (loops) in series. Each turn is made up of two conductors connected to one end by an end connection. In other words, each side of a turn is called a **conductor**.

In a lap winding there are always as many paths in parallel through the armature winding as there are number of poles (or brushes). Each path is made up of a series connection between a number of terminal coils that is approximately equal to the total number of armature coils divided by the number of poles. In such a lap winding the current in each armature coil is equal to the armature terminal current divided by the number of poles.

In a wave winding there are always two paths in parallel through the armature winding from one terminal to the other. At any given time, each path is made up of a series connection of approximately one-half of the total armature coils between the terminals. The current in each armature coil is one-half of the armature terminal current.

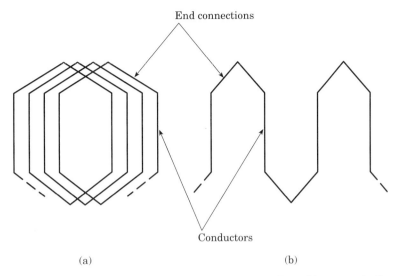

Figure 8.8   Basic armature winding types: (a) lap winding; (b) wave winding.

The internal generated voltages in each of the parallel paths of a lap winding are equal if the machine is geometrically and magnetically symmetrical. If complete symmetry does not exist, the internal generated voltage of the different paths will not be exactly the same. Therefore, there will be a circulating current flowing between brush sets of the same polarity. To prevent this, equalizers are used. **Equalizers** are bars located on the rotor of a lap-wound dc machine that short together the points in the winding at the same voltage level in different paths. Figure 8.7 shows a close-up view of the commutator of a dc motor. Notice that the equalizers are mounted in the small ring just in front of the commutator segments.

8.4   AN ELEMENTARY DC MACHINE

Figure 8.9 shows an elementary two-pole dc generator. The armature winding consists of a single coil of N turns. The voltage induced in this rotating armature is alternating. However, by using a **commutator** this voltage is rectified mechanically into dc voltage for the external circuit . Here, the commutator has two half rings which are made up of two copper segments insulated from each other and from the shaft. Each end of the armature coil is connected to a segment. Stationary carbon brushes held against the commutator surface connect the coil to the external armature terminals. Since the brushes remain in the same position as the coil rotates, each fixed terminal is always connected to the side of the coil where the relative motion between the coil side and the field is the same. In other words, the action of the commutator is to reverse the armature coil connections to the external circuit when the current reverses in the armature coil. Therefore, the commutator at all times connects the coil side under the south pole to the positive brush and the one under the north pole to

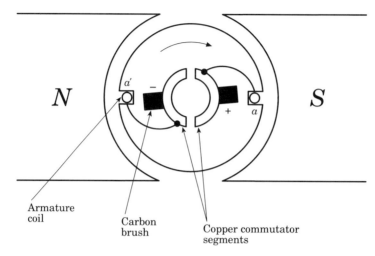

Figure 8.9   Simple representation of a dc machine.

the negative pole. Thus, the polarity of the voltage difference between the two fixed brushes is always the same and the voltage is now unidirectional. However, a pulsating direct current, like the one produced by this type of single-coil generator, is not suitable for most commercial uses. As shown in Figure 8.10, the total internal generated voltage between brushes (i.e., simply the brush voltage) can be made practically constant by using a large number of coils and commutator segments with the coils evenly distributed around the armature surface.

8.5 ARMATURE VOLTAGE

In a dc machine the armature voltage is the internal generated voltage. By applying Faraday's law of electromagnetic induction, the armature voltage[3] can be expressed as

$$E_a = \left(\frac{Zp}{2\pi a} \right) \Phi_d \omega_m \tag{8.1}$$

where Z = total number of conductors in the armature winding
p – number of poles (of field or stator)
a = number of parallel paths in the armature winding
Φ_d = direct-axis air-gap flux per pole in webers
ω_m = angular velocity of the armature (or shaft) in mechanical radians per second

---

[3] It is also known as the **speed voltage**. Some authors define this voltage as the **internal source voltage** when the machine is operating as a generator, and as the **countervoltage** (or **back emf**) when the machine is operating as a motor.

The armature voltage can also be expressed as

$$E_a = K_a \Phi_d \omega_m \qquad (8.2)$$

where

$$K_a = \frac{Zp}{2\pi a} \qquad (8.3)$$

and is called the **armature constant**. The speed of the machine may be given in revolutions per minute (rpm) rather than in radians per second. Since

$$\omega_m = \left(\frac{2\pi}{60}\right) n_m \qquad (8.4)$$

the armature voltage can be expressed as

$$E_a = K_{a1} \Phi_d n_m \qquad (8.5)$$

where

$$K_{a1} = \frac{Zp}{60a} \qquad (8.6)$$

Therefore, the armature voltage is a function of the flux in the machine, the speed of its rotor, and a constant that depends on the machine.

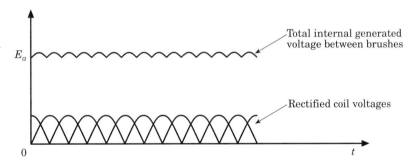

Figure 8.10   The total internal generated voltage between brushes in a dc machine as a function of time.

The armature voltage, or more precisely the internal generated voltage, is not the terminal voltage. Consider the circuit representation of a separately excited dc generator and motor as shown in Figures 8.11(a) and 8.11(b), respectively. The armature voltage E_a can be expressed as

$$E_a = V_t \pm I_a R_a \qquad (8.7)$$

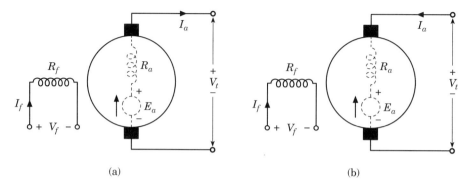

(a)   (b)

Figure 8.11   Simple representation of a dc machine: (a) circuit representation of a dc generator; (b) circuit representation of a dc motor.

where the plus sign is used for a generator and the minus sign for a motor. V_t is the terminal voltage and R_a is the armature resistance.[4] Therefore, in the case of a generator, the armature voltage is always greater than the terminal voltage. In a motor, the armature voltage is less than the terminal voltage. Regardless of whether the machine is used as a generator or as a motor, a **brush- contact voltage drop,** usually assumed to be 2 V, exists due to the resistive voltage drop between the brushes and commutator. Also, the term **armature winding circuit resistance** can include not only the resistance of the armature winding R_a, but also the resistances of the series field winding R_{se}, commutating winding R_{cw}, compensating winding R_{cp}, as well as the resistance of any external wires (used in laboratories to make the necessary connections) $R_{a, ext}$. Therefore, the general expression for the armature voltage becomes

$$E_a = V_t \pm I_a(\Sigma R_a) \pm 2.0 \tag{8.8}$$

where

$$\Sigma R_a = R_a + R_{se} + R_{cw} + R_{cp} + R_{a, ext} \tag{8.9}$$

and represents the total armature winding circuit resistance. In equation (8.8) the plus sign is used for a generator and the minus sign for a motor. Note that the voltage polarity of the brushes is a function of the rotational direction and the magnetic polarity of the stator field poles.

Example 8.1

Assume that a six-pole dc machine has an armature with a radius of 15 cm and an effective length of 30 cm, and that the poles cover 70 percent of the

---

[4] Even though the armature resistance R_a actually exists between the brushes in the armature, in some books it is not explicitly represented in the armature circuits. Such representation agrees

armature periphery. The armature winding has 50 coils of five turns each and is wave-wound with $a = 2$ paths. Assume that the average flux density in the air gap under the pole faces is 0.7 T and determine the following:

(a) The total number of conductors in the armature winding.
(b) The flux per pole.
(c) The armature constant K_a.
(d) The induced armature voltage if the speed of the armature is 900 rpm.
(e) The current in each coil if the armature current is 200 A.

SOLUTION:

(a) The total number of conductors in the armature winding is

$$Z = (2 \text{ conductors/turn})(5 \text{ turns/coil})(50 \text{ coils}) = 500$$

(b) Since the pole area is

$$A_p = \frac{2\pi(0.15 \text{ m})(0.30 \text{ m})(0.7 \text{ T})}{4} = 0.0495 \text{ m}^2$$

the flux per pole is

$$\Phi_d = A_p B = (0.0495 \text{ m}^2)(0.7 \text{ Wb/m}^2) = 0.0346 \text{ Wb}$$

(c) The armature constant is

$$K_a = \frac{Zp}{2\pi a} = \frac{(500)(6)}{2\pi(2)} = 238.73$$

(d) The speed of the armature is

$$\omega_m = n_m \left(\frac{2\pi}{60}\right) = (900 \text{ rpm})\left(\frac{2\pi}{60}\right) = 94.25 \text{ rad/s}$$

Therefore, the induced armature voltage is

$$E_a = K_a \Phi_d \omega_m = (238.73)(0.0346 \text{ Wb})(94.25 \text{ rad/s}) = 779.32 \text{ V}$$

(e) The current in each coil is

$$I_{coil} = \frac{I_a}{a} = \frac{200 \text{ A}}{2} = 100 \text{ A}$$

---

with the dc machine panels of an electromechanical laboratory since only armature terminals A_1 and A_2 can possibly be reached, as shown in Figure 8.12. However, in the rest of the book, the armature resistance is explicitly represented to avoid confusing the beginner.

Example 8.2

Assume that a separately excited shunt dc machine has a rated terminal voltage of 230 V and a rated armature current of 100 A. Its armature winding resistance, commutating winding resistance, and compensating winding resistance are 0.08, 0.01, and 0.008 Ω, respectively. The resistance of external wires (i.e., $R_{a,\ ext}$) is 0.002 Ω. Determine the following:

(a) The induced armature voltage if the machine is operating as a generator at full load.

(b) The induced armature voltage if the machine is operating as a motor at full load.

SOLUTION:

(a) Since the total armature winding circuit resistance is

$$\Sigma R_a = R_a + R_{se} + R_{cw} + R_{cp} + R_{a,\ ext}$$

$$= 0.08 + 0 + 0.01 + 0.008 + 0.002 = 0.10\ \Omega$$

When the machine is operating as a generator its induced armature voltage is

$$E_a = V_t + I_a(\Sigma R_a) + 2.0$$

$$= 230 + (100\ \text{A})(0.10\ \Omega) + 2.0 = 242\ \text{V}$$

(b) When the machine is operating as a motor its induced armature voltage is

$$E_a = V_t - I_a(\Sigma R_a) - 2.0$$

$$= 230 - (100\ \text{A})(0.10\ \Omega) - 2.0 = 218\ \text{V}$$

8.6 METHODS OF FIELD EXCITATION

The field circuit and the armature circuit of a dc machine can be interconnected in several different ways to produce various operating characteristics. There are basically two types of field windings, namely **shunt field winding** and **series field winding**. The shunt windings have a great many turns and are built from thinner wires. Therefore, the required field current is a very small portion (less than 5 percent) of the rated armature current. On the other hand, the series windings have relatively fewer number of turns and are built from thicker wires. The series windings are connected in series with the armature and therefore, its field current is the armature current. The shunt winding can be separately excited from a separate source. In a **separately excited** machine there is no electrical interconnection between the field and the armature windings, as shown in Figure 8.12(a). When the field is interconnected with the armature winding the machine is said to be **self-excited**. The self-excited machines may be shunt, series, or compound, as shown in Figures 8.12(b) through 8.12(g). Notice that a compound machine has both shunt and series field windings in addition to the armature winding. If the relative polarities of the shunt and series field windings are additive, the machine is called **cumulative compound**. If they oppose each other the machine is called **differential compound**. A compound machine may be connected **short-shunt** with the shunt field in parallel with the armature only or **long-shunt** with the shunt field in

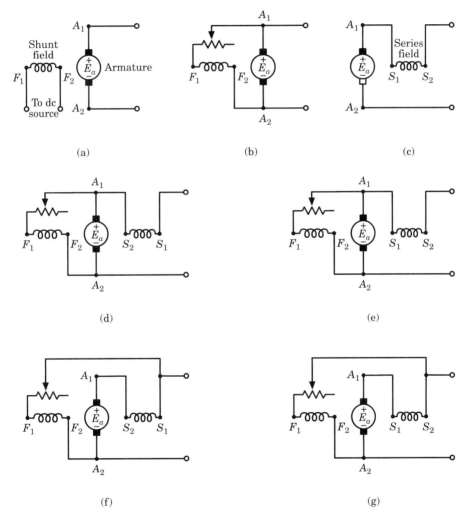

Figure 8.12   Typical field-excitation methods for a dc machine: (a) separately excited; (b) shunt; (c) series; (d) short-shunt connection for cumulative compound motor or differential compound generator; (e) short-shunt connection for cumulative compound generator or differential compound motor; (f) long-shunt connection for cumulative compound motor or differential compound generator; (g) long-shunt connection for cumulative compound generator or differential compound motor.

parallel with both the armature and series field. The circuits shown in Figure 8.12 are labeled according to the NEMA standards.[5]

8.7 ARMATURE REACTION

Armature reaction is defined as the effect of the armature mmf field upon the flux distribution of the machine. Figure 8.13(a) shows the main field flux Φ_f that is established by the mmf produced by the field current when there is no current flowing in the armature. Figure 8.13(b) shows the armature flux Φ_a that is established by the armature mmf produced by the current flowing in the armature when there is no current flowing in the field winding of the machine.

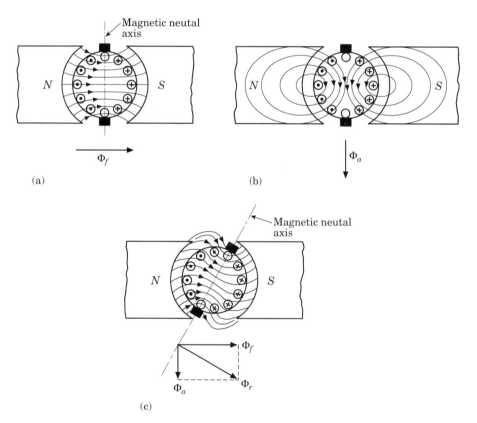

Figure 8.13   The effect of armature reaction in a dc machine: (a) the effect of the air gap on the pole flux distribution; (b) the armature flux alone; (c) resultant distortion of the field flux produced by the armature flux.

---

[5] Notice that currents flowing into terminals F_1 and S_1 result in a cumulative compound effect; whereas currents flowing into terminal F_1 and out of terminal S_1 result in a differential compound effect. Also notice that if F_1 were connected to A_1 in each case, it would become a **short-shunt** connection.

The brushes are located on the magnetic neutral axis. Figure 8.13(c) shows the situation when both the main field flux and the armature flux exist at the same time. It is clear that the armature flux causes a distortion in the distribution of the main field flux. As shown in the figure, the phasor sum of the two mmfs produces a resultant flux Φ_r. Notice that the flux produced by the armature mmf opposes the flux under one half of the pole and aids under the other half of the pole. As a result, flux density under the pole increases in one half of the pole and decreases under the other half of the pole. As shown in Figure 8.14(d) the magnetic neutral axis is shifted from the geometric neutral axis. The shift is forward in the direction of rotation for a generator and backward against rotation for a motor. The magnitude of flux shift is a function of saturation in the pole tips and the amount of armature (load) current. If there is no saturation, the increase of flux in one pole tip is canceled by a corresponding decrease in the other.[6] With saturation, on the other hand, there is a net decrease in total flux,[7] causing a decrease in the terminal voltage of the generator and an increase in the speed of the motor.

8.8 COMMUTATION

Commutation is the process of reversing the direction of the current in an armature coil as the commutator segments to which the coil is connected move from the magnetic field of one polarity to the influence of the magnetic field of the opposite polarity, as shown in Figure 8.15. The time interval required for this reversal is called the **commutation period**. As the commutator sweeps past the brushes, any given coil connected to one of the segments has current in a particular direction. The current in the coil is reversed as the commutator segment approaches and passes the brush. Consider the coil b connected to commutator segments 2 and 3, as shown at the top of Figure 8.15, and notice that the current in the coil is flowing from left to right. The middle diagram of Figure 8.15 shows that there is no current flow in the coil since it is short-circuited by the brush. The bottom of Figure 8.15 shows the moment at which brush contact with commutator segment 3 is interrupted. Notice that the direction of current flow in coil b is reversed and that it is now flowing from right to left.

As shown in Figure 8.16, during the commutation period Δt, the commutator segments to which a coil is connected are passing under the brush; the current I has to be totally reversed as the commutator segments pass from under the brush to prevent the formation of an arc. The reversal of the coil induces a self-inductance voltage which opposes the change of current that can cause a spark to appear at the trailing edge of the brush. Figure 8.17 shows the ideal process of commutation as well as undercommutation[8] due

---

[6] It is called the **cross-magnetizing armature reaction**.
[7] It is called the **demagnetizing effect of the cross-magnetizing armature reaction**.
[8] It is also called the **incomplete** or **delayed commutation**.

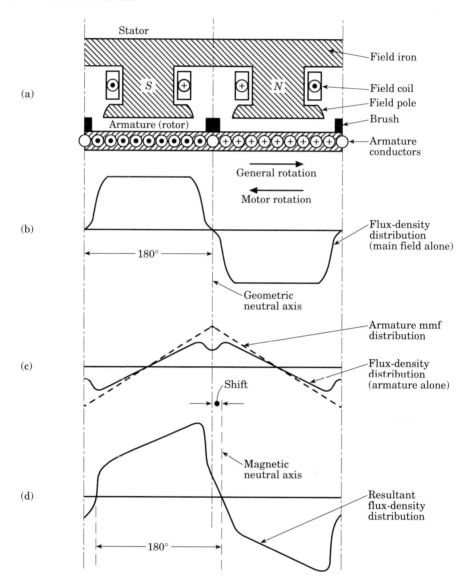

Figure 8.14 Main field, armature, and resultant flux-density distributions with brushes on the geometric neutral axes: (a) linear representation of stator and armature-magnetic circuits; (b) flux-density distribution due to main field alone; (c) flux-density distribution due to armature mmf alone; (d) resultant flux-density distribution.

to the reactance voltage. If the reactance voltage is large enough it may cause sparking at the trailing edge of the brush. Excessive sparking burns the brushes and the commutator surface, but sparking can be prevented by inducing a voltage in the coil undergoing commutation. This can be accomplished by the use of commutating poles, which are small poles placed be-

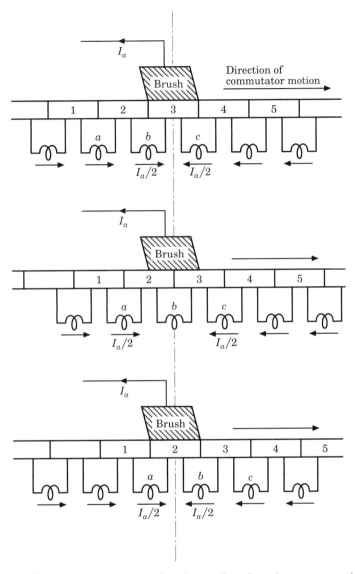

Figure 8.15   The reversal of current flow in a coil undergoing commutation.

tween the main poles of a dc machine. The commutating windings are interconnected in a such way that they have the same polarity as the following main pole in the direction of rotation. Almost all dc machines of more than 1 hp are furnished with commutating poles (or interpoles). As previously stated, the commutating windings are permanently connected in series with the armature and their leads are not brought out to the terminal box. In small machines (with 1 hp or less) the commutation can be improved by shifting the brushes. As a result, the coils undergoing commutation can have current reversals supported by flux from the main poles.

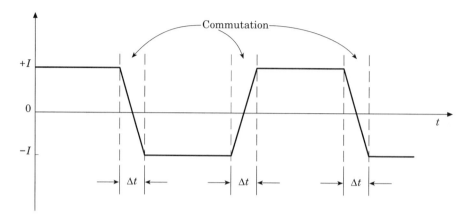

Figure 8.16 Waveform of a current in an armature coil during linear commutation.

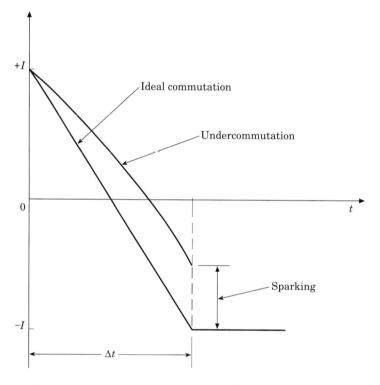

Figure 8.17 Coil current as a function of time during ideal commutation and undercommutation (or delayed commutation.)

8.9 COMPENSATING WINDINGS

As the armature current increases due to the armature reaction, the corresponding flux density distortion also increases, which in turn causes the commutator flashover probability to increase. The commutating windings located on the commutating poles can neutralize the effect of armature reaction in the interpolar areas. However, they cannot stop the flux distortions in the air gaps over the pole faces. These flux distortions can be eliminated by placing compensating windings in slots distributed along the pole faces, as shown in Figure 8.2. Each compensating winding has a polarity opposite that of the adjoining armature winding. By allowing armature current to flow through such a pole-face winding, the armature reaction can be completely neutralized by a proper amount of ampere-turns. The only disadvantage of compensating windings is that they are very expensive. For this reason they are only used in large dc machines that handle heavy overloads, or suddenly changing loads or in motors subject to high acceleration and sudden reversals in rotational directions.

8.10 MAGNETIZATION CURVE

The magnetization curve of a dc machine can be obtained by running the machine as a generator at its rated speed[9] with no load and varying its field current. The internal generated voltage of the machine is

$$E_a = K_a \Phi_d \omega_m \tag{8.2}$$

and if the speed of the armature is kept constant, then E_a will be proportional to the flux set up by the field winding. The magnetization curve (mc) is usually obtained by exciting the field winding separately, regardless of whether the machine is normally used as a generator or as a motor. Note that the armature (or load) current is zero since the machine is running without a load, and that the terminal voltage is equal to the internal generated voltage E_a. As the magnetizing flux per pole is increased by raising the current in the field winding, the voltage E_a also increases. However, above a certain point called the **saturation point**, it becomes increasingly difficult to further magnetize the core. Due to the saturation of the magnetic core above the saturation point (i.e., the knee of the curve), the relationship between the voltage E_a and the field current I_f becomes nonlinear, as shown in Figures 8.18(a) and 8.18(b). The resultant mc curve shows the relationship between the voltage E_a and the field mmf or the field current I_f. However, the shape of the curve is determined mainly by the characteristics of the magnetic circuit. Because of a small residual magnetism that exists in the field structure, the voltage does not start from zero, except in a new machine.[10] As shown in Figure 8.18(a), if the field current is increased from zero to a value that yields an armature voltage well above the rated voltage of the machine, the resultant curve is the **ascending curve**. If the

---

[9] The **rated speed** is the speed at which the machine is designed to operate to produce the rated voltage.

[10] Just as a new machine may not have residual magnetic flux, in any machine it may be lost as a result of conditions such as mechanical jarring during transportation, excessive vibrations, inactivity for long periods of time, extreme heat, or having its alternating current un-

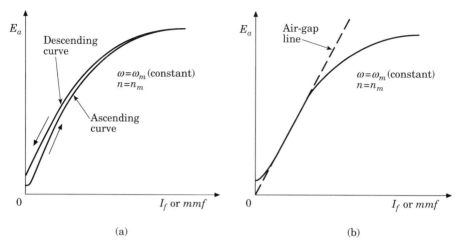

(a)   (b)

Figure 8.18   Magnetization curve: (a) explicitly showing the effect of hysteresis; (b) average magnetization curve.

field current I_f is progressively decreased to zero again, this curve is the **descending curve**. The reason that the descending curve is above the ascending curve can be explained by hysteresis. The magnetization curve[11] shown in Figure 8.18(b) is the average of the two curves. Note that ω_m and n_m represent the rated speed[12] of the machine at which the magnetization curve is developed given in rad/s and rpm, respectively.

Consider the magnetization curves shown in Figure 8.19. Assume that the top and the bottom curves are obtained at constant speeds of ω_{m1} and ω_{m2}, respectively. Therefore, the associated internal generated voltages are

$$E_{a1} = K_a \Phi_d \omega_{m1} \tag{8.10}$$

and

$$E_{a2} = K_a \Phi_d \omega_{m2} \tag{8.11}$$

Since

$$K_a \Phi_d = \frac{E_{a1}}{\omega_{m1}} \tag{8.12}$$

and

---

intentionally connected across the field winding. In such cases, the field winding must receive an initial dc excitation to provide the machine with a suitable level of residual flux. This process is known as **flashing the field.**

[11] It is also called the **saturation curve**, the open-circuit characteristic, or the **no-load characteristic.**

[12] Hence it is called the **magnetization-curve speed**.

$$K_a\Phi_d = \frac{E_{a2}}{\omega_{m2}} \qquad (8.13)$$

then

$$E_{a2} = \left(\frac{\omega_{m2}}{\omega_{m1}}\right)E_{a1} \qquad (8.14)$$

where E_{a1} and E_{a2} are generated at constant speeds ω_{m1} and ω_{m2}, respectively. Alternatively, if the constant speeds are given in rpm, then

$$E_{a2} = \left(\frac{n_{m2}}{n_{m1}}\right)E_{a1} \qquad (8.15)$$

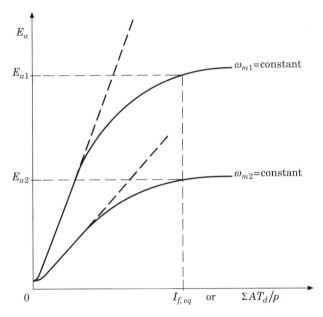

Figure 8.19 Magnetization curves developed at two separate speeds with the excitation held constant.

Up to now, it has been assumed that the machine involved is a separately excited shunt machine. However, it is possible to apply this approach to other types of dc machines. This can be accomplished by considering the direct-axis air-gap flux produced by the combined mmf. For example, suppose that the machine involved is a compound machine. Its net excitation per pole on the d-axis (i.e., the total ampere turns per pole on the d-axis) can be expressed as

$$\Sigma AT_d/p = N_f I_f \pm N_{se}I_a \qquad (8.16)$$

where N_f = number of turns per pole of the shunt field winding
I_f = the current in the shunt field winding
N_{se} = the number of turns per pole of the series field winding
I_a = the current in the series field winding

This equation can be modified so that the effects of the armature reaction can be taken into account. Thus,

$$\Sigma AT_d/p = N_f I_f \pm N_{se} I_{se} - K_d I_a \tag{8.17}$$

where the term $K_d I_a$ is a simplified linear approximation to account for the demagnetization of the d-axis (i.e., the armature reaction) caused by the armature mmf. In equations (8.16) and (8.17) the plus sign is used for a cumulative compounded machine and the minus sign is used for a differential compounded machine. The K_d is the armature reaction constant for the machine involved. Hence equation (8.17) gives the total effective mmf per pole.

If the machine has a self-excited shunt field, the net excitation per pole on the d-axis can be expressed as

$$\Sigma AT_d/p = \frac{V_t N_f}{\Sigma R_f} \pm N_{se} I_a - K_d I_a \tag{8.18}$$

where the total shunt field circuit resistance is

$$\Sigma R_f = R_f + R_{rheo} \tag{8.19}$$

where V_t = the terminal voltage
R_f = the resistance of the shunt field winding
R_{rheo} = the resistance of the shunt field rheostat[13]

However, if the machine has a separately excited shunt field, the net excitation per pole on the d-axis can be expressed as

$$\Sigma AT_d/p = \frac{V_f N_f}{\Sigma R_f} \pm N_{se} I_a - K_d I_a \tag{8.20}$$

where V_f is the voltage across the shunt field winding.

If the magnetization curve is given in terms of E_a versus I_f, it is necessary to define an equivalent I_f that would produce the same voltage E_a as the combination of all the mmfs in the machine.[14] Since

---

[13] Usually, a rheostat is included in the circuit of the shunt winding to control the field current and to vary the shunt field mmf.

[14] In other words, it is as if the machine were replaced by an equivalent machine with a shunt field in order to find the corresponding equivalent I_f current.

$$\Sigma AT_d/p = N_f I_{f, \, eq} \tag{8.14}$$

such equivalent shunt field current can be found from

$$I_{f, \, eq} = \frac{\Sigma AT_d/p}{N_f} \tag{8.21}$$

Once the equivalent shunt field current is found, the corresponding voltage E_a can be found from the magnetization curve. Some magnetization curves are plotted in terms of E_a versus $\Sigma AT_d/p$ or $I_{f, \, eq}$, as shown in Figure 8.19. As can be observed from equation (8.20), when the I_a is zero the net excitation per pole on the d-axis is produced only by the shunt field winding. Since

$$\Sigma At_d/p = N_f I_f \tag{8.22}$$

then

$$I_f = \frac{\Sigma AT_d/p}{N_f} \tag{8.23}$$

8.11 DC GENERATORS

As stated in Section 8.1, dc generators are generally not used to produce bulk power today. Instead, solid-state-controlled rectifiers are preferred for many applications. However, there are still some dc generators that are used to provide dc power to isolated loads and for special applications. In such use, the dc machine operating as a generator is driven by a prime mover at a constant speed with the armature terminals connected to the load. It can be used as a separately excited generator, a self-excited shunt generator, a series generator, or a compound generator.

8.12 THE SEPARATELY EXCITED GENERATOR

In the separately excited dc generator the shunt field winding is supplied by a separate external dc power source, as shown in Figure 8.20. The external dc power source can be a small dc generator, a solid-state dc power supply, or a battery. In the equivalent circuit, E_a is the internal generated voltage, V_t is the terminal voltage, I_a is the armature current, which is also the load current, I_f is the field current, R_a is the resistance of the armature winding, R_f is the resistance of the field winding, R_{rheo} is the resistance of the shunt field rheostat, and V_f is the voltage of a separate source. According to Kirchhoff's voltage law the terminal voltage is

$$V_t = E_a - I_a R_a - 2.0 \tag{8.24}$$

and the field voltage is

$$V_f = I_f R_f \tag{8.25}$$

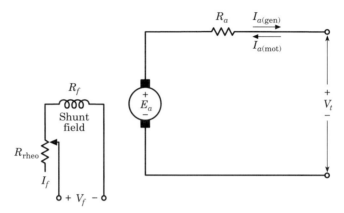

Figure 8.20   The separately excited shunt generator or motor schematic diagram with current directions.

Equation (8.24) represents the **terminal characteristic**[15] of the separately excited dc generator as shown in Figure 8.21. Notice that the terminal voltage differs from the no-load voltage by the three voltage drops representing the armature resistance voltage drop, the brush-contact voltage drop, and the armature reaction voltage drop. The **load characteristic** is determined by

$$V_t = I_a R_L$$

where R_L represents the load resistance. As shown in Figure 8.21, the intersection of the terminal characteristic and the load characteristic is the **operating point** for the generator. The operation of the separately excited generator is stable with any field excitation. Therefore, a wide range of output voltages are available.

8.13 THE SELF-EXCITED SHUNT GENERATOR

As shown in Figure 8.22(a), in the self-excited shunt generator the field winding is connected directly across the armature winding. Therefore, the armature voltage can provide the field current. However, any change in the armature current results in a change in the $I_a R_a$ voltage drop. Because of this both the terminal voltage and the field current[16] must vary. Thus, the internal generated voltage E_a in a self-excited generator is a function of the armature current I_a. Accordingly the terminal voltage V_t changes as the I_a changes. Hence,

$$V_t = E_a - I_a R_a - 2.0 \qquad (8.26)$$

---

[15] It is also called the **external characteristic**.
[16] Typically, the field current I_f is about 5% of the rated armature current.

and the field voltage is

$$V_t = I_f R_f \tag{8.27}$$

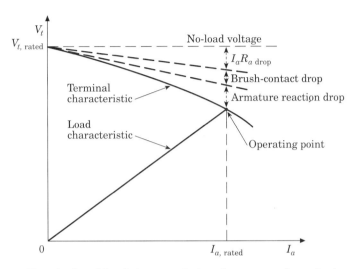

Figure 8.21   Terminal and load characteristics of a separately excited generator.

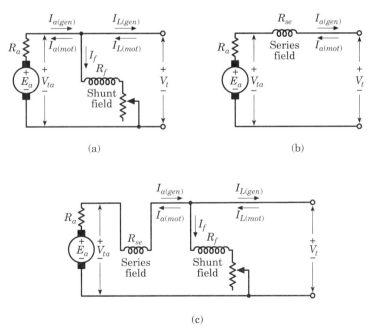

Figure 8.22   Motor or generator connection diagram with current directions of a dc machine having: (a) shunt winding; (b) series winding; (c) compound (long shunt) winding.

The terminal characteristic of a self-excited generator is similar to that of the separately excited generator except that its terminal voltage falls off faster as the load current I_L increases. Notice that

$$I_a = I_L + I_f \tag{8.28}$$

and

$$I_f = \frac{V_f}{R_f} \tag{8.29}$$

The decrease in the terminal voltage, due to increased $I_a R_a$ and armature reaction voltage drops, also causes the field current I_f to decrease and the terminal voltage to drop further. The field current can be adjusted by using the shunt-field rheostat, which is connected in series with the shunt field winding. Under no-load conditions the armature current is equal to the field current.

The operation of the self-excited generator depends on the existence of some residual magnetism in its magnetic circuit. As shown in Figure 8.23, when such a generator is brought up to its constant speed because of the presence of a **residual flux** $\Phi_{d,\ res}$ in its field poles, there will be a small internal generated voltage $E_{a,\ res}$ even at zero field current since

$$E_{a,\ res} = K_a \Phi_{d,\ res} \omega_m \tag{8.30}$$

Once this voltage appears at the terminals, there will be a small amount of field current flowing in the shunt field winding since it is connected directly across the brushes. This field current produces an mmf in the field, causing the flux to increase and inducing a higher E_a, which in turn causes more current to flow through the field windings. This **voltage buildup** is depicted on the mc curve in Figure 8.23. Notice that at no-load field current $I_{f,\ nl}$ the corresponding no-load terminal voltage is $V_{t,\ nl}$. Also shown in this figure is the field resistance line which is a plot of $R_f I_f$ versus I_f. Such a **field resistance line** is governed by

$$(8.31)$$

As long as there is some residual magnetism in the magnetic poles of the

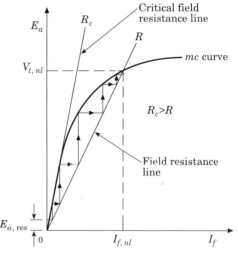

Figure 8.23   Voltage built up in a self-excited shunt dc generator.

$$R_f = \frac{V_f}{I_f} \qquad (8.31)$$

gener-

ator, the voltage buildup will take place if the following conditions are satisfied: (1) there must be a residual flux in the magnetic circuit; (2) the field winding mmf must act to aid this residual flux; (3) the total field-circuit resistance must be less than the critical field circuit resistance. As shown in Figure 8.23, the **critical field resistance** is the value that makes the resistance line coincide with the linear portion of the mc curve. In other words, it represents the resistance value of the shunt-field circuit below which the voltage buildup takes place. Therefore, the slope of the mc curve in the linear region is the **critical resistance**. Since a shunt generator maintains approximately constant voltage on load, it is widely used as an exciter to provide the field current for a large generator.

8.14 THE SERIES GENERATOR

The series generator is a self-excited generator that has its field winding connected in series with its armature, as shown in Figure 8.22(b). Therefore, its armature current I_a, field current I_{se}, and load current I_L are all equal to each other. Thus,

$$(8.32)$$

$$I_a = I_{se} = I_L \qquad \text{and}$$

$$(8.33)$$

$$V_t = E_a - I_a(R_a + R_{se})$$

where R_{se} is the resistance of the series-field winding. Under no-load conditions the internal generated voltage is due to the residual magnetism. As the load increases so do the field current I_a and the voltage E_a as shown in Figure 8.24. The terminal voltage continues to increase until the magnetic circuit of the machine becomes saturated. Series generators are used as *voltage boosters* and as *constant-current generators* in arc welding.

8.15 THE COMPOUND GENERATOR

As shown in Figure 8.22(c), a compound generator has both series and shunt field windings. If the mmf of the series field aids the mmf of the shunt field, it is called a **cumulative compound generator**. On the other hand, if the mmf of the series field opposes the mmf of the shunt field, it is called a **differential compound generator**. If the shunt field winding is connected across the armature, as shown in Figures 8.12(d) and 8.12(e), this type of compound generator is called a **short-shunt generator**. If the shunt field winding is connected across the series combination of armature and series windings,

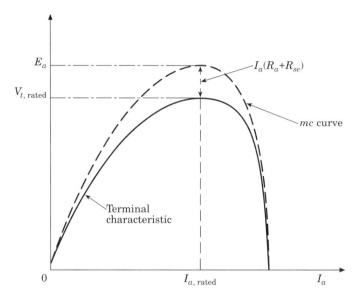

Figure 8.24 Terminal characteristic of a series generator.

as shown in Figure 8.22(g), it is called a **long-shunt generator**. For the short-shunt compound generator,

(8.34)

$$V_t = E_a - I_a R_a - I_L R_{se} \qquad \text{and}$$

(8.35)

$$I_a = I_f + I_L \qquad \text{where}$$

R_{se} is the resistance of the series field winding and I_L is the load current. For the long-shunt compound generator,

$$V_t = E_a - I_a(R_a + R_{se}) \tag{8.36}$$

and

$$I_a = I_f + I_L \tag{8.37}$$

where

$$I_f = \frac{V_t}{R_f} = \frac{V_t}{R_a + R_{se}} \tag{8.38}$$

As shown in Figure 8.25, a *cumulatively compounded generator* may be *flat-compounded*, *overcompounded*, or *undercompounded*, depending on the strength of the series field. If the terminal voltage at the rated load (i.e., full load) is equal to the rated voltage (i.e., no-load voltage), the generator is called a **flat-compounded generator**. If the terminal voltage at the rated load is greater than the no load voltage, the generator is called an **overcompounded generator**. In the event that the terminal voltage at the rated load is less than the no load voltage, the generator is called an **undercompounded generator**. Over-compounding can be used to compensate for line drop when the load served is located far from the generator. It may also be used to counteract the effect of a drop in the prime-mover speed as the load increases.

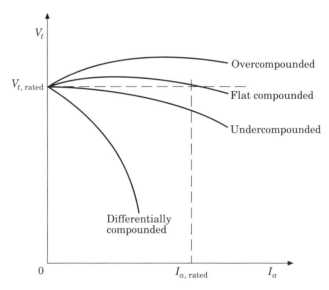

Figure 8.25 Terminal characteristics of compound generators operating at constant speed.

8.16 VOLTAGE REGULATION

The terminal voltage of a generator normally changes as the load current changes. This voltage variation is described by voltage regulation. The percent voltage regulation (% *V Reg*) of a generator is defined as

$$\% \, V \, Reg = \frac{V_{t,nl} - V_{t,fl}}{V_{t,fl}} \times 100 \tag{8.39}$$

where $V_{t,\,nl}$ = no-load terminal voltage
$V_{t,\,fl}$ = full-load terminal voltage

Also,

$$\% \ V \ Reg = \frac{E_a - V_{t,fl}}{V_{t,fl}} \times 100 \qquad (8.40)$$

since E_a is equal to $V_{t,\ nl}$.

Example 8.3

A dc machine has a rated terminal voltage of 250 V and a rated (full-load) armature current of 100 A. Its armature circuit resistance (i.e., armature winding resistance plus commutating field resistance) is 0.10 Ω. Its shunt field winding has a 100 Ω resistance and 1000 turns per pole. The total brush-contact voltage drop is 2 V and the demagnetization of the d-axis by armature mmf is neglected. The magnetization curve data of the dc machine for the rated speed n_{mc} of 1200 rpm is tabulated in Table 8.1 and the curve is plotted in Figure 8.26.

TABLE 8.1 The Magnetization Curve Data for n_{mc} = 1200 rpm

$E_{a(mc)}$	70 V	140 V	195 V	235 V	260 V	276 V
I_f	0.5 A	1.0 A	1.5 A	2.0 A	2.5 A	3.0 A

Assume that the machine is operated as a separately excited shunt generator and is driven at a constant speed of 1200 rev/min.
(a) Determine the values of the terminal voltage at no load and full load as the field current is set at 1.0, 1.5, and then 2.5 A, separately.
(b) Plot the found values of the terminal voltage V_t versus the armature current I_a.
(c) Determine the percent voltage regulation at each field current setting.

SOLUTION:
(a) When the field current is set at 1.0 A at no load, the terminal voltage $V_{t(0\ A)}$ is equal to the internal generated voltage $E_{a(mc)}$ found from the magnetization curve given in Table 8.1. Thus,

$$V_{t(0\ A)} = E_{a(mc)} = E_a = 140 \ V$$

At full load the internal voltage drop due to the armature circuit resistance is

$$I_a(\Sigma R_a) = (100 \ A)(0.1 \ \Omega) = 10 \ V$$

Hence, at full load the terminal voltage of the generator is

$$V_{t(100\ A)} = E_a - I_a(\Sigma R_a) - 2.0 = 140 - (100 \ A)(0.1 \ \Omega) - 2.0 = 128 \ V$$

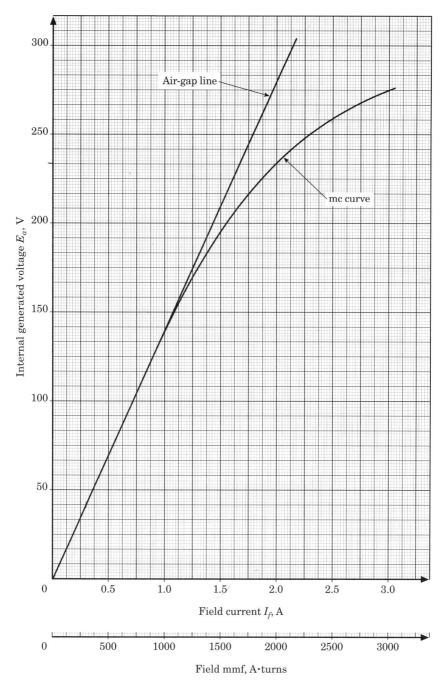

Figure 8.26   Magnetization curve for a dc machine at 1200 rev/min for Example 8.2

Similarly, when the field current is set at 1.5 and then at 2.5 A, the corresponding full load terminal voltages can be found in the same way. The results are presented in Table 8.2. Figure 8.27 shows the plot of terminal voltage at no load versus the field current I_f.

TABLE 8.2

I_f	At $I_a = 0$ $V_{t(0\,A)} = E_{a(mc)}$ $= E_a$	At $I_a = 100\,A$ $I_a(\Sigma R_a)$	$V_{t(100\,A)}$
1.0 A	140 V	10 A	128 V
1.5	195	10	183
2.5	260	10	248

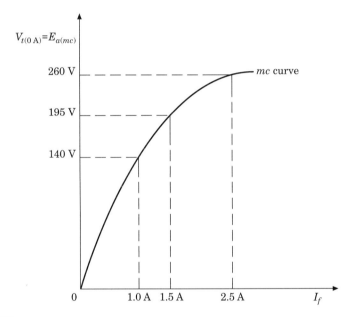

Figure 8.27

(b) The plot of the terminal voltage V_t versus the armature current I_a is shown in Figure 8.28.

(c) At $I_f = 1.0$ A,

$$\% \, V \, Reg = \frac{V_{t,\,nl} - V_{t,\,fl}}{V_{t,\,fl}} \times 100$$

$$= \frac{140 - 128}{128} \times 100 = 9.4$$

At $I_f = 1.5$ A,

$$\% V\,Reg = \frac{195 - 183}{183} \times 100 = 6.6$$

At $I_f = 2.5$ A,

$$\% V\,Reg = \frac{260 - 248}{248} \times 100 = 4.8$$

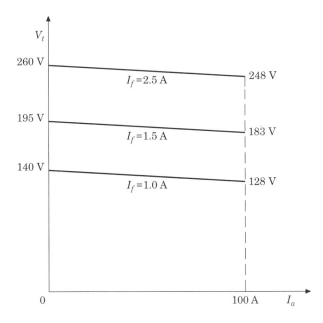

Figure 8.28

8.17 DEVELOPED POWER

As previously stated, a dc machine is a versatile machine, which can be used as a generator or a motor. In the generator mode the input is the mechanical power provided by a prime mover (a diesel engine, a gas turbine, or an electrical motor) and the output is the electrical power. Conversely, in the motor mode the input is the electrical power and the output is the mechanical power. However, as illustrated in Figure 8.29 in both modes of operation a dc excitation current I_f must be provided to establish the magnetic field. The developed power of a separately excited dc machine is

$$P_d = E_a \times I_a \tag{8.41}$$
$$E_a I_a$$

The power at the armature terminals is

$$P_{ta} = \cancel{V_{ta} + I_a} \qquad (8.42)$$

$$V_{ta} I_a$$

where V_{ta} is the voltage at the armature terminals. Thus, from equation (8.8),

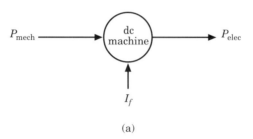

(a)

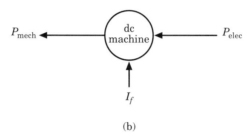

(b)

Figure 8.29   Block diagram of a dc machine operating in: (a) generator mode; (b) motor mode.

$$V_{ta} = E_a \pm I_a(\Sigma R_a) \pm 2.0 \qquad (8.43)$$

where the total armature winding circuit resistance is given by equation (8.9) as

$$\Sigma R_a = R_a + R_{se} + R_{cp} + R_{cw} + R_{a,\,ext} \qquad (8.9)$$

Observe that in equation (8.43) the minus sign is used for a generator and the plus sign for a motor. By substituting equation (8.43) into equation (8.42) the power at the armature terminals can be expressed as

$$P_{ta} - \left[E_a \mp I_a^2(\Sigma R_a) \mp 2.0 \right] \qquad (8.44)$$
$$I_a \qquad\qquad I_a$$

or

$$P_{ta} = E_a I_a \mp I_a^2(\Sigma R_a) \mp 2.0 I_a \qquad (8.45)$$

or

$$P_{ta} = P_d \mp P_{cu} \mp P_{brush} \qquad (8.46)$$

where P_d is the developed power, P_{cu} is the armature-circuit copper losses, and P_{brush} is the brush-contact loss. Therefore,

$$P_{cu} = I_a^2 (\Sigma R_a) \tag{8.47}$$

and

$$P_{brush} = 2.0 I_a \tag{8.48}$$

The developed power P_d is greater than the armature terminal power P_{ta} for a generator, but it is less than P_{ta} for a motor.

The **shaft power**[17] can be expressed as

$$P_{shaft} = P_d \pm P_{rot} \tag{8.49}$$

where P_{rot} represents the rotational losses and is the sum of the friction and windage losses P_{FW} and the core losses P_{core}. Thus,

$$P_{rot} = P_{FW} + P_{core} \tag{8.50}$$

In equation (8.49) the plus sign is used for a generator and the minus sign for a motor because the shaft power P_{shaft} is greater than the developed power P_d for a generator, but it is less than P_d for a motor.

In addition to the rotational losses there may be a **stray-load loss** P_{stray} for those losses that cannot be easily accounted for. It is usually ignored in small machines, but in large machines above 100 horsepower it is generally assumed to be about 1 percent of the output power.

8.18 DEVELOPED TORQUE

Assume that a dc machine has an armature voltage of E_a and armature current of I_a. Its developed power can be expressed as

$$P_d = E_a I_a \tag{8.51}$$

and its developed torque[18] is found from

$$T_d = \frac{P_d}{\omega_m} \tag{8.52}$$

or

---

[17] It is also called the **mechanical power**. It is the **input power** for a generator and the **output power** for a motor.

[18] Some authors prefer to call the torque of a dc motor the **developed torque** and the torque of a dc generator the **countertorque** or the **induced torque** since it opposes the torque applied to the shaft by the prime mover.

$$T_d = \frac{E_a I_a}{\omega_m} \tag{8.53}$$

Using equation (8.2),

$$T_d = \frac{(K_a \Phi_d \omega_m) I_a}{\omega_m} \tag{8.54}$$

or

$$T_d = K_a \Phi_d I_a \tag{8.55}$$

where

$$K_a = \frac{Zp}{2\pi a} \tag{8.3}$$

and is defined as the **winding constant** since it is fixed by the design of the winding. Hence, the developed torque of a dc machine is a function of the flux in the machine, the armature current in the rotor, and a constant that depends on the machine. In the above equations the torque is in N·m. If it is requested in English units, it must be multiplied by 0.7373. Note that the developed torque can also be determined from

$$T_d = \frac{P_d}{\omega_m} = \frac{E_a I_a}{\omega_m}$$

Since

$$\omega_m = \frac{2\pi n_m}{60}$$

then

$$T_d = \frac{E_a I_a}{2\pi n_m / 60}$$

where the speed n_m is in rpm.

8.19 POWER-FLOW AND EFFICIENCY

Consider the equivalent-circuit diagram of the self-excited compound machine shown in Figure 8.22(c). Figure 8.30(a) shows the power flow of this type of dc generator. The input power is the mechanical power or the shaft power. Depending on the machine size, the rotational losses are between 3 to 15 percent, the stray-load loss is about 1 percent for machines larger than 100 hp (otherwise, it is usually ignored), the armature-circuit copper loss is between 3 to 6 percent, the shunt-field loss is between 1 to 5 percent, and the brush-contact loss is about $2I_a$ as explained before. If the shunt field is separately excited, its losses are not supplied by the prime mover through the shaft and therefore, must be handled separately. Figure 8.30(b) shows the power flow of a self-excited compound motor. Notice that here, the input is in electrical power and the output is in mechanical power. The efficiency[19] of a dc machine can be determined from

$$Efficiency = \frac{P_{out}}{P_{in}} \tag{8.57}$$

but since

$$P_{out} = P_{in} - \Sigma P_{loss} \tag{8.58}$$

then

$$Efficiency = 1 - \frac{\Sigma P_{loss}}{P_{in}} \tag{8.59}$$

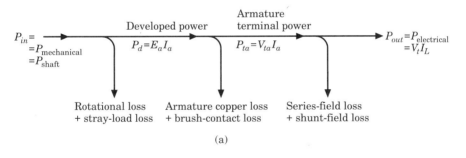

(a)

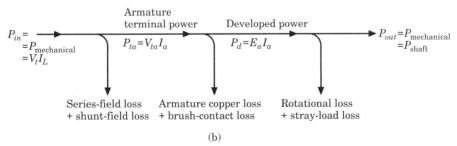

(b)

Figure 8.30   Power flow in a dc machine: (a) generator; (b) motor.

---

[19] The efficiency of a dc machine can be determined more accurately by a test using the **Kapp-Hopkinson method**. In such a test two similar machines are mechanically coupled and electrically connected back-to-back. For further information see Daniels [17].

Thus, the percent efficiency can be expressed as

$$\% \ \eta = \left(1 - \frac{\Sigma P_{loss}}{P_{in}}\right) \times 100 \tag{8.60}$$

The maximum efficiency at a given constant speed is obtained when the sum of the rotational loss and the shunt-field copper loss is equal to the armature copper loss. The percent efficiency can also be found from

$$\% \ \eta = \frac{P_{out}}{P_{out} + \Sigma P_{loss}} \times 100 \tag{8.61}$$

Example 8.4

Assume that a separately excited shunt motor operating at 1000 rev/min has a load current of 100 A and a terminal voltage of 240 V. If the armature winding resistance is 0.1 Ω, determine:
(a) The developed torque.
(b) The shaft speed and the load current if the torque is doubled at the same excitation.

SOLUTION:
(a) From equation (8.8),

$$E_a = V_t - I_a \left(\Sigma R_a\right) - 2.0$$

$$= 240 - (100)(0.1) - 2.0 = 228 \ V$$

at a speed of

$$\omega_m = \left(1000 \ \text{rev/min}\right)\left(\frac{2\pi}{60}\right) = 104.72 \ \text{rad/s}$$

From equation (8.2),

$$E_a = K_a \Phi_d \omega_m$$

or

$$K_a \Phi_d = \frac{E_a}{\omega_m} = \frac{228 \ V}{104.72 \ \text{rad/s}} = 2.1772$$

Thus, the developed torque is

$$T_d = K_a \Phi_d I_a$$

$$= (2.1772)(100) = 217.72 \ \text{N} \cdot \text{m}$$

(b) When $T_d = 2(217.72) = 435.44$ N·m

$$I_a = \frac{T_d}{K_a \Phi_d} = \frac{435.44}{2.1772} = 200 \text{ A}$$

Therefore, the corresponding speed is

$$\omega_m = \frac{E_a}{K_a \Phi_d} = \frac{V_t - I_a R_a}{K_a \Phi_d}$$

$$= \frac{240 - (200)(0.1)}{2.1772} = 101.05 \text{ rad/s}$$

or

$$n = (101.05 \text{ rad/s})\left(\frac{60}{2\pi}\right) = 964.96 \text{ rev/min}$$

Example 8.5

Assume that a 25-hp 250-V self-excited shunt motor is supplied by a full-load line current of 83 A. The armature and field resistances are 0.1 Ω and 108 Ω, respectively. If the total brush-contact voltage drop is 2 V and the friction and core losses are 650 W, determine the following:
(a) The shunt-field winding loss.
(b) The armature winding loss.
(c) The total loss of the motor.
(d) The percent efficiency of the motor.

SOLUTION:
The input power is

$$P_{in} = V_t I_L = (250 \text{ V})(83 \text{ A}) = 20{,}750 \text{ W}$$

(a) The shunt-field winding loss is

$$P_f = I_f^2 R_f = V_f I_f = \frac{V_f^2}{R_f} = \frac{(250 \text{ V})^2}{108 \text{ Ω}} = 579 \text{ W}$$

Thus, the field current is

$$I_f = \frac{P_f}{V_f} = \frac{579 \text{ W}}{250 \text{ V}} = 2.316 \text{ A}$$

(b) Since the full-load armature current is

$$I_a = I_L - I_f = 83 - 2.316 = 80.684 \text{ A}$$

the armature winding loss is

$$P_a = I_a^2 R_a = (80.684 \text{ A})^2 (0.1 \text{ }\Omega) = 651 \text{ W}$$

and the brush-contact loss is

$$P_{brush} = 2I_a = 2(80.684 \text{ A}) = 161.4 \text{ W}$$

(c) Since the rotational loss is given as 650 W, the total power loss is

$$\Sigma P_{loss} = P_f + P_a + P_{brush} + P_{rot}$$

$$= 579 + 651 + 161.4 + 650 = 2{,}041.4 \text{ W}$$

(d) The percent efficiency of the motor is

$$\% \text{ } \eta = \left(1 - \frac{\Sigma P_{loss}}{P_{in}}\right) \times 100$$

$$= \left(1 - \frac{2{,}041.4 \text{ W}}{20{,}750 \text{ W}}\right) 100 = 90.16\%$$

8.20 DC MOTOR CHARACTERISTICS

As previously stated, a dc machine can be used both as a generator and as a motor. In fact, in certain applications dc machines operate alternately as a motor and as a generator. However, in general there are some design differences, depending on whether the dc machine is intended for operation as a motor or as a generator. Unlike dc generators, dc motors are still very much in use in many industrial applications because of their attractive performance characteristics.

8.20.1 SPEED REGULATION

The dc motors excel in speed control applications, where they are compared by their speed regulations. The **speed regulation** (*Speed Reg*) of any motor is determined from

$$Speed \text{ } Reg = \frac{\omega_{nl} - \omega_{fl}}{\omega_{fl}} \tag{8.62}$$

Thus, the percent speed regulation is

$$\% \ Speed \ Reg \ = \left(\frac{\omega_{nl} - \omega_{fl}}{\omega_{fl}}\right) \times 100 \qquad (8.63)$$

where ω_{nl} = no-load angular speed in rad/s
ω_{fl} = full-load angular speed in rad/s

Alternatively, if the speeds are given in rpm then the percent speed regulation of a motor is found from

$$\% \ Speed \ Reg = \left(\frac{n_{nl} - n_{fl}}{n_{fl}}\right) \times 100 \qquad (8.64)$$

The magnitude of the speed regulation indicates the steepness of the slope of the torque-speed characteristic.

8.20.2 THE SPEED-CURRENT CHARACTERISTIC

In general, motors are designed to provide a rated horsepower at a rated speed. Since the internal generated voltage is a function of the angular velocity, then

$$E_a = K_a \Phi_d \omega_m \qquad (8.65)$$

from which the angular velocity ω_m can be expressed as

$$\omega_m = \frac{E_a}{K_a \Phi_d} \qquad (8.66)$$

By substituting equation (8.65) into equation (8.66) the angular velocity or shaft speed can be found as

$$\omega_m = \frac{V_t - I_a R_a}{K_a \Phi_d} \qquad (8.67)$$

This equation is called the **motor speed equation**. Notice that the speed of a dc motor depends on the applied terminal voltage V_t, the armature current I_a, the resistance R_a, and the field flux per pole Φ_d. The K_a is a design constant and cannot be changed to control the speed. Equation (8.67) can be expressed as

$$\omega_m = \frac{V_t}{K_a \Phi_d} - \frac{I_a R_a}{K_a \Phi_d} \qquad (8.68)$$

Since at no load the second term becomes zero,[20] the speed is

---

[20] In reality, however, the armature current I_a at no-load is not zero, but is about 5 percent of the full-load current due to rotational losses.

$$\omega_{nl} = \frac{V_t}{K_a \Phi_d} \tag{8.69}$$

For a given armature current the speed of a dc motor will decrease as the armature-circuit resistance increases. As can be observed in Equation (8.67), such a speed adjustment can be made by weakening the field flux by inserting resistance in the field circuit using a field rheostat. Such speed control is a smooth and efficient means of changing the motor speed from basic speed to maximum speed. However, if the field circuit is opened accidentally, the field flux will suddenly decrease to its relatively small residual value. If the armature circuit is not opened immediately, the motor speed will increase to dangerously high values and the motor will destroy itself in a few seconds either by the windings being forced from the slots or the commutator segments being thrown out by centrifugal force. Since the sudden decrease in the field flux reduces the counter voltage to a very small amount, the armature current of the motor will increase to a very high value. This will take place before the motor starts to rotate at a high speed. A properly sized circuit breaker inserted into the armature circuit can prevent such a disaster. Succinctly put, the **field circuit of a shunt motor must never be opened if the motor is running. Otherwise, the motor will 'run away' and will destroy itself in a few seconds!**

Similarly, as the load is removed from a series motor, its field flux will decrease. If all the mechanical load is removed from its shaft the field flux will decrease to almost zero, and the motor speed will increase to a dangerously high level until it destroys itself. **Thus, a series motor must never be run without a load!** Nor should it be connected to a mechanical load by a belt since it can break or slip. Figure 8.31 shows the speed-current characteristics of shunt, series, and cumulative compound motors.

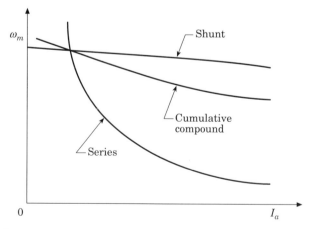

Figure 8.31   Speed-current characteristics.

8.20.3 THE SPEED-TORQUE CHARACTERISTIC

Figure 8.32 shows the speed-torque characteristics of shunt, series, and cumulative compound motors. Notice that the speed of a shunt motor changes very little (in fact, less than 5 percent in large motors or less than 8 percent in small motors). Because of this, shunt motors are classified as **constant-speed motors**. The speed of a series motor changes drastically as the load or the load torque changes. However, the cumulative-compound motor combines the operating characteristics of the shunt and series motors. Unlike a series motor, a compound motor has a definite no-load speed and can be safely operated at no load. As the load or load torque is increased, the growth in the field flux decreases the speed more in a series motor than in a shunt motor.

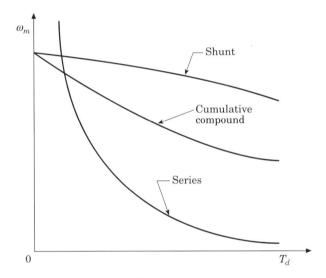

Figure 8.32   Speed-torque characteristics.

8.20.4 THE TORQUE-CURRENT CHARACTERISTIC

Figure 8.33 shows the torque-load current characteristics of shunt, series, and cumulative compound motors. Notice that except for lighter loads the series motor has a much greater torque than the shunt or the cumulative compound motor for a given armature current . Since a series motor has a much greater starting torque, it is exceptionally well suited for starting heavy loads at a reduced speed.

8.20.5 THE INTERNAL GENERATED VOLTAGE-CURRENT CHARACTERISTIC

Figure 8.34 shows the internal generated voltage-armature current characteristics of shunt, series, and cumulative compound motors. In a motor the

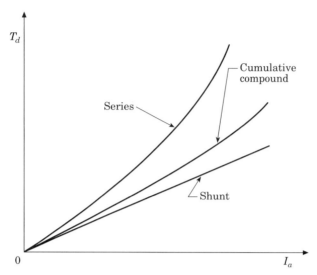

Figure 8.33   Torque-load current characteristics.

internal generated voltage is called the **countervoltage** (i.e., the **counter emf** or **back emf**). Notice that the generated voltage decreases as the load increases because of the increased voltage drop $I_a \Sigma R_a$ in the armature circuit. In series and compound motors, the armature-circuit resistance ΣR_a is the sum of the armature winding resistance R_a and the series-field winding resistance R_{se}. Thus,

$$\Sigma R_a = R_a + R_{se} \tag{8.70}$$

However, in the shunt motor it is equal to the armature winding resistance R_a. That is,

$$\Sigma R_a = R_a \tag{8.71}$$

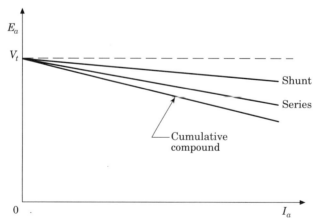

Figure 8.34   Internal generated voltage armature current characteristics.

Example 8.6

Assume that the separately excited dc machine given in Example 8.3 is being used as a shunt motor to drive a mechanical load. Its field current is kept constant at 2 A. As before the full-load current of the machine is 100 A. The terminal voltage of the motor is kept variable at 150 V, 200 V, and 250 V by using a *control rectifier*. Determine the following:
(a) The developed torque at full load of 100 A.
(b) The ideal no-load speed in rpm.
(c) The full-load speed in rpm.
(d) A sketch of the torque-current characteristic based on the results found in Part (a).
(e) A sketch of the speed-current characteristics based on the results found in Parts (b) and (c).

SOLUTION:
(a) Since the magnetization curve is drawn at the rated and constant speed of $n_{mc} = 1200$ rpm at $I_f = 2$ A, the corresponding voltage from the magnetization curve given in Table 8.1 is found as $E_{a(mc)} = 235$ V. However, at a different speed n_m the corresponding voltage would be E_a. Therefore, from equation (8.14),

$$\frac{E_a}{E_{a(mc)}} = \frac{\omega_m}{\omega_{mc}}$$

or

$$\frac{E_a}{\omega_m} = \frac{E_{a(mc)}}{\omega_{mc}}$$

The right side of this equation represents *constant excitation on the d-axis.* Since

$$E_a = K_a \Phi_d \omega_m$$

from which

$$K_a \Phi_d = \frac{E_a}{\omega_m}$$

Therefore, the developed torque is

$$T_d = K_a \Phi_d I_a$$

$$= \left(\frac{E_a}{\omega_m} \right) I_a$$

$$= \left(\frac{E_{a(mc)}}{\omega_{mc}}\right) I_a$$

Hence, at full load

$$T_d = \frac{(235 \text{ V})(100 \text{ A})}{(1200 \text{ rpm})\left(\dfrac{2\pi}{60}\right)} = 187 \text{ N} \cdot \text{m}$$

(b) Since

$$\frac{E_a}{E_{a(mc)}} = \frac{\omega_m}{\omega_{mc}}$$

or

$$\frac{E_a}{E_{a(mc)}} = \frac{n_m}{n_{mc}}$$

then

$$n_m = \left(\frac{E_a}{E_{a(mc)}}\right) n_{mc}$$

where $n_{mc} = 1200$ rpm and $E_{a(mc)} = 235$ V from the mc curve as long as $I_f = 2$ A is kept constant. Since at no load the terminal voltage V_t and the internal generated voltage E_a are the same at $E_a = V_t = 150$ V, the ideal no-load speed is

$$n_{nl} = \left(\frac{150 \text{ V}}{235 \text{ V}}\right)(1200 \text{ rpm}) = 766 \text{ rpm}$$

The other ideal no-load speeds that correspond to the terminal voltages of 200 V and 250 V are given in Table 8.3.

(c) Similarly, the speed at full load is

$$n_{fl} = \left(\frac{E_{a, fl}}{E_{a(mc)}}\right) n_{mc}$$

where the internal generated voltage is

$$E_{a, fl} = V_t - \Sigma I_a R_a - 2.0$$

If the applied terminal voltage V_t is 150 V, then the internal generated voltage is

$$E_{a,fl} = (150 \text{ V}) - (100 \text{ A})(0.10 \text{ }\Omega) - 2.0 = 138 \text{ V}$$

Therefore, the corresponding speed at full load is

$$n_{fl} = \left(\frac{138 \text{ V}}{235 \text{ V}}\right)(1200 \text{ rpm}) = 705 \text{ rpm}$$

The other full-load speeds that correspond to the terminal voltages of 200 V and 250 V are given in Table 8.3.

TABLE 8.3

V_t	At $I_a = 100$ A $\Sigma R_a = 0.10$ Ω $\Sigma I_a R_a$	At I_a $= 0$ A $E_{a, nl}$	At I_a $= 100$ A $E_{a, fl}$	At I_f $= 2.0$ A $E_{a(mc)}$	At I_a $= 0$ A n_{nl}	At I_a $= 100$ A n_{fl}
150 V	10 V	150 V	138 V	235 V	766 rpm	705 rpm
200	10	200	188	235	1021	960
250	10	250	238	235	1277	1215

(d) Figure 8.35 shows the torque-current characteristic based on the results found in Part (a). Note that if the excitation current is increased, the slope of the characteristic increases and the motor provides more torque. Similarly, if the excitation current is decreased, the slope of the characteristic decreases and the motor provides less torque.

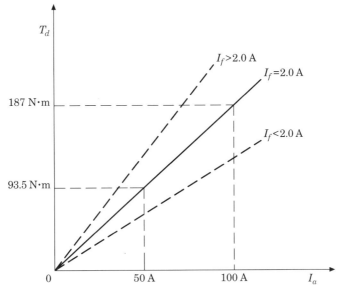

Figure 8.35   Torque-current characteristics.

(e) Figure 8.36 shows the speed-current characteristics based on the results found in Parts (b) and (c).

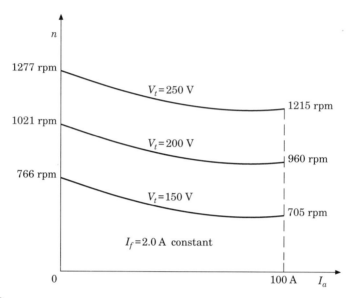

Figure 8.36

For the sake of comparison, Figure 8.37 shows the magnetization curves corresponding to the no-load speed of 1277 rpm, the full-load speed of 1215 rpm and also the mc curve drawn at 1200 rpm.

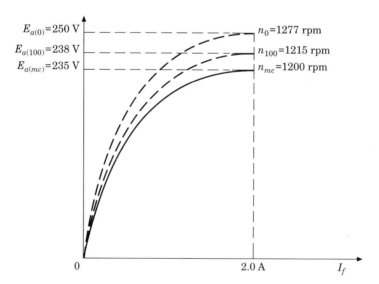

Figure 8.37

8.21 CONTROL OF DC MOTORS

The speed of a dc motor can be controlled with relative ease over a wide range above and below the base (rated) speed.[21] Speed control methods for dc motors are simpler and less expensive than those for ac motors. As can be observed from equation (8.67), the speed of a dc motor can be changed by using the following methods:

(a) Field control method.
(b) Armature resistance control method.
(c) Armature voltage control method.

The **field control method** is the simplest, cheapest, and is most applicable to shunt motors. In this method, the armature circuit resistance R_a and the terminal voltage V_t are kept constant and the speed is controlled by changing the field current I_f. As the value of the rheostat resistance in the shunt field circuit is increased, the flux Φ_d decreases and the speed increases. Thus, the internal generated voltage does not change considerably as the speed is increased. However, the torque of the motor decreases as the field flux decreases.[22] This speed control method is also called a **constant-horsepower drive** and it is well suited for drives requiring increased torque at low speeds. If the motor has a series field, speed control above the base speed can be obtained by inserting a diverter-resistance in parallel with the series winding to make the field current less than the armature current. If the shunt field is separately excited, a solid-state control can be used without a significant change in motor losses.

In the **armature resistance control** method, the armature terminal voltage V_t and the field current I_f (and therefore, the field) are maintained constant at their rated values. The speed of the motor is controlled by varying the resistance of the armature circuit by inserting an external resistance in series with the armature.[23] Even though it can also be applied to compound and series motors,[24] it is more easily applied to shunt motors. The armature resistance control method is simple to perform and requires a small initial investment, but has the disadvantage of considerable power loss and low overall efficiency with the full armature current passing through the external resistance connected in series. Today this speed control method is still used in various transit system vehicles. The same armature rheo-

---

[21] The **base speed** is the speed obtained with rated armature voltage, normal field flux, and normal armature resistance.

[22] The process of decreasing the field current is also known as the **field weakening**. By inserting external resistance in series with the motor field, the *speed of a motor can only be increased* from a minimum speed, that is, the **base speed**.

[23] By inserting external resistance in series with the armature, the *speed of a motor can only be decreased* from a maximum speed, that is, the **base speed**.

[24] In shunt and compound motors the external resistor must be connected between the shunt field winding and the armature, not between the line and the motor.

stat control can be used for both starting and speed control. The speed range begins at zero speed.

In the **armature voltage control method**, the armature circuit resistance R_a and the field current I_f are kept constant,[25] and the speed is controlled by varying the armature terminal voltage V_t. This is the most flexible method of speed control and avoids the disadvantages of poor speed regulation and low efficiency that are characteristic of the armature-resistance method. It can be applied to shunt, series, and compound motors. The speed is easily controlled from zero to a maximum safe speed in either forward or backward directions. The controlled-voltage source may be a dc machine or a solid-state controlled rectifier. If a dc machine is used, the speed control system is called a **Ward-Leonard system**.

Figure 8.38 shows a Ward-Leonard system for dc motor speed control. In such a system a three-phase induction motor or a three-phase synchronous motor drives a separately excited dc generator, with its armature connected directly to the armature of the separately excited dc motor that drives a mechanical load. Note that the motor of the motor-generator set operates at a constant speed. The armature voltage of the dc drive motor can be controlled by changing the field current of the dc generator. Thus, such control of the armature voltage allows for smooth control of the motor's speed from a very small value up to the base speed.[26] The motor speed for speeds above the base speed can be controlled by reducing the field current I_{fm} in the motor.

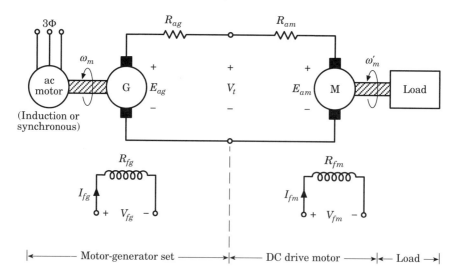

Figure 8.38   Ward-Leonard system for dc motor speed control.

---

[25] The field current is usually kept constant at its rated value.
[26] Here, the **base speed** is the rated speed of the motor. The rated speed is the speed that can be obtained when the motor is operating at its rated terminal voltage, power, and field current.

Figure 8.39 shows the torque and power limits as a function of speed for a shunt motor that has a combined armature voltage and field resistance control. Notice that the range below base speed is a constant-torque drive since the flux and permissible armature current are almost constant. The range above base speed represents a constant-horsepower drive. In the event that the field current I_{fg} of the generator is reversed, the polarity of the generator armature voltage is also reversed and so is the direction of rotation. A Ward-Leonard system is capable of providing a wide variation of speed in both forward and reverse directions. Increasing the field resistance R_{fg} of the generator decreases its field current I_{fg} and its internal generated voltage E_{ag}. Hence, the speed of the dc drive motor will decrease. The opposite will be true if the field resistance R_{fg} of the generator is decreased. Increasing the field resistance R_{fm} of the motor will decrease its field current I_{fm} and increase the speed of the motor. Decreasing the field resistance R_{fm} of the motor will result in a decrease in speed. Furthermore, the Ward-Leonard system has the ability to "regenerate," that is, to return stored energy in the machine to the supply lines. For example, when a heavy load is first lifted and then lowered by the dc drive motor of a Ward-Leonard system, the dc motor starts to act as a generator as the load is coming down, using its countertorque as a brake. Under these conditions, the "generator" itself starts to operate in the motor mode, driving the synchronous machine as a generator and supplying power back to the ac system. This process is known as **regeneration**.

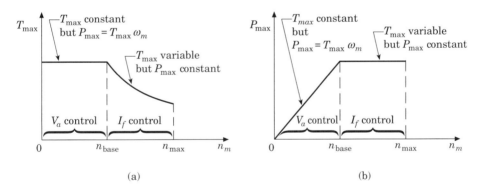

Figure 8.39   (a) Torque and (b) power limits as a function of speed for a shunt motor having combined armature voltage and field resistance control.

While the dc drive motor is operating, if the voltage V_t is suddenly decreased to a value below the counter emf of the motor, the armature current is reversed with the motor acting as a generator, driving the dc generator as a motor. This establishes **dynamic braking** which brings the motor to a quick stop.

In summary, the main advantage of the Ward-Leonard system is that the speed is adjustable over a wide range without large power losses which results in high efficiency at all speeds. This system is satisfactory for a maximum-to-minimum speed range of 40 to 1, but must be modified for greater speed ranges. The disadvantage is that a special motor-generator set is needed for each dc drive motor. Instead of one machine, three machines of basically equal ratios must be purchased and therefore a greater initial investment is required. If there are long periods when the motor is operating under light load the losses in the motor-generator set are high. It is relatively inefficient since several energy transformations are involved. In recent years the application of silicon-controlled rectifiers (SCR) has resulted in solid-state dc motor drives with precise speed control. However, they will not be presented here.

8.22 DC MOTOR STARTING

Only small dc motors of one horsepower or less can be connected directly to a line of rated voltage safely, but they must have very small moments of inertia. Any dc motor larger than 1 hp requires a starting device to protect the armature from excessive current during the starting operation. Typically, the armature resistance of a dc motor is about 0.05 per unit. If such a motor is connected directly to a line of rated voltage $V_{a,B}$, the armature current on starting is

$$I_{a, start} = \frac{V_a}{R_a} = \frac{V_{a,B}}{R_a} \tag{8.72}$$

since

$$R_{a, pu} = \frac{R_a}{R_{a,B}} \tag{8.73}$$

then

$$R_a = R_{a, pu} \times R_{a, B} \tag{8.74}$$

Therefore,

$$I_{a, start} = \frac{V_{a,B}}{R_{a, pu} \times R_{a, B}} \tag{8.75}$$

or

$$I_{a, start} = \frac{I_{a,B}}{R_{a, pu}} \tag{8.76}$$

where $I_{a,B}$ is the base or rated armature current and $R_{a, pu}$ is the armature resistance per unit. Since the typical $R_{a, pu}$ is about 0.05 per unit, then

$$I_{a, start} = \frac{I_{a, B}}{0.05} = 20I_{a, B} \tag{8.77}$$

In other words, **the armature current of a dc motor on starting is about 20 times its base** or **rated armature current!** This is due to the fact that the motor is at a standstill on starting and the counter emf is zero. Therefore, the armature starting is limited only by the resistance of the armature circuit.

All except very small dc motors are started with variable external resistance in series with their armatures to limit the starting current to the value (about 1.5 to 2 times rated value) that the motor can commutate without any damage. Such starting resistance is taken out of the circuit either manually or automatically as the motor comes to speed. To develop maximum starting torque, the shunt and compound motors are normally started with full field excitation (i.e., full line voltages are applied across the field circuits with their field rheostat resistances set at zero). Of course the series motor is always started under load. Figure 8.40(a) illustrates how to start a shunt motor with a dc motor starter. Figure 8.40(b) illustrates how to start a dc motor with starting resistors and accelerating contactors. Note that the starting resistor in each case consists of a series of pieces, each of which is cut out from the circuit[27] in succession as the speed of the motor increases. This limits the armature current of the motor to a safe value and for rapid acceleration it does not allow the current to decrease to a value that is too low, as shown in Figure 8.41. As the armature resistance is reduced at each step the motor accelerates, but requires less time to reach its asymptotic speed (zero acceleration) and reduces its current to approximately the rated load. The number of acceleration steps is almost a function of the horsepower capacity of the motor. Accordingly, larger motors with more inertia require more steps and a longer time interval to attain a given asymptotic speed.

Figure 8.42 shows a simplified diagram of an automatic starter[28] using "counter-emf" relays for a dc motor. It is based on the principle that as the motor speeds up, its countervoltage E_a starts to increase from an initial value of zero and causes a reduction in the armature current. The segments of the starting resistor are then cut out in steps as the counter emf E_a increases. When the E_a has increased to an adequately high value the starting resistor is short-circuited entirely, and the motor is then connected di-

---

[27] If the motor starting is achieved by using a manual dc motor starter, at start-up the handle is moved to position 1 at start up so that all the resistances R_1, R_2, R_3, R_4, and R_5 are in series with the armature to reduce the starting current. As the motor speed increases, the handle is moved to positions 2, 3, 4, 5 and finally the "run" position. At the "run" position all the resistances in the starter are cut out of the armature circuit. If the motor starting is achieved by using starting resistors and accelerating contactors, the individual segments of the resistor are cut out of the circuit by closing the 1A, 2A, and 3A contactors.

[28] Such starters are also called **controllers**. An electric controller is defined as a device, or group of devices, which governs in some predetermined fashion the electric power delivered to the apparatus to which it is connected. Therefore, an electric **starter** is defined as a controller whose main task is to start and accelerate a motor.

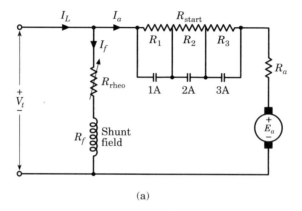

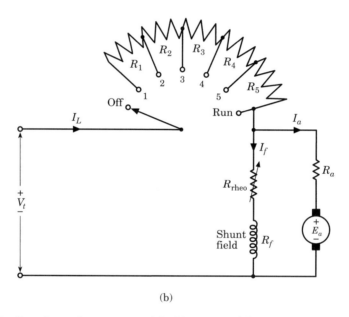

Figure 8.40   Starting a shunt motor with: (a) a manual dc motor starter; (b) starting resistors and accelerating contactors.

rectly to the line. Figure 8.43 shows some typical symbols used in automatic starter circuits. An automatic starter such as the one shown in Figure 8.42 is operated simply by pushing a button. Note that the field circuit has a relay labeled FL known as the **field loss-relay**. In the event that the field current is lost, this field loss relay is de-energized and cuts off the power to the M relay by deactivating the normally open FL contacts even after the start button is pushed. It also causes the normally open M contacts in the armature circuit to open as well. The three relays that are located across the armature are called the **accelerating relays** (AR). They are fast-acting relays with their contacts located in control lines 2, 3, and 4, respectively, as shown in Figure 8.43. The

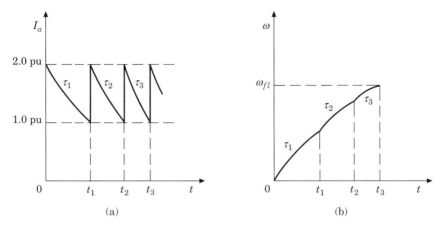

Figure 8.41 (a) Armature current versus time; (b) speed versus time during the start-ing of a dc motor.

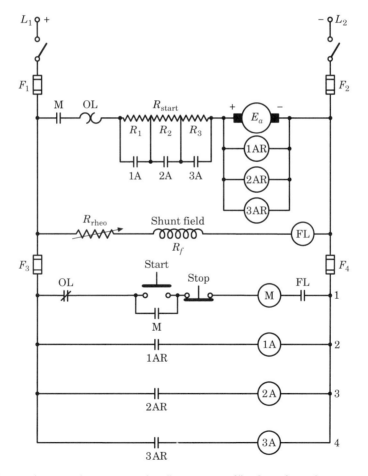

Figure 8.42 Automatic starter using "counter-emf" relays for a dc motor.

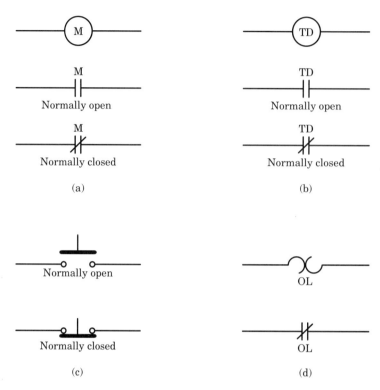

Figure 8.43   Typical symbols used in automatic starter diagrams: (a) a relay coil and its contacts; (b) a time delay relay coil and its contacts; (c) normally open and normally closed push-button switches; (d) a thermal overload device and its normally closed contacts.

control relays are labeled as 1A, 2A, and 3A and are located on control lines 2, 3, and 4. They short out the segments of the starting resistance as the counter emf of the motor increases. If the overload device OL that is located in the armature circuit heats up excessively due to excessive power demands on the motor, its normally closed contact located on the control line 1 will open and de-energize the main relay M. When the relay M is de-energized its contacts will open and disconnect the motor from the line.

Figure 8.44 shows a simplified diagram of an automatic starter for a dc motor using time delay relays. The operation is based on series time delay relays adjusted in a predetermined manner to close their individual contacts and short out each section of the starting resistor at the proper time intervals.

Example 8.7

Assume that a 100-hp 240-V dc motor has a full-load current of 343 A and an armature resistance of 0.05 per unit. Determine the following:

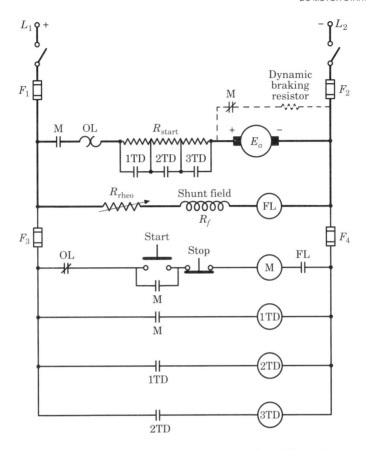

Figure 8.44   Automatic starter for a dc motor using time delay relays.

(a) The base value of the armature resistance in ohms.
(b) The value of the armature resistance in ohms.
(c) The value of the armature current on starting in amps.
(d) The value of the external resistance required if twice the full-load current
     is permitted to flow through the armature at the time of starting.

SOLUTION:
(a) The base value of the armature resistance is

$$R_{a,B} = \frac{V_{a,B}}{I_{a,B}}$$

$$= \frac{240 \text{ V}}{343 \text{ A}} = 0.6997 \ \Omega$$

(b) Since

$$R_{a,pu} = \frac{R_a}{R_{a,B}}$$

then the value of the armature resistance in ohms is

$$R_a = R_{a,pu} \times R_{a,B}$$

$$= (0.05 \text{ pu})(0.6997 \text{ } \Omega) = 0.035 \text{ } \Omega$$

(c) The value of the armature current on starting is

$$I_{a,start} = \frac{V_t - E_a}{R_a}$$

$$= \frac{240 \text{ V} - 0 \text{ V}}{0.035 \text{ } \Omega} \cong 6860 \text{ A}$$

or by using equation (8.75) directly,

$$I_{a,start} = 20 \text{ } I_{a,B}$$

$$= 20(343 \text{ A}) = 6860 \text{ A}$$

(d) The value of the external resistance required is

$$R_{ext} = \frac{V_t}{2I_{a,fl}} - R_a$$

$$= \frac{240 \text{ V}}{2(343 \text{ A})} - 0.035 = 0.3149 \text{ } \Omega$$

8.23 DC MOTOR BRAKING

By using automatic motor starters a number of additional control actions can be accomplished. These actions may include dynamic braking, reversing, jogging, plugging, and regenerative braking.[29]

When the stop button is pushed the power supply of the motor is cut off and the motor coasts to a stop. Since the only braking effect is mechanical friction it will take some time for the motor to come to rest. The time that it

---

[29] In addition, **eddy-current braking** can be used. An eddy-current brake is a disc of conducting material affected by the magnetic field of a coil. This disc rotates with the shaft of the motor. When the motor is turned off the coil is energized. As the shaft continues to rotate the eddy currents produced in the disc develop torque in the opposite direction of the rotation and stop the motor.

takes for the motor to stop completely is a function of the kinetic energy that is stored in the motor armature and the attached mechanical load.

As briefly discussed in Section 8.22, a motor can be stopped quickly by the use of the **dynamic braking** technique. In dynamic braking the shunt field of the motor is left connected to the supply after the armature is disconnected by the opening of the main (M) contactor. When the M contactor opens, a resistor called a **dynamic braking resistor** is connected across the armature terminals. With its shunt field energized, the dc machine behaves like a generator and produces a counter torque that quickly slows the armature by releasing the stored kinetic energy in the resistor as heat. During this braking operation a current flows in the armature winding in a direction opposite to that of the motor mode of operation. A shunt motor operation can be smoothly converted from a motor mode to a generator mode without changing the field winding connections. In a series machine either the series winding connections have to be reversed or it has to be connected to a separate voltage source to achieve good braking. In a compound machine the series winding is left disconnected during the braking operation. In dynamic braking the amount of braking effort is a function of the motor speed, the motor-field strength, and the value of the resistance. It is used extensively in the control of elevators and hoists and in other applications in which motors have to be started, stopped, and reversed frequently.

In some motor applications it may be necessary to quickly reverse the direction of rotation. This can be achieved by using dynamic braking to stop the motor quickly and then reversing the voltage applied to the armature. The operation of running a motor for only a fraction of a revolution or a few revolutions without going through the starting sequence is called **jogging**. It is often used for positioning applications.

Plugging is used when a motor has to be brought to a stop quickly or when a fast reversal of the direction of rotation is needed. It can be used in some motor applications where there is a sudden reversal in direction at full speed. This is done by reversing the armature connections by leaving the field winding connections undisturbed to maintain the magnetic field direction the same. Because the armature winding resistance is very small, the counter emf is almost equal and opposite to the applied voltage. However, the counter emf and the applied voltage are in the same direction at the time of plugging. Therefore, the total voltage in the armature circuit is almost twice that of the applied voltage. To protect the motor from the sudden increase in armature current, an external resistance (known as the **plugging resistance**) must be inserted in the armature circuit. The armature current reverses its direction and develops a force that tends to rotate the armature in a direction opposite to that of its initial rotation and brings the motor to a stop. The kinetic energy of the armature and mechanical load is hence being dissipated as heat in the plugging resistor.

In general, **regenerative braking** is used in motor applications where the motor speed is likely to increase from its normal speed. Such applications include electric railway locomotives, elevators, cranes, and hoists. The speed of motors driving such loads can be reduced significantly without mechanical braking by using regenerative braking to feed electrical energy back into the electrical system. Note that as the speed increases, so does the counter emf in the motor. When the counter emf becomes greater than the supply voltage, the current in the armature winding reverses its direction, causing the motor to operate as a generator. Regenerative braking can be used to maintain safe speeds but cannot be used to stop a mechanical load. For this action, dynamic braking, plugging, or mechanical braking[30] are required.

Example 8.8

Assume that a 240-V self-excited shunt motor is supplied by a line current of 102.4 A when it is loaded with a full load at a speed of 1000 rev/min. The armature-circuit resistance and the shunt-field circuit resistance of the motor are 0.1 Ω and 100 Ω, respectively. Assume that a braking resistor of 1.05 Ω is used for *dynamic braking* and determine the following:
(a) The value of the counter emf.
(b) The value of the armature winding current at the time of initial braking.
(c) The full-load torque.
(d) The value of the initial dynamic braking torque.

SOLUTION:
(a) The field current of the motor is
    Therefore, the armature winding current is

$$I_f = \frac{V_f}{R_f} = \frac{240 \text{ V}}{100 \ \Omega} = 2.4 \text{ A}$$

Hence, the value of the counter emf generated is

$$I_a = I_L - I_f = 102.4 \text{ A} - 2.4 \text{ A} = 100 \text{ A}$$

(b) At the time of the initial braking, since E_a, speed, and flux have not changed,

$$E_a = V_t - I_a R_a = 240 \text{ V} - (100 \text{ A})(0.1 \ \Omega) = 230 \text{ V}$$

then the value of the armature winding current is

$$I_{a, \, brake} = \frac{E_a}{R_a + R_{brake}} = \frac{230 \text{ V}}{0.1 \ \Omega + 1.05 \ \Omega} = 200 \text{ A}$$

Note that during the braking the value of the armature winding current has increased about twice.
(c) The full-load torque of the motor is

---

[30] This mechanical braking may be operated by a magnetic solenoid.

$$T_d = \frac{E_a I_a}{\omega_m} = \frac{E_a I_a}{2\pi n_m / 60}$$

$$= \frac{(230 \text{ V})(100 \text{ A})}{2\pi(1000 \text{ rev/min})/60} = 219.6 \text{ N} \cdot \text{m}$$

(d) Therefore, the value of the initial dynamic braking torque is

$$T_d = (219.6 \text{ N} \cdot \text{m})2 = 439.2 \text{ N} \cdot \text{m}$$

PROBLEMS

8.1   An eight-pole 500-V 500-kW dc generator has a lap winding with 640 armature conductors. If the generator has 6 commutating poles, determine the following:
(a)   The number of turns in the commutating winding if the mmf of the commutating poles is 1.4 times that of the armature.
(b)   The number of conductors for the compensating winding in each pole-face if the generator has a compensating winding and the pole face covers 80 percent of the pole-span.
(c)   The number of turns per pole in the commutating winding when the compensating winding is in the circuit.

8.2   Resolve Example 8.1 assuming that the winding of the armature is lap-wound.

8.3   Assume that the armature of a dc machine operating at 1,800 rpm is lap-wound with 720 conductors and that the machine has four poles. If the flux per pole is 0.05 Wb, determine the following:
(a)   The induced armature voltage.
(b)   The induced armature voltage if the armature is wave-wound.

8.4   Suppose that a separately excited shunt dc machine has a rated terminal voltage of 250 V and a rated armature current of 100 A. Its armature winding, commutating winding, and compensating winding resistances are 0.1, 0.02, and 0.009 Ω, respectively. Determine the following:
(a)   The induced armature voltage if the machine is operating as a generator at full load.
(b)   The induced armature voltage if the machine is operating as a motor at full load.

8.5   Suppose that a four-pole wave-wound dc machine is operating at 1,050 rpm at a terminal voltage of 250 V and that the resistance of the winding between terminals is 0.15 Ω. The armature winding has 100 coils of three turns each. If the cross-sectional area of each pole face is

150 cm^2 and the average flux density in the air gap under the pole faces is 0.75 T, determine the following:
(a) The total number of conductors in the armature winding.
(b) The flux per pole.
(c) The armature constant K_a.
(d) The induced armature voltage.
(e) Is the machine operating as a motor or a generator?
(f) The armature current.
(g) The developed power.
(h) The developed torque and its direction with respect to the direction of rotation.

8.6 A shunt motor operating at 1200 rev/min has an armature current of 38 A from a 240-V source when providing 8,398 W of mechanical power. If the armature winding resistance is 0.3 Ω, determine the following:
(a) The loss torque of the motor at the given speed.
(b) The required armature current to provide half the mechanical (shaft) power at the same speed.

8.7 Consider the shunt motor in Example 8.4 and determine the following:
(a) The developed torque if the load current is 125 A.
(b) The developed torque if a 25 percent increase in full-load current results in a 12 percent increase in the flux due to the demagnetizing effect of the armature reaction.

8.8 A series motor operating at a full-load speed of 1200 rev/min has a terminal voltage of 240 V and a rated load current of 74 A. The speed at which its magnetization curve has been developed is 1200 rev/min. From its magnetization curve its no-load voltages (i.e., internal generated voltages) are given as 230 V and 263.5 V at the series-field currents of 74 A and 100 A, respectively. The armature winding and the series field winding resistances of the motor are 0.085 Ω and 0.05 Ω, respectively. Ignore the effect of armature reaction and determine the torque, speed, and power output of the motor when the line current is 100 A. The developed torque is 125.66 N·m at the rated load current of 74 A.

8.9 Consider the shunt generator of Example 8.3. Assume that it is used as a separately excited cumulative compound generator by the addition of a series winding of 5 turns per pole. The resistance of the series winding is 0.03 Ω. The applied voltage at the terminals of the shunt field is 250 V and the resistance of the shunt field rheostat setting is 25 Ω. The generator is driven at a speed of 1200 rpm at no load and at 1150 rpm at a full load of 100 A. As before, its armature circuit resistance is 0.1 Ω and its shunt field winding has a 100 Ω resistance and is made up of 1,000 turns per pole. The total brush-contact voltage drop is 2 V and the demagnetization of the d-axis by the armature mmf is neglected. Use the magnetization curve data given in Table 8.1 and the magneti-

zation curve plotted in Figure 8.26. Determine:
(a) The excitation current, the terminal voltage, and the developed torque at no load.
(b) The excitation current, the terminal voltage, and the developed torque at a full load of 100 A.

8.10 Consider the speed-current characteristics developed in Example 8.6 and describe how and why they would be altered if the demagnetization of the d-axis (i.e., the armature reaction) by the armature mmf (due to the load current) were accounted for.

8.11 Assume that a 100-kW 250-V long-shunt compound generator is driven at its rated speed of 1800 rpm. Its armature winding resistance, the series winding resistance, and the interpole winding resistance are given as 0.018 Ω, 0.006 Ω, and 0.006 Ω, respectively. Its shunt-field current is 3 A. Its no-load rotational loss is 4,500 Ω. Assume that its brush contact voltage drop is 2 V and that its stray-load loss is 1 percent of the machine output. Determine the following:
(a) The total armature-circuit resistance excluding the brush-contact resistance.
(b) The armature current.
(c) The total losses.
(d) The efficiency at the rated load.

8.12 Assume that a 125-kW 250-V long-shunt compound generator is driven at its rated speed of 1,000 rpm. Its armature winding resistance, the series winding resistance, and the shunt winding resistance are given as 0.03 Ω, 0.01 Ω, and 35 Ω, respectively. Its stray-load loss at the rated voltage and speed is 1,250 W. Its rated field current is 4 A. If its rotational losses are 1,250 W, determine the following:
(a) The shunt-field copper loss.
(b) The series-field copper loss.
(c) The total losses.
(d) The percent efficiency of the machine.
(e) The maximum percent efficiency at its rated speed.

8.13 Consider the results of Example 8.5 and determine the following for the shunt motor:
(a) The armature terminal power.
(b) The developed power.
(c) The shaft power.

8.14 A 20-kW 250-V self-excited dc shunt generator is driven at its rated speed of 1,200 rpm. Its armature circuit resistance is 0.15 Ω and the field current is 2 A when the terminal voltage is 250 V at rated load. If its rotational loss is given as 1,000 W determine the following:
(a) The internal generated voltage.

(b) The developed torque.

(c) The percent efficiency of the generator.

8.15 Assume that the separately excited dc machine given in Example 8.3 is being used as a separately excited shunt motor to drive a mechanical load. Its terminal voltage is kept constant at 200 V. The full-load current of the machine is 100 A as before. Its field current is kept variable at 1.0 A, 1.5 A, and 2.5 A by using the field rheostat. For each value of the field current determine the following:

(a) The developed torque at a full load of 100 A.

(b) The developed power at a full load of 100 A.

(c) The ideal no-load speed in rpm.

(d) The full-load speed in rpm.

(e) Sketch the torque-current characteristic based on the results found in Part (a).

(f) Sketch the speed-current characteristic based on the results found in Parts (c) and (d).

8.16 Assume that the dc machine given in Example 8.3 is being used as a self-excited cumulative compound motor to drive a mechanical load. Its series winding has 5 turns per pole and a resistance of 0.03 Ω. Its shunt field current is kept constant at 2 A by using a 25 Ω shunt field rheostat setting. If the applied terminal voltage is 250 V determine the following:

(a) The short-shunt connection diagram of the motor.

(b) The ideal no-load speed in rpm.

(c) The full-load speed in rpm.

(d) The developed torque at a full load of 100 A.

(e) Sketch the speed-current characteristic of the machine and compare it with the one given in Example 8.3. Explain the difference in performance.

(f) Sketch the torque-current characteristic of the machine and compare it with the one given in Example 8.3. Explain the difference in performance.

8.17 Consider the solution of Problem 8.16 and account approximately for the demagnetization of the d-axis by the armature mmf on the q-axis. This demagnetization is due to the nonlinearity of the magnetization curve. Assume that the armature reaction constant K_d is 2.0 ampere turns per pole and that the machine in Problem 8.16 has no compensating winding. Determine the *modified value* of the number of turns of the series winding N_{se} (i.e., $N_{se} \neq 5$ turns) that will give approximately the same performance found in Problem 8.16 when the demagnetization due to the armature mmf was ignored.

8.18 Explain the effect that demagnetization due to armature mmf has on ideal (i.e., with no demagnetization) dc generator external characteristics.

8.19 Explain the effect of demagnetization due to armature mmf on the developed torque and the speed of the motor found in Problem 8.17.

8.20 Assume that the dc machine given in Example 8.3 is being used as a series motor to drive a mechanical load. Its series winding has 25 turns per pole and a resistance of 0.08 Ω. Its shunt field is totally disconnected. If the applied terminal voltage is 250 V, determine the following:
(a) The speed in rpm when the load current is 20 A.
(b) The developed torque when the load current is 20 A.

8.21 Consider the dc machine of Problem 8.20 and determine the following:
(a) The speed in rpm when the load current is 100 A.
(b) The developed torque when the load current is 100 A.

8.22 Consider the solutions of Problem 8.20 and Problem 8.21 and sketch the following characteristics:
(a) The torque-speed characteristic.
(b) The speed-current characteristic.
(c) The torque-current characteristic.

8.23 Assume that the dc machine given in Example 8.3 is being used as a self-excited shunt motor. The machine is being considered for an application that requires the motor to have a developed torque of 375 N·m at the start. The armature current at starting is desired to be as small as possible but not to be greater than 200 percent of the rated full-load armature current. The motor starter to be designed will have a connector to short circuit the field rheostat at starting. The supply line voltage is maintained at 250 V. Determine the following:
(a) The armature current at starting if there is no starting resistance connected.
(b) The value of the field current at starting.
(c) The value of $K_a\Phi_d$ of the motor at starting.
(d) The armature current at starting if the 375 N·m starting torque is to be developed.
(e) The value of the starting resistance.

8.24 Suppose that the dc machine given in Example 8.3 is being used as a self-excited shunt motor and that its speed control is achieved by inserting an external armature circuit resistance of 1.10 Ω into its circuit. Assume that the supply line voltage is maintained at 250 V and that the field current is 2.0 A. Also assume that the core loss and the friction and windage losses of the machine are one percent of the machine's rating. Determine the following:
(a) The developed torque at full load.
(b) The ideal no-load speed in rpm.
(c) The full-load speed in rpm.

    (d)   The full-load (overall) efficiency of the motor.
    (e)   The conversion efficiency of the motor.

8.25   Assume that the dc machine given in Example 8.3 is being used as a self-excited shunt motor and that its speed control is achieved by using the variable voltage method. The thyristor equipment used in such an application is made up of two solid state controlled rectifiers. Both rectifiers are continuously adjustable from 250 V terminal dc voltage; one of them has 3 A continuous rating, the other 100 A. The current and the resistance of the shunt field are 2 A and 100 Ω, respectively. Assume that the core loss and the friction and windage losses of the machine are one percent of the machine's rating. To achieve the maximum efficiency and minimum losses, the field rheostat resistance is set at zero. Determine the following:
    (a)   The voltages V_f and V_t to be set to duplicate the performance of Problem 8.24 at full load of 100 A armature current and also to achieve the maximum possible efficiency.
    (b)   The full-load (overall) efficiency of the motor.
    (c)   The conversion efficiency of the motor.

8.26   A 200-hp 240-V dc motor has a full-load current of 675 A and an armature resistance of 0.05 per unit. Determine the following:
    (a)   The base value of the armature resistance in ohms.
    (b)   The value of the armature resistance in ohms.
    (c)   The value of the armature current on starting in amps.

8.27   A 50-hp 240-V dc motor has a full-load current of 173 A and an armature resistance of 0.05 per unit. Determine the following:
    (a)   The base value of the armature resistance in ohms.
    (b)   The value of the armature resistance in ohms.
    (c)   The value of the armature current on starting in amps.
    (d)   The value of the external resistance needed if twice the full-load current is permitted to flow through the armature at the time of starting.

8.28   A 250-V self-excited dc shunt motor draws 102 A when it is loaded with a full load at a speed of 15 rev/s. Its armature circuit resistance is 0.15 Ω and the shunt field current is 2 A when the terminal voltage is 250 V at the rated load. An external plugging resistance is inserted into the armature circuit so that the armature current does not exceed 150 percent of its rated load value when the motor is plugged. Determine the following:
    (a)   The value of the plugging resistance.
    (b)   The braking torque at the instant of plugging.
    (c)   The braking torque when the motor reaches zero speed.

8.29   A 250-V self-excited shunt motor is supplied by a line current of 102.5 A when it is loaded with a full load at a speed of 1200 rev/min. The armature-circuit resistance and the shunt-field circuit resistance of the motor are 0.1 Ω and 100 Ω, respectively. If a braking resistor of 1.1 Ω is used for dynamic braking, determine the following:

(a)   The value of the counter emf.
(b)   The value of the armature winding current at the time of initial braking.
(c)   The full-load torque of the motor.
(d)   The value of the initial dynamic braking torque.

8.30   A 20-kW 250-V self-excited generator supplying the rated load has an armature circuit voltage drop of 4 percent of the terminal voltage and a shunt-field current equal to 4 percent of rated load current. Determine the resistance of the armature circuit and that of the field circuit.

CHAPTER 9

SINGLE-PHASE AND SPECIAL PURPOSE MOTORS

The conscious mind allows itself to be trained like a parrot, but the unconscious does not-which is why St. Augustine thanked God for not making him responsible for his dreams.
Carl G. Jung, *Psychology and Alchemy, 1952*

Mind moves matter.
Virgil

What is mind? No matter. What is matter? Never mind.
Thomas H. Key

9.1 INTRODUCTION

Today it can be said without exaggeration that about ninety percent of all motors manufactured are the single-phase type. They are used extensively in homes, businesses, farms, and small industries. Most of them are built as fractional-horsepower (hp) or subfractional-horsepower motors (one horsepower is equal to 746 watts). Standard ratings for **fractional-horsepower motors** range from 1/20 to 1 hp. The small motors rated for less than 1/20 hp, called **subfractional-horsepower motors**, are rated in millihorsepower (mhp) and range from 1 to 35 mhp. The single-phase motors manufactured in standard integral horsepower sizes are in the 1.5, 2, 3, 5, 7.5, to 10 hp range. However, special integral horsepower sizes can range from several hundred up to a few thousand e.g., in locomotive service using single-phase ac series motors. They can also be designed for very rugged use in cranes and hoists. Unlike **integral-horsepower motors**, small single-phase motors are manufactured in many different types of designs with different characteristics. This is especially true of subfractional-horsepower motors. There are three basic types of single-phase ac motors: single-phase induction motors, universal motors, and single-phase synchronous motors.

9.2 SINGLE-PHASE INDUCTION MOTORS

Single-phase induction motors generally have a distributed stator winding and a squirrel-cage rotor. Figure 9.1 shows a schematic diagram of a single-phase induction motor. The ac supply voltage is applied to the stator winding, which in turn creates a nonrotating (i.e., stationary in position and pulsating with time) magnetic field.[1] As shown in Figure 9.1(a), currents are induced in the squirrel-cage rotor windings by transformer action. These currents produce an mmf opposing the stator mmf. Since the axis of the rotor-mmf wave coincides with that of the stator field, the torque angle is zero and no starting torque develops. At standstill, therefore, the motor behaves like a single-phase nonrotating transformer with a short-circuited secondary. Hence, a single-phase induction motor is not self-starting. However, if the rotor of a single-phase induction motor is given a spin or started by auxiliary means, it will continue to run and develop power.[2] As shown in Figure 9.1(a), the single-phase induction motor can develop torque when it is running. This phenomenon can be explained by the **double-revolving field theory**.[3]

According to the double-revolving field theory, a pulsating mmf (or flux) field can be replaced by two rotating fields half the magnitude but rotating at the same speed in opposite directions. For a sinusoidally distributed stator winding, the mmf along a position θ can be expressed as

$$\mathscr{F}(\theta) = Ni\cos\theta \tag{9.1}$$

where N is the effective number of turns of the stator winding and i is the instantaneous value of the current in the stator winding. Thus,

$$i = I_{max}\cos\omega t \tag{9.2}$$

Hence, the mmf can be written as a function of space and time as

$$\mathscr{F}(\theta, t) = NI_{max}\cos\theta\cos\omega t \tag{9.3a}$$

$$= \frac{NI_{max}}{2}\cos(\omega t - \theta) + \frac{NI_{max}}{2}\cos(\omega t + \theta) \tag{9.3b}$$

---

[1] This type of mmf field is sometimes referred to as a **breathing field** because it expands and contracts in the same place on the stator. In other words, the stator winding does not provide a rotating mmf field for the rotor mmf to chase.

[2] Historically, the first single-phase induction motors were started by wrapping a rope or a strap around the shaft and pulling to spin the rotor. Fortunately, today the necessity for manual starting can be overcome by various relatively simple methods.

[3] It can also be explained by the **cross-field theory**. For a discussion of the cross-field theory, see references [14] and [100].

$$= \mathscr{F}_f + \mathscr{F}_b \tag{9.3c}$$

In other words, the stator mmf is the sum of a positive- and a negative-traveling mmfs in the direction of θ. $\mathscr{F}_f$ *is the rotating mmf in the direction of θ only it represents the* **forward rotating field**); $\mathscr{F}_b$ is the rotating mmf in the opposite direction and represents the **backward rotating field**. Here, it is assumed that the rotational direction of the forward rotating field is the same as the rotational direction of the rotor.[4]

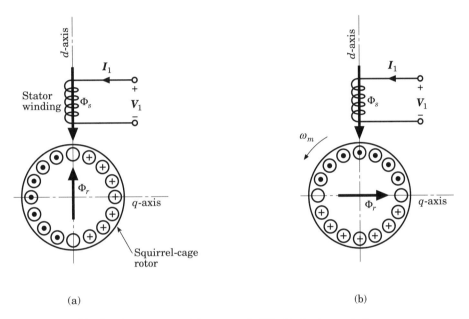

(a)   (b)

Figure 9.1   Single-phase motor: (a) at standstill; (b) during rotation.

As shown in Figure 9.2, the forward-rotating mmfs and backward-rotating mmfs both produce induction motor action; that is, they both produce a torque on the rotor, though in opposite directions. Notice that at standstill, the torques caused by the fields are equal in magnitude and the resultant starting torque is zero. At any other speed, however, the two torques are not equal and therefore the resultant torque causes the motor to rotate in the rotational direction of the motor.

Assume that the rotor is made to rotate at a speed of n_m rpm in the forward direction and that the synchronous speed is n_s rpm. The slip with respect to the forward rotating field is

---

[4] Since by definition the forward direction is that direction in which the motor is initially started, it is also referred to as the **positive sequence**. Similarly, the backward direction is also referred to as the **negative sequence**.

$$s_f = s = \frac{n_s - n_m}{n_s} = 1 - \frac{n_m}{n_s} \tag{9.4}$$

However, because the direction of rotation is opposite that of the backward-rotating field, the slip with respect to the backward field is

$$s_b = \frac{n_s - (-n_m)}{n_s} = \frac{n_s + n_m}{n_s} = 1 + \frac{n_m}{n_s} = 2 - s \tag{9.5}$$

or

$$s_b = 2 - s_f \tag{9.6}$$

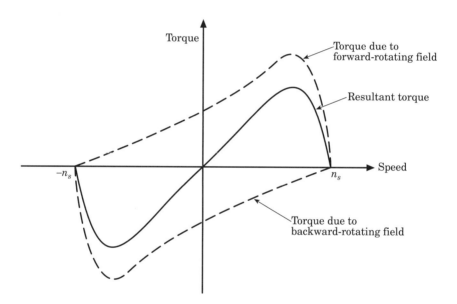

Figure 9.2   Torque-speed characteristics of a single-phase induction motor based on the revolving field theory.

9.2.1 EQUIVALENT CIRCUIT

At standstill, a single-phase induction motor behaves like a transformer with its secondary short-circuited. Figure 9.3(a) shows the corresponding equivalent circuit where R_1 and X_1 are the resistance and reactance of the stator winding, respectively. Here, X_m is the magnetizing reactance, and R_2' and X_2' are the standstill values of the rotor resistance and reactance referred to the stator winding by use of the appropriate turns ratio. The core losses of the motor are not shown, but are included in the rotational losses along with the mechanical and stray losses. Based on the double-revolving field theory, the equivalent circuit can be modified to include the effects of the two counterrotating fields of constant magnitude. At standstill, the magnitudes of the forward and backward resultant mmf fields are both equal to half the magnitude of the pulsating field. Therefore, the rotor-equivalent circuit can be split into equal sections. The equivalent circuit of a single-phase induction motor, then, consists of the series connection of a forward- and a backward-rotating field equivalent circuits, as shown in Figure 9.3(b).

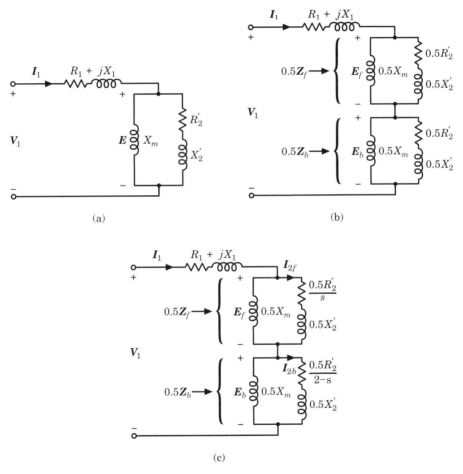

Figure 9.3 Equivalent circuit of a single-phase induction motor based on the revolving-field theory: (a) conventional configuration at standstill; (b) modified configuration at standstill; (c) typical torque-speed characteristic.

After the motor has been brought up to speed by the use of an auxiliary winding (which is switched out again after obtaining the proper speed) and is running in the direction of the forward-rotating field at a slip s, its equivalent circuit has to be modified, as shown in Figure 9.3(c). Therefore, the rotor resistance in the forward equivalent circuit is $0.5 R_2'/s$. Also, since the rotor is rotating at a speed that is s less than the forward-rotating field, the difference[5] in speed between the rotor and the backward-rotating field is $2 - s$. Hence, the rotor resistance in the equivalent backward circuit is represented by $0.5 R'_2/(2 - s)$.

To simplify the calculations the impedances shown in Figure 9.3(c) corresponding to the forward and backward fields are defined, respectively, as

$$Z_f = R_f + jX_f = \frac{jX_m(R'_2/s + jX'_2)}{jX_m + (R'_2/s + jX'_2)} \tag{9.7}$$

and

$$Z_b = R_b + jX_b = \frac{jX_m[R'_2/(2-s) + jX'_2]}{jX_m + [R'_2/(2-s) + jX'_2]} \tag{9.8}$$

These impedances which represent the reactions of the forward- and backward-rotating fields with respect to the single-phase stator winding are $0.5Z_f$ and $0.5Z_b$, respectively.

After the motor is started, the forward air-gap flux wave increases and the backward wave decreases. This is due to the fact that during normal operation the slip is very small. Because of this, the rotor resistance in the forward field, $0.5 R'_2/s$, is much greater than its standstill value, whereas the resistance in the backward field, $0.5 R'_2/(2 - s)$, is smaller. As a result of this, the forward field impedance Z_f is greater than its standstill value, while the backward field impedance Z_b is smaller. This also means that Z_f is much greater than Z_b, during the normal operation of the motor. Consequently, since each of these impedances carries the same current, the magnitude of the voltage E_f, is much greater than that of the voltage E_b. Therefore, the magnitude of the forward field Φ_f which produces E_f is much greater than that of the backward field Φ_b which produces E_b.

9.2.2 PERFORMANCE ANALYSIS

Based on the equivalent circuit shown in Figure 9.3(c), the input current can be found from

$$I_1 = \frac{V_1}{R_1 + jX_1 + 0.5Z_f + 0.5Z_b} \tag{9.9}$$

Therefore, the air-gap powers developed by the forward and backward fields respectively are

$$P_{g,f} = I_1^2(0.5R_f) \tag{9.10}$$

---

[5] The total difference in speed between the forward and backward-rotating fields is 2.

and

$$P_{g,b} = I_1^2(0.5R_b) \tag{9.11}$$

Hence, the total air-gap power is

$$P_g = P_{g,f} - P_{g,b} \tag{9.12}$$

Thus, the developed torques due to the forward and backward fields, respectively, are

$$T_{d,f} = \frac{P_{g,f}}{\omega_s} \tag{9.13}$$

and

$$T_{d,b} = \frac{P_{g,b}}{\omega_s} \tag{9.14}$$

The total developed torque is

$$T_d = \frac{P_g}{\omega_s} \tag{9.15a}$$

$$= \frac{P_{g,f} - P_{g,b}}{\omega_s} \tag{9.15b}$$

$$= T_{d,f} - T_{d,b} \tag{9.15c}$$

Since the rotor currents produced by the two component air-gap fields are different frequencies, the total rotor copper loss is the sum of the rotor copper losses caused by each field. These rotor copper losses of the forward and backward fields, respectively, are

$$P_{2,cu,f} = sP_{g,f} \tag{9.16}$$

and

$$P_{2,cu,b} = (2-s)P_{g,b} \tag{9.17}$$

Therefore, the total rotor copper loss is

$$P_{2,cu} = P_{2,cu,f} + P_{2,cu,b} \tag{9.18}$$

The mechanical power developed in the motor can be found from

$$P_d = P_{mech} = T_d \omega_m \tag{9.19a}$$

$$= T_d \omega_s (1 - s) \tag{9.19b}$$

$$= (1 - s)P_g \tag{9.19c}$$

$$= (1 - s)(P_{g,\, f} - P_{g,\, b}) \tag{9.19d}$$

$$= 0.5 I_1^2 (R_f - R_b)(1 - s) \tag{9.19e}$$

Hence, the output power is

$$P_{out} = P_d - P_{rot} \tag{9.20a}$$

$$= P_d - (P_{core} + P_{FW} + P_{stray}) \tag{9.20b}$$

Example 9.1

A 1/4-hp single-phase 120-V 60-Hz two-pole induction motor has the following constants in ohms referred to the stator:

$$R_1 = 2.0\,\Omega \qquad\qquad R_2' = 4.1\,\Omega$$
$$X_1 = 2.5\,\Omega \qquad\qquad X_2' = 2.2\,\Omega$$
$$X_m = 51\,\Omega$$

The core losses of the motor are 30 W; and the friction, windage, and stray losses are 15 W. The motor is operating at the rated voltage and frequency with its starting winding open. For a slip of 5 percent, determine:
(a) The shaft speed in rpm.
(b) The forward and backward impedances of the motor.
(c) The input current.
(d) The power factor.
(e) The input power.
(f) The total air-gap power.
(g) The developed power.
(h) The output power.
(i) The developed torque.
(j) The output torque.
(k) The efficiency of the motor.

SOLUTION:
(a) The synchronous speed is

$$n_s = \frac{120 f_1}{p} = \frac{120\,(60\text{ Hz})}{2} = 3600\text{ rev/min}$$

Thus, the rotor's mechanical shaft speed is

$$n_m = (1-s)n_s = (1-0.05)(3600 \text{ rev/min}) = 3420 \text{ rev/min}$$

(b) The forward impedance of the motor is

$$Z_f = R_f + jX_f = \frac{jX_m(R_2'/s + jX_2')}{jX_m + (R_2'/s + jX_2')}$$

$$= \frac{j51(4.1/0.05 + j2.2)}{j51 + 4.1/0.05 + j2.2}$$

$$= 42.8\angle 58.56° = 22.32 + j2.185 \ \Omega$$

Similarly, the backward impedance of the motor is

$$Z_b = R_b + jX_b = \frac{jX_m[R_2'/(2-s) + jX_2']}{jX_m + [R_2'/(2-s) + jX_2']}$$

$$= \frac{j51[4.1/(2-0.05) + j2.2]}{j51 + [4.1/(2-0.05) + j2.2]}$$

$$= 2.915\angle 48.56° = 1.929 + j2.185 \ \Omega$$

(c) The stator input current of the motor is

$$I_1 = \frac{V_1}{R_1 + jX_1 + 0.5Z_f + 0.5Z_b}$$

$$= \frac{120\angle 0°}{2.0 + j2.5 + 0.5(22.32 + j36.51) + 0.5(1.929 + j2.185)}$$

$$= \frac{120\angle 0°}{14.1245 + j21.8475} = 4.61\angle -57.12° \ A$$

(d) The stator power factor of the motor is

$$PF = \cos 57.12° = 0.543 \text{ lagging}$$

(e) The input power of the motor is

$$P_{in} = VI\cos\theta = (120)(4.61)\cos 57.12° = 300.4 \text{ W}$$

(f) The air-gap power due to the forward field is

$$P_{g,f} = I_1^2(0.5R_f) = (4.61)^2(0.5)(22.32) = 237.17 \text{ W}$$

and the air-gap power due to the backward-field is

$$P_{g,b} = I_1^2(0.5R_b) = (4.61)^2(0.5)(1.929) = 20.5 \text{ W}$$

Therefore, the total air-gap power of the motor is

$$P_g = P_{g,f} - P_{g,b} = 237.17 - 20.5 = 216.67 \text{ W}$$

(g) The developed mechanical power is

$$P_d = P_{mech} = (1-s)P_g = (1-0.05)216.67 = 205.84 \text{ W}$$

(h) The output (shaft) power is

$$P_{out} = P_d - (P_{core} + P_{FW} + P_{stray})$$

$$= 205.84 - (30 + 15) = 160.84 \text{ W}$$

(i) The developed torque is

$$T_d = \frac{P_g}{\omega_s}$$

$$= \frac{216.67 \text{ W}}{(3600 \text{ rev/min})(1 \text{ min/60 s})(2\pi \text{ rad/rev})} = 0.575 \text{ N} \cdot \text{m}$$

(j) The output torque is

$$T_{out} = T_{load} = \frac{P_{out}}{\omega_m}$$

$$= \frac{160.84 \text{ W}}{(3420 \text{ rev/min})(1 \text{ min/60 s})(2\pi \text{ rad/rev})} = 0.45 \text{ N} \cdot \text{m}$$

(k) The efficiency of the motor is

$$\eta = \frac{P_{out}}{P_{in}} \times 100 = \frac{160.84 \text{ W}}{300.4 \text{ W}} \times 100 = 53.5\,\%$$

9.3 STARTING OF SINGLE-PHASE INDUCTION MOTORS

As stated previously, a single-phase induction motor cannot be started by its main winding alone, but must be started by an auxiliary (starting) winding or some other means. The auxiliary winding may be disconnected automatically by the operation of a centrifugal switch at about 75 percent of synchronous speed. Once the motor is started, it continues to run in the same direction. A single-phase motor is designed so that the current in its auxiliary winding leads that of the main winding by 90 electrical degrees.[6] Accordingly, the field of its auxiliary winding builds up first. The direction of rotation of the motor can be reversed by reversing the connections of the main or the auxiliary winding. However, reversing the connections of both the main and auxiliary windings will not reverse the direction of rotation.

Consider the phasor diagram of a motor at starting shown in Figure 9.4(b). The phase angle α between the two currents I_m and I_a is about 30° to 45°. Therefore, the starting torque can be expressed as

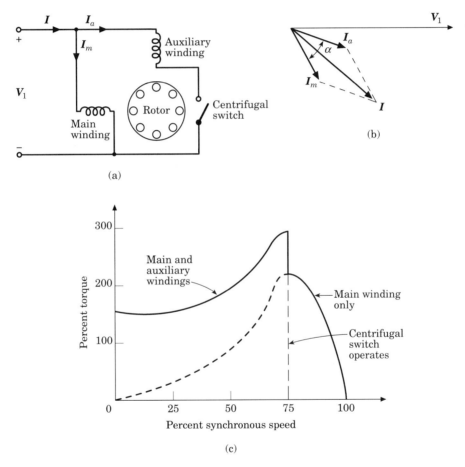

Figure 9.4   Split-phase induction motor: (a) schematic diagram; (b) phasor diagram at starting; (c) typical torque-speed characteristic.

---

[6] Therefore, the operation of a single-phase induction motor is very similar to that of two-phase motors. The two windings of a single-phase induction motor are placed in the stator with their axes displaced 90 electrical degrees in space.

$$T_{start} \propto I_m I_a \sin \alpha \qquad (9.21)$$

or

$$T_{start} = K I_m I_a \sin \alpha \qquad (9.22)$$

where K is a constant. Thus, the starting torque is a function of the magnitudes of the currents in the main and auxiliary windings and the phase difference between these two currents.

9.4 CLASSIFICATION OF SINGLE-PHASE INDUCTION MOTORS

Single-phase induction motors are categorized based on the methods used to start them. Each starting method differs in cost and in the amount of starting torque it produces.

9.4.1 SPLIT-PHASE MOTORS

A split-phase motor is a single-phase induction motor with two stator windings: a main (stator) winding, m, and an auxiliary (starting) winding, a, as shown in Figure 9.4(a). The axes of these two windings are displaced 90 electrical degrees in space and somewhat less than 90° in time.[7] As shown in Figure 9.4(b), the auxiliary winding has a higher resistance-to-reactance ratio than the main winding, so that its current *leads* the current in the main winding. The most common way to obtain this higher R/X ratio is to use smaller wire for the auxiliary winding. This is acceptable since the auxiliary winding is in the circuit only during the starting period. The auxiliary winding is disconnected by a centrifugal switch or relay when the speed of the motor reaches about 75 percent of the synchronous speed. The rotational direction of the motor can be reversed by switching the connections of the auxiliary winding while the connections of the main winding remain the same.[8] A typical torque-speed characteristic of the split-phase motor is shown in Figure 9.4(c). A higher starting torque can be obtained by inserting a series resistance in the auxiliary winding. Alternatively, a series inductive reactance can be inserted into the main winding to achieve the same

---

[7] Note that when two identical motor stator windings spaced 90 electrical degrees apart are connected in parallel to a single-phase source, the currents through the two windings lag the applied voltage by the same angle. Connecting a resistance in series with one winding causes the current in that winding to be more nearly in phase with the applied voltage. Since the current in the first winding is not affected by the added resistance, the currents in the two windings are displaced in time phase. This is the required condition to produce a revolving field. A motor using this method of phase splitting is called a **resistance-start motor**, a **resistance split-phase motor**, or simply a **split-phase motor**.

[8] However, such reversal (**plugging**) can never be done under running conditions even though it is sometimes done with polyphase induction motors.

result. Split-phase motors which are rated up to 1/2 hp are relatively less costly than other motors and are used to drive easily-started loads, such as fans, blowers, saws, pumps, and grinders.

When the motor is at standstill, the impedances of the main and the auxiliary windings, respectively, are

$$\mathbf{Z}_m = R_m + jX_m \tag{9.23}$$

$$\mathbf{Z}_a = R_a + jX_a \tag{9.24}$$

Thus, the magnitude of the auxiliary (starting) winding current can be determined from

$$I_a = \frac{V_1}{(R_a^2 + X_a^2)^{1/2}} \tag{9.25}$$

where

$$R_a = \frac{X_a}{X_m}(R_m + Z_m) \tag{9.26}$$

or

$$R_a = \left(\frac{N_a}{N_m}\right)^2 (R_m + Z_m) \tag{9.27}$$

and

$$X_a = \left(\frac{N_a}{N_m}\right)^2 X_m \tag{9.28}$$

Hence, for *design purposes,* it is easier to assume a number of turns for the auxiliary winding N_a to determine the value of R_a for maximum starting torque and the current of the auxiliary winding. If the optimum values of starting torque and current are not achieved, the process can be repeated until the proper design is found.

9.4.2 CAPACITOR-START MOTORS

A capacitor-start motor is also a split-phase motor. As shown in Figure 9.5(a), a capacitor is connected in series with the auxiliary winding of the motor. By selecting the proper capacitor size, the current in the auxiliary winding can be made to *lead* the voltage V_1 and to bring about a 90° time displacement between the phasors of currents I_m and I_a, as shown in Figure 9.5(b). This produces a much greater starting torque than resistance split-phase starting, as shown in Figure 9.5(c). The auxiliary winding is

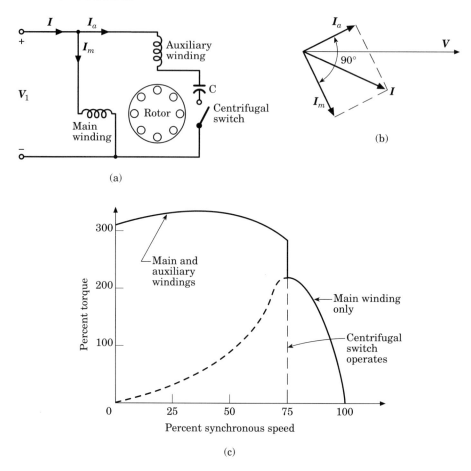

Figure 9.5   Capacitor-start induction motor: (a) schematic diagram; (b) phasor diagram at starting; (c) typical torque-speed characteristic.

disconnected by a centrifugal switch when the speed of the motor reaches about 75 percent of the synchronous speed. In contrast to the split-phase motor, the capacitor-start motor is a **reversible** motor. To reverse direction of the motor it is temporarily disconnected and its speed is allowed to drop to a slip of 20 percent (about four times the rated slip of 5 percent). At the same time, its centrifugal switch is closed over a reversely connected (with respect to the main winding) auxiliary winding. These two simultaneous actions reverse the rotational direction of the motor. The cost of the capacitor is an added cost and makes these motors more expensive[9] than split-phase motors. They are used in applications that require high-starting torques, such as compressors, pumps, air conditioners, conveyors, larger washing machines, and other hard-to-start loads.

---

[9] However, since the capacitor is in the circuit only during the relatively short starting period, it can be an inexpensive ac electrolytic type.

For design purposes, the value of the capacitive reactance which is connected in series with the auxiliary winding and provides the maximum starting torque [84] can be expressed as

$$X_c = X_a + \frac{R_a R_m}{X_m + Z_m} \tag{9.29}$$

The value of this capacitance can be found from

$$C = \frac{1}{\omega X_c} \tag{9.30}$$

or

$$C = \frac{1}{\omega \left(X_a + \dfrac{R_a R_m}{X_m + Z_m} \right)} \tag{9.31}$$

However, as suggested by Sen [84], the best design for the motor may be found by maximizing the starting torque per ampere of starting current rather than by maximizing the starting torque alone. The value of this capacitive reactance can be determined from

$$X_c = X_a + \{-X_m R_a + [R_a(R_a + R_m)]^{1/2} Z_m\}/R_m \tag{9.32}$$

The value of the capacitance is determined from

$$C = \frac{1}{\omega X_c} \tag{9.30}$$

9.4.3 CAPACITOR-RUN MOTORS

The **capacitor-split-capacitor motor** is also called the **permanent-split-capacitor motor** or simply the **capacitor motor**, because it is designed to operate with its auxiliary winding and its series capacitor permanently connected, as shown in Figure 9.6(a). It is simpler than the capacitor-start motor since there is no need for any centrifugal switch. Its torque,[10] efficiency, and power factor are also better since the motor runs effectively as a two-phase motor. In this motor the value of the capacitor is based on its optimum **running** rather than its starting characteristic. Since at starting the current in the capacitive branch is very low, the capacitor motor has a very low starting torque, as shown in Figure 9.6(b). The **reversible** operation is not only possible, but also more easily done than in other motors. Its speed can be controlled by vary-

---

[10] It produces a constant torque, not a pulsating torque as in other single-phase motors. Therefore, its operation is smooth and quiet.

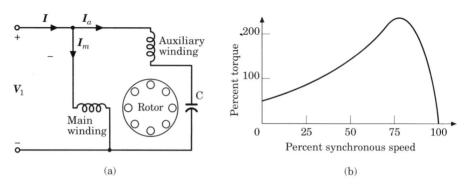

Figure 9.6 Capacitor-run induction motor: (a) schematic diagram; (b) typical torque-speed characteristic.

ing its stator voltage using various methods. Capacitor-run motors are used for fans, air conditioners, and refrigerators. Since at starting, slip s is unity and R_f is equal to R_b, the starting torque of a capacitor-run induction motor is determined from

$$T_{start} = \frac{2aI_aI_m(R_f + R_b)}{\omega_s}\sin(\theta_a + \theta_m) \tag{9.33}$$

or

$$T_{start} = KI_aI_m\sin\alpha \tag{9.22}$$

as before. Here, a is the turns ratio of the auxiliary and main windings, and θ_a and θ_m are the impedance angles of the auxiliary and main windings, respectively.

9.4.4 CAPACITOR-START CAPACITOR-RUN MOTORS

The **capacitor-start capacitor-run motor** is also called the **two-value capacitor motor**. In this motor, the high starting torque of the capacitor-start motor is combined with the good running performance of the capacitor-run motor, as shown in Figure 9.7(b). This is achieved by using two capacitors, as shown in Figure 9.7(a). Both the auxiliary winding capacitor and the capacitor C_{run} are usually the electrolytic type and are connected in parallel at starting. Since the running capacitor C_{run} must have a continuous rating, this motor is expensive but provides the best performance.

9.4.5 SHADED-POLE MOTORS

The shaded-pole induction motor is used widely in applications that require 1/20 hp or less. As shown in Figure 9.8(a), the motor has a salient-pole

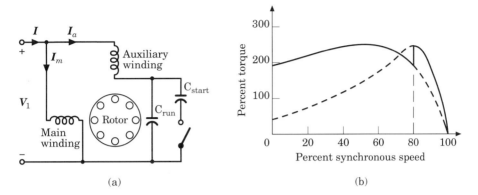

(a)   (b)

Figure 9.7   Capacitor-start capacitor-run induction motor: (a) schematic diagram; (b) typical torque-speed characteristic.

construction, with one-coil-per-pole main windings, and a squirrel-cage rotor. One portion of each pole has a **shading band** or **coil**. The shading band is simply a short-circuited copper strap (or single-turn solid copper ring) wound around the smaller segment of the pole piece. The purpose of the shading band is to retard, in time, the portion of the flux passing through it in relation to the flux coming out of the rest of the pole face. In other words, the current induced in the shading band causes the flux in the shaded portion of the pole to lag the flux in the unshaded portion of the pole. Therefore, the flux in the unshaded portion reaches its maximum before the flux in the shaded portion. The result is like a rotating field moving from the unshaded to the shaded portion of the pole, and causing the motor to produce a slow starting torque. The shaded-pole motor is rugged, cheap, small in size and needs minimum maintenance. It has very low starting torque, efficiency, and power factor and is used in turntables, motion-picture projectors, small fans, and vending machines.

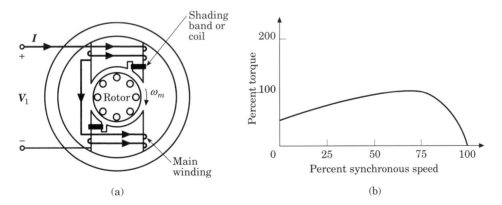

(a)   (b)

Figure 9.8   Shaded-pole induction motor: (a) schematic diagram; (b) typical torque-speed characteristic.

Example 9.2

Assume that a single-phase 120-V 60-Hz two-pole induction motor has the following *standstill* impedances when tested at rated frequency:

Main winding:	$\boldsymbol{Z}_m = 1.6 + j4.2$ W
Auxiliary winding:	$\boldsymbol{Z}_a = 3.2 + j6.5$ Ω

Determine the following:

(a) The value of external resistance that needs to be connected in series with the auxiliary winding to have maximum starting torque, if the motor is operated as a resistance split-phase motor.

(b) The value of the capacitor to be connected in series with the auxiliary winding to have maximum starting torque, if the motor is to be operated as a capacitor-start motor.

(c) The value of the capacitor that needs to be connected in series with the auxiliary winding to have maximum starting torque per ampere of the starting current as a capacitor-start motor.

SOLUTION:

(a) The value of the external resistance that needs to be connected in series with the auxiliary winding is found from equation (9.26) as

$$R_a = \frac{X_a}{X_m}(R_m + Z_m)$$

$$= \frac{6.5}{4.2}(1.6 + 4.495) = 12.5\ \Omega$$

where

$$\boldsymbol{Z}_m = Z_m\angle\theta_m = 4.495\angle 69.15°\ \Omega$$

Therefore, the value of external resistance is found as

$$R_{ext} = 12.5 - 6.5 = 6\ \Omega$$

(b) The value of the capacitive reactance that needs to be connected in series with the auxiliary winding is found from equation (9.29) as

$$X_c = X_a + \frac{R_a R_m}{X_m + Z_m}$$

$$= 6.5 + \frac{(3.2)(1.6)}{4.2 + 4.495} = 6.7712\ \Omega$$

Thus, the value of the capacitance is found as

$$C = \frac{1}{\omega X_c} = \frac{10^6}{2\pi 60(6.7712)} = 391.75 \ \mu F$$

(c) To have maximum starting torque per ampere of the starting current, the value of the capacitive reactance is found from equation (9.32) as

$$X_c = X_a + \{-X_m R_a + [R_a(R_a + R_m)]Z_m\}/R_m$$

$$= 6.5 + \{-4.2(3.2) + [3.2(3.2 + 1.6)]^{1/2} 4.495\}/1.6 = 9.11 \ \Omega$$

Hence, the value of the capacitance is found as

$$C = \frac{1}{\omega X_c} = \frac{10^6}{2\pi 60(9.11)} = 291.17 \ \mu F$$

9.5 UNIVERSAL MOTORS

A universal motor is a single-phase **series** motor that can operate on either alternating or direct current with similar characteristics as long as both the stator and the rotor cores are completely laminated. It is basically a series dc motor with laminated stator and rotor cores without laminated cores, the core losses would be tremendous if the motor were supplied by an ac power source. Since such a motor can run from either an ac (at any frequency up to design frequency) or a dc (zero frequency) power source, it is often called a **universal motor**. Here, the main field and armature field are in phase, because the same current flows through the field and armature.[11] When it is supplied by an ac power source both the main field and armature field will reverse at the same time, but the torque and the rotational direction will always be in the same direction. Like all series motors, the **no-load speed** of the universal motor is usually high, often in the range of 1500 to 20,000 rpm, and is limited by windage and friction. It is typically used in fractional horsepower ratings (1/20 hp or less) in many commercial appliances[12] such as electric shavers, portable tools, sewing machines, mixers, vacuum cleaners, small hand-held hair dryers, drills, routers, and hand-held grinders. In such applications, it is always directly loaded with little danger of motor runaway. The best way to control the speed and torque of the universal motor is to vary its input voltage by using a solid-state device (a SCR or TRIAC).

---

[11] Therefore, a shunt dc motor cannot operate on an ac power source due to the fact that the shunt field is highly inductive, and the armature is basically highly resistive. Hence, the armature and the field are not in phase. The high inductance of the field winding causes the field current to lag the armature current by such a large angle that a very low net torque is produced.

[12] Such applications require a motor with relatively high starting torques and speeds that exceed the maximum synchronous speed of 3600 rpm at 60 Hz. They are built for voltages from 1.5 V to 250 V. Therefore, universal motors are ideal for such applications.

There are also large (in the range of 500 hp) single-phase series ac motors that are still extensively used for traction applications such as electric locomotives.

Under dc excitation, the developed torque and induced voltage of a universal motor can be expressed, respectively, as

$$T_d = K_a \Phi_{d(dc)} I_a \tag{9.34}$$

$$E_a = K_a \Phi_{d(dc)} \omega_{m(dc)} \tag{9.35}$$

If magnetic linearity can be assumed, then the developed torque and induced voltage can be expressed, respectively, as

$$T_d = KI_a^2 \tag{9.36}$$

$$E_a = KI_a \omega_{m(dc)} \tag{9.37}$$

Under ac excitation, the average developed torque and the rms value of the induced voltage of a universal motor can be expressed, respectively, as

$$T_d = K_a \Phi_{d(ac)} I_a \tag{9.38}$$

$$E_a = K_a \Phi_{d(ac)} \omega_{m(ac)} \tag{9.39}$$

where $\Phi_{d(ac)}$ is the rms value of the d-axis flux and I_a is the rms value of the motor current. If magnetic linearity can be assumed, then the average developed torque and the rms value of the induced voltage can be expressed, respectively, as

$$T_d = KI_a^2 \tag{9.40}$$

$$E_a = KI_a \omega_{m(ac)} \tag{9.41}$$

Since the developed mechanical power is

$$P_d = P_{mech(ac)} = E_a I_a \tag{9.42}$$

then

$$T_d = \frac{P_d}{\omega_{m(ac)}} = \frac{E_a I_a}{\omega_{m(ac)}} \tag{9.43}$$

Notice that the terminal voltage (under ac excitation) is

$$V_1 = E_a + I_a Z_a + I_a Z_{se} \tag{9.44}$$

where

$$Z_a = R_a + jX_a \tag{9.45}$$

$$Z_{se} = R_{se} + jX_{se} \tag{9.46}$$

Therefore, the input voltage is

$$V_1 = E_a + I_a(R_a + jX_a) + I_a(R_{se} + jX_{se}) \tag{9.47}$$

or

$$V_1 = E_a + I_a(R_a + R_{se}) + jI_a(X_a + X_{se}) \tag{9.48}$$

Hence, the induced voltage can be expressed as

$$E_a = V_1 - I_a(R_a + R_{se}) - jI_a(X_a + X_{se}) \tag{9.49}$$

Based on the assumption that the armature current under dc excitation and the rms value of the armature current under ac excitation are the same, it can be shown that

$$\frac{E_{a(dc)}}{E_{a(ac)}} = \frac{K_a \Phi_{d(dc)} \omega_{m(dc)}}{K_a \Phi_{d(ac)} \omega_{m(ac)}} \cong \frac{\omega_{m(dc)}}{\omega_{m(ac)}} \tag{9.50}$$

Furthermore, if saturation takes place while the motor is under ac excitation, then the flux under $\Phi_{d(ac)}$ is a little less than the flux under dc excitation $\Phi_{d(dc)}$. Thus, the ratio of the induced voltage becomes

$$\frac{E_{a(dc)}}{E_{a(ac)}} \cong \frac{1}{\cos \theta} \rangle 1.0 \tag{9.51}$$

which is greater than unity. When the terminal voltage, armature current, and torque are constant, the speed of a universal motor is lower under ac excitation than under dc excitation. In summary, under ac excitation, the universal motor produces a lower speed, a poorer power factor, and a pulsating torque.

Example 9.3

A single-phase 120-V 60-Hz universal motor is operating at 1800 rpm and its armature current is 0.5 A when it is supplied by a 120-V dc source. Its resistance and reactance are 22 Ω and 100 Ω, respectively. If the motor is supplied by ac power, determine the following:
(a) The speed of the motor when it is connected to an ac source.
(b) The power factor of the motor when it is connected to an ac voltage source.
(c) The developed torque of the motor when it is connected to an ac voltage source.

SOLUTION:

(a) When the motor is supplied by the dc source:

$$E_{a(dc)} = V_1 - I_a R_a = (120 \text{ V}) - (0.5 \text{ A})(22 \text{ W}) = 109 \text{ V}$$

When the motor is supplied by the ac source:

$$E_{a(ac)} + I_a R_a = [V_1^2 - (I_a X)^2]^{1/2}$$

or

$$E_{a(ac)} = [V_1^2 - (I_a X)^2]^{1/2} - I_a R$$

$$= \{(120 \text{ V})^2 - [(0.5 \text{ A})(100 \text{ } \Omega)]^2\}^{1/2} - (0.5 \text{ A})(22 \text{ A}) = 98.09 \text{ V}$$

By assuming the same flux for the same current under the dc and ac operation from equation (9.50),

$$\frac{E_{a(dc)}}{E_{a(ac)}} = \frac{n_{dc}}{n_{ac}}$$

Thus, the speed of the motor when it is connected to an ac source is

$$n_{ac} = \frac{n_{dc} E_{a(ac)}}{E_{a(dc)}} = \frac{(1800 \text{ rpm})(98.09 \text{ V})}{109 \text{ V}} = 1,619.83 \text{ rpm}$$

(b) The power factor of the rotor is found as

$$\cos\theta = \frac{E_a + I_a R_a}{V_1} = \frac{(98.09 \text{ V}) + (0.5 \text{ A})(22 \text{ } \Omega)}{120 \text{ V}} = 0.91 \text{ lagging}$$

(c) The developed (mechanical) power of the motor is

$$P_d = P_{mech} = E_a I_a = (98.09 \text{ V})(0.5 \text{ A}) = 49 \text{ W}$$

Therefore, the developed torque of the motor is

$$T_d = \frac{P_d}{\omega_m} = \frac{P_d}{n_m\left(\frac{2\pi}{60}\right)} = \frac{49 \text{ W}}{(1619.83 \text{ rpm})\left(\frac{2\pi}{60}\right)} = 0.289 \text{ N} \cdot \text{m}$$

9.6 SINGLE-PHASE SYNCHRONOUS MOTORS

Single-phase synchronous motors are used for applications that require precise speed. They include the reluctance motor, the hysteresis motor, and the stepper motor. Reluctance and hysteresis motors are used in electrical clocks, timers, and turntables. Stepper motors are used in electrical typewriters, printers, computer disk drives, VCRs, and other electronic equipment.

9.6.1 RELUCTANCE MOTORS

A **reluctance motor**[13] is a salient-pole synchronous machine with no field excitation. The operation of this type of motor depends on reluctance torque that tends to align the rotor under the nearest pole of the stator and defines the direction of rotation. The torque applied to the rotor of the motor is proportional to sin 2δ, where δ is defined as the electrical angle between the rotor and stator magnetic fields. Hence, the reluctance torque of the motor becomes maximum when the angle between the rotor and stator magnetic fields is 45°. In general, any induction motor can be modified into a self-starting reluctance type synchronous motor. This can be done by altering the rotor so that the laminations have salient rotor poles, as shown in Figure 9.9(a). Notice that the saliency is introduced by removing some rotor teeth from the proper sections to make a four-pole rotor structure. This rotor structure can then be used for a four-pole reluctance motor. The reluctance of the air-gap flux path will be far greater at the places where there are no rotor teeth. Thus, the reluctance motor can start as an induction motor as long as the squirrel-cage bars and end rings are left in place. This motor, coming up to speed as an induction motor, will be pulled into synchronism by the pulsating ac single-phase field due to a reluctance torque produced by the salient iron poles with lower-reluctance air gaps.

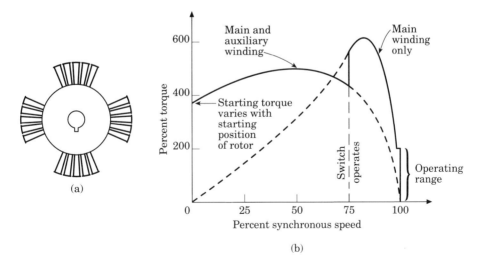

Figure 9.9   Reluctance motor: (a) rotor design; (b) typical torque-speed characteristic.

---

[13] It is also referred to as a **single-phase salient-pole synchronous-induction motor** or simply as a **synchronous motor**.

In summary, the torque develops because of the tendency of the rotor to align itself with the rotating field so that a reluctance motor starts as an induction motor, but continues to operate as a synchronous motor.

There are two stator windings, namely, a main winding and an auxiliary winding. When the motor starts as an induction motor it has both windings energized. At a speed of approximately 75 percent of the synchronous speed a centrifugal switch disconnects the auxiliary winding so that the speed of the motor increases to almost the synchronous speed. At that time, as a result of the reluctance torque, the rotor snaps into synchronism and continues to rotate at synchronous speed.[14] Figure 9.9(b) shows the torque-speed characteristic of a typical single-phase reluctance motor. Note that the value of the starting torque depends on the position of the unsymmetrical rotor with respect to the field winding. Also, since there is no dc excitation in the rotor of a reluctance motor, it develops less torque than an excited synchronous motor of the same size. Since the volume of a machine is approximately proportional to the torque, the reluctance motor is about three times larger than a synchronous motor with the same torque and speed .

9.6.2 HYSTERESIS MOTORS

These motors use the phenomenon of hysteresis to develop a mechanical torque. The rotor of a hysteresis motor is a smooth cylinder made up of a special magnetic material such as hard steel, chrome, or cobalt and has no teeth, laminations, or windings. The stator windings are made up of distributed windings in order to have a sinusoidal space distribution of flux. The stator windings can be either single- or three-phase. In single-phase motors, the stator windings are customarily permanent-split-capacitor type, as shown in Figure 9.6(a). If the stator windings are energized, a revolving magnetic field is developed, rotating at synchronous speed. This rotating field magnetizes the metal of the rotor and induces **eddy currents**. Due to the hysteresis, the magnetization of the rotor lags with respect to the inducing revolving field, as shown in Figure 9.10(a). The lag angle δ exists because the metal of the rotor has a large hysteresis loss. The angle by which the rotor magnetic field lags the stator magnetic field depends on the hysteresis loss of the rotor. At synchronous speed, the stator flux stops to sweep across the rotor, causing the eddy currents to disappear, and the rotor behaves like a permanent magnet. At that time, the developed torque in the motor is proportional to the angle δ between the rotor and stator magnetic fields, which is dictated by the hysteresis of the motor. Consequently, a constant torque [indicated as the hysteresis torque in Figure 9.10(b)] exists from zero up to and including synchronous speed. As indicated in the figure, a hysteresis motor, whose rotor is round and not laminated, has an induction torque which is added to the hysteresis torque until synchronous speed is reached. Hysteresis motors are self-starting and are manufactured up to

---

[14] If the load of this type of reluctance motor increases significantly, the motor will slip out of synchronism. However, it will continue to run with some slip just like an induction motor.

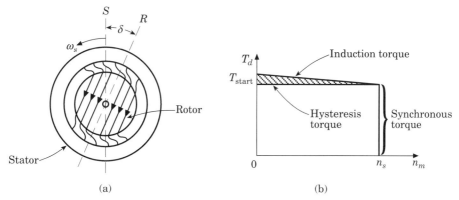

Figure 9.10   Hysteresis motor: (a) stator and rotor field; (b) torque-speed characteristic.

about 200 W for use in precise-speed drives. The applications include clocks,
record players, compact disc players, and servomechanisms.

9.6.3 STEPPER MOTORS

These motors are also referred to as **stepping** or **step motors**. Basically, a
stepper motor is a type of ac motor that is built to rotate a specific number of
degrees in response to a digital input in the form of a pulse. Step sizes typically
vary from 1°, 2°, 2.5°, 5°, 7.5°, 15°, or more for each electrical pulse. Stepper
motors are often used in digital control systems, where the motor is given open-
loop commands in the form of a train of pulses and the controller directs pulses
sequentially to the motor windings to turn a shaft or move an object a specified
distance. They are excellent devices for accurate speed control or precise position control without any feedback. In such usage, the axis of the motor's magnetic field *steps* around the air gap at a speed that is based on the frequency of
pulses. The rotor inclines to align itself with the axis of the magnetic field.
Therefore, the rotor *steps* in synchronism with the motion of the magnetic field.
Because of this, the motor is referred to as a **stepper motor**. These motors are
relatively simple in construction and can be controlled to step in equal incre-
ments in either direction. They are increasingly used in digital electronic systems because they do not need a position sensor or a feedback system to make
the output response follow the input command. Figure 9.11 illustrates a prim-
itive form of control implementation in a stepper motor. Notice that a train of f
pulses per second is furnished to the digital driver circuit and that the input of
the controller is divided so that the output is sent in sequence to one phase
winding at a time. In the event that $2p$ is the number of phases and k is the
number of teeth, then the rotor angular motion per pulse is a *step* of π/kp radians. Accordingly, the rotor moves n steps per second. Hence, the angular speed
is exactly $\pi n/kp$ rad/s.

Stepper motors are classified according to the type of motor used. If a permanent-magnet motor is used it is called a **permanent-magnet stepper motor**.

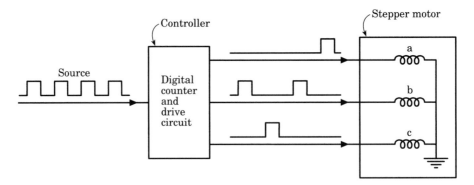

Figure 9.11   Driver for a stepper motor.

If a variable-reluctance motor is used, it is called a **variable-reluctance stepper motor**. Permanent-magnet stepper motors have a higher inertia and thus a slower acceleration than variable reluctance stepper motors. For example, the maximum step rate for permanent-magnet stepper motors is 300 pulses per second, but it can be as high as 1200 pulses per second for variable-reluctance stepper motors. The permanent-magnet stepper motor develops more torque per ampere stator current than the variable-reluctance stepper motor. There is also a **hybrid stepper motor** that has a rotor with an axial permanent magnet in the middle and ferromagnetic teeth at the outer sections. The hybrid stepper motor combines the characteristics of the variable-reluctance and permanent-magnet stepper motors. A variable-reluctance stepper motor can be the single-stack type or the multiple-stack type. The latter one is used to provide smaller step sizes. Its motor is segmented along its axial length into magnetically isolated sections which are called **stacks** which are excited by a separate winding called a **phase**. Even though variable-reluctance stepper motors with up to seven stacks and phases are used, three-phase arrangements are more often used.

Figure 9.12(b) shows a variable-reluctance stepper motor that has a rotor with eight poles and three separate eight-pole stators arranged along the rotor. If phase-a poles of a stator are energized by a set of series-connected coils with current i_a, the rotor poles align with the stator poles of phase-a. As can be observed in Figure 9.12(a), the phase-b stator is the same as the phase-a stator except that its poles are displaced by 15° in a counterclockwise direction. Similarly, the phase-c stator is displaced from the phase-b stator by 15° in the counterclockwise direction. When the flow of the current i_a in phase a is interrupted and phase-b is energized, the motor will develop a torque, rotating its rotor by 15° in the counterclockwise direction. Similarly, when the flow of the current i_b in phase b is interrupted and phase c is energized, the motor will rotate another 15° in the counterclockwise direction. Finally, when the flow of the current i_c in phase c is interrupted and phase a is energized, the motor will rotate another 15° in the counterclockwise direction, completing a one step (i.e., 45°) rotation in the counterclockwise direction. Therefore, additional current pulses in the abc sequence will develop additional counterclockwise stepping motions. Reversing

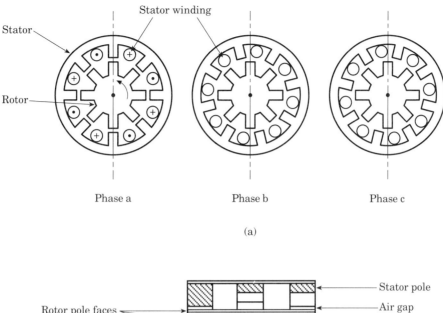

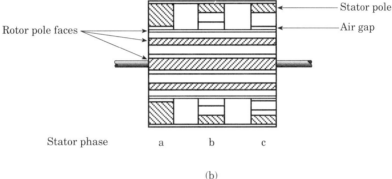

Figure 9.12   A variable-reluctance stepper motor having a rotor with eight poles and three separate eight-pole stators arranged along the rotor.

the current-pulse sequence to *abc* will develop reversed rotation. For an *n*-stack motor, the rotor or stator (but not both) on each stack is displaced by 1/*n* times the pole-pitch angle. Permanent-magnet stepper motors require two phases and current polarity is important. The hybrid stepper motor varies significantly from a multistack variable-reluctance stepper motor in that the stator pole structure is continuous along the length of the rotor.

9.7 SUBSYNCHRONOUS MOTORS

A subsynchronous motor has a rotor with an overall cylindrical outline and yet it is as toothed as a many-pole salient-pole rotor. For example, a typical motor may have 16 teeth or poles, and in combination with a 16-pole stator will normally rotate at a synchronous speed of 450 rpm when operated at 60 Hz. The motor starts as a hysteresis motor. At synchronous speed, the rotor poles induced in a hysteresis rotor stay at fixed spots on the rotor surface as the rotor rotates into synchronism with the rotating magnetic field of the stator. The hysteresis torque is in effect when the rotor rotates at less than synchronous speed. Subsynchronous motors, which are self-starters, start and accelerate with hysteresis torque just as the hysteresis synchronous motor does. There is no equivalent induction-motor torque like the one found in reluctance motors. This type of motor has a higher starting torque but less synchronous speed torque than reluctance torque. If such a motor operating at 450 rpm were temporarily overloaded, it would drop out of synchronism. As the speed drops down toward the maximum torque point, the motor will again lock into synchronism at a submultiple speed of 225 rpm. For this reason it is called a **subsynchronous motor**.

9.8 PERMANENT-MAGNET DC MOTORS

A permanent-magnet motor is a motor that has poles made up of permanent magnets. Even though most permanent-magnet machines are used as dc machines, they are occasionally built to operate as synchronous machines with the rotating field winding replaced by a permanent magnet. The permanent-magnet ac motor operation resembles that of the permanent-magnet stepper motor. Just as in the stepper motor, the frequency of the excitation dictates the motor speed, and the angular position between the rotor magnetic axis and a particular phase when it is energized affects the developed torque. Often, a permanent-magnet ac motor is called a **brushless motor** or **brushless dc motor**.

Permanent-magnet dc motors are widely used in automobiles to drive air conditioners, heater blowers, windshield wipers and washers, power seats and power windows, tape decks, and radio antennas. They are used in the home to operate electric shavers, electric toothbrushes, carving knives, vacuum cleaners, power tools, miniature motors in many toys, lawn mowers, and other equipment that uses batteries. They are used as starter motors for outdoor motors. In computers, they are used for capstan and tape drives. They can also be used in control systems such as dc servomotors and tape drives. In these applications, they are often used as fractional-horsepower motors for economic reasons. However, they can also be built in sizes greater than 200 hp.

Since there is no field winding in a permanent-magnet dc motor it has a smooth stator structure on which a cylindrical shell made up of a permanent magnet is mounted. Hence, the magnetic field is produced by the permanent magnet. The rotor of this permanent-magnet motor is a wound armature. The dc power supply is connected directly to the armature conductors through a brush/commutator assembly.

In these motors, there are basically three types of permanent magnets, namely, Alnico magnets, ceramic (or ferrite) magnets, and rare-earth magnets (samarium-cobalt magnets). Ceramic magnets are usually used for low-horsepower slow-speed motors. They are most economical in fractional horsepower motors and are also less expensive than Alnico in motors up to 10 hp. The rare-earth magnets are very expensive; however, they have proven to be the most cost effective in very small motors. In general, Alnico magnets are used in very large motors up to 200 hp. It is also possible to use special combinations of magnets and ferromagnetic materials to achieve high performance (i.e., high torque, high efficiency, low volume) at a low cost.

Figure 9.13 shows a cutaway view of a permanent-magnet dc motor. A permanent-magnet dc motor is basically a shunt dc motor with its field circuit replaced by permanent magnets. Since the flux of the permanent magnet cannot be changed, its speed can only be controlled by varying its armature voltage and armature circuit resistance. Therefore, the equivalent circuit of a permanent-magnet dc motor is made up of an armature connected in series with the armature circuit resistance R_a. Hence, the internal generated voltage can be determined from

$$E_a = K_a \Phi_d \omega_m \qquad (9.52)$$

where K_a is the armature constant, and Φ_d is the net flux per pole. In a permanent-magnet dc machine Φ_d is constant, thus

Figure 9.13   Cutaway view of a permanent-magnet dc motor. (Courtesy of General Electric.)

$$E_a = K\omega_m \tag{9.53}$$

where

$$K = K_a \Phi_d \tag{9.54}$$

and is called the **torque constant of the motor**. It is determined by the armature geometry and the properties of the permanent magnet used. The developed torque of the motor is found from

$$T_d = \frac{E_a I_a}{\omega_m} = K I_a \tag{9.55}$$

Figure 9.14(a) and Figure 9.14(b) show typical current-torque and speed-torque characteristics of a permanent-magnet dc motor respectively. Varying terminal voltage V_t of the motor changes the no-load speed of the motor, but the slope of the curves remains constant, as shown in Figure 9.15(a). However, varying the armature-circuit resistance R_a changes the speed-torque characteristic, but does not affect the no-load speed ω_o of the motor, as shown in Figure 9.15(b).

Example 9.4

Assume that the armature resistance of a permanent-magnet dc motor is 1.2 W. When it is operated from a dc source of 60 V, it has a no-load speed of 1950 rpm and is supplied by 1.5 A at no load. Determine the following:
(a) The torque constant.
(b) The no-load rotational losses.
(c) The output in horsepower if it is operating at 1500 rpm from a 50-V source.

SOLUTION:
(a) The internal generated voltage of the motor is

$$E_a = V_t - I_a R_a$$

$$= (60\ \text{V}) - (1.5\ \text{A})(1.2\ \Omega) = 58.2\ \text{V}$$

At speeds of 1950 rpm, its speed is

$$K = \frac{E_a}{\omega_m} = \frac{58.2\ \text{V}}{204.2\ \text{rad/s}} = 0.285\ \text{V/(rad/s)}$$

(b) Since all the power supplied at no load is used for rotational losses of the motor,

$$P_{rot} = E_a I_a = (58.2\ \text{V})(1.5\ \text{A}) = 87.3\ \text{W}$$

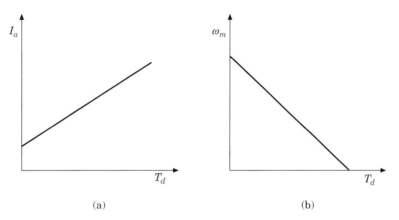

Figure 9.14   For a permanent-magnet dc motor: (a) typical current-torque characteristic; (b) typical speed-torque characteristic.

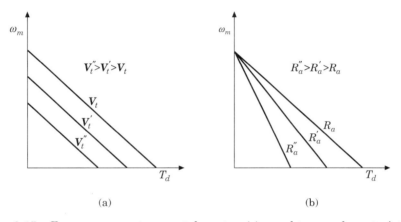

Figure 9.15   For a permanent-magnet dc motor: (a) speed-torque characteristics for
different supply voltages; (b) speed-torque characteristics for different
armature circuit resistances.

(c) At 1500 rpm,

$$\omega_m = 1500\left(\frac{2\pi}{60}\right) = 157.01 \text{ rad/s}$$

hence

$$E_a = K\omega_m = (0.285 \text{ V/(rad/s)})(157.01 \text{ rad/s}) = 44.75 \text{ V}$$

Therefore, the input power is

$$P_{shaft} = P_d = E_a I_a = (44.73 \text{ V})(4.38 \text{ A}) = 196 \text{ W}$$

Since the rotational losses are approximately constant, the output power of the motor is

$$P_{out} = P_{shaft} - P_{rot} = 196 - 87.3 = 108.7 \text{ W}$$

or in horsepower,

$$P_{out} = \frac{108.7 \text{ W}}{746 \text{ W/hp}} = 0.314 \text{ hp}$$

PROBLEMS

9.1 A 1/2-hp single-phase 120-V four-pole induction motor has the following constants in ohms referred to the stator:

$$R_1 = 1.9 \; \Omega \qquad\qquad R'_2 = 3.5 \; \Omega$$
$$X_1 = 2.5 \; \Omega \qquad\qquad X'_2 = 2.5 \; \Omega$$
$$X_m = 62 \; \Omega$$

The core losses of the motor are 35 W, and the friction, windage, and stray losses are 14 W. The motor is operating at the rated voltage and frequency with its starting winding open. For a slip of 4 percent, determine:

(a) The shaft speed in rpm.
(b) The forward and backward impedances.
(c) The input current.
(d) The power factor.
(e) The input power.
(f) The total air-gap power.
(g) The developed power.
(h) The output power.
(i) The developed torque.
(j) The output torque.
(k) The input power.
(l) The efficiency of the motor.

9.2 A 1/4-hp single-phase 120-V 60-Hz four-pole induction motor has the following constants in ohms referred to the stator:

$$R_1 = 2.5 \; \Omega \qquad\qquad R'_2 = 3.8 \; \Omega$$
$$X_1 = 2.2 \; \Omega \qquad\qquad X'_2 = 1.9 \; \Omega$$
$$X_m = 59 \; \Omega$$

The core losses of the motor are 30 W, the friction and windage losses are 10 W, and the stray losses are 4 W. The motor is operating at the rated voltage and frequency with its starting windings open. For a slip of 5 percent, determine:

(a) The forward and backward impedances.
(b) The input current.
(c) The input power.

(d) The total air-gap power.
(e) The developed torque.
(f) The developed power.
(g) The output power in watts and horsepower.
(h) The output torque.
(i) The efficiency of the motor.

9.3   Use the data given in Problem 9.1 and determine the following for the single-phase induction motor:
(a) The air-gap power due to the forward field.
(b) The air-gap power due to the backward field.
(c) The rotor copper loss due to the forward field.
(d) The rotor copper loss due to the backward field.
(e) The total rotor copper loss.

9.4   Use the data given in Problem 9.2 and determine:
(a) The air-gap power due to the forward field.
(b) The air-gap power due to the backward field
(c) The rotor copper loss due to the forward field.
(d) The rotor copper loss due to the backward field.
(e) The total rotor copper loss.

9.5   Determine the developed torque given in Problem 9.1, if it is operating at 4 percent slip and its terminal voltage is
(a) 208 V.
(b) 240 V.

9.6   Determine the developed torque in the motor given in Problem 9.2, if it is operating at 5 percent slip and its terminal voltage is
(a) 208 V.
(b) 240 V.

9.7   Assume that the currents in the main and the auxiliary windings of a single-phase induction motor are given respectively as
$$i_m = \sqrt{2}I_m \cos \omega t \qquad \text{and} \qquad i_a = \sqrt{2}I_a \cos (\omega t + \theta_a)$$
and that the windings are located in quadrature with respect to each other. If the effective number of turns for the main and auxiliary windings are N_m and N_a, determine:
(a) A mathematical expression for the rotating mmf wave of the stator.
(b) The magnitude and the phase angle of the auxiliary winding current needed to produce a balanced two-phase system.

9.8   Consider the solution of Example 9.2 and develop a table to compare the starting torques and starting currents in part (a), (b), and (c) expressed as per unit of the starting torque without any external element in the auxiliary circuit, when connected to a supply of 120 V at 60 Hz.

9.9   Assume that the impedances of the main and auxiliary windings of a single-phase 120-V 60-Hz, capacitor-start induction motor are

$$\boldsymbol{Z}_m = 4.2 + j3.4 \; \Omega \qquad \text{and} \qquad \boldsymbol{Z}_a = 9 + j3 \; \Omega$$

Determine the value of the starting capacitance that will cause the main and auxiliary winding currents to be in quadrature at starting.

9.10   Assume that the armature resistance of a permanent-magnet dc motor is 1.4 Ω. When it is operating from a dc source of 75 V, it has a no-load speed of 2,200 rpm and is supplied by 1.7 A at no load. Determine the following:
(a)   The torque constant.
(b)   The no-load rotational losses.
(c)   The output in horsepower if it is operating at 1800 rpm from a 70 V source.

9.11   Determine the best motor selection for the following applications and explain the reasoning: (a) electric drill, (b) electric clock, (c) refrigerator, (d) vacuum cleaner, (e) air conditioner fan, (f) air conditioner compressor, (g) electric sewing machine, (h) electric shaver, and (i) electric toothbrush.

9.12   Assume that a permanent-magnet dc motor is operating with a magnetic flux of 5 mWb, that its armature resistance is 0.7 Ω and the supply voltage is 30 V. If the motor load is 2 N•m and its armature constant is 110, determine the following:
(a)   The operating speed of the motor.
(b)   The developed torque under a blocked-rotor condition.

CHAPTER 10

TRANSIENTS AND DYNAMICS OF ELECTRIC MACHINES

Our understanding of a phenomenon becomes complete when, and only when, we can measure the quantities and factors involved and can assign numbers to them and can tie these numbers together with mathematical equations.

William Thomson (Lord Kelvin)

10.1 INTRODUCTION

Steady-state operation and behavior of ac and dc electromechanical machines have been reviewed in previous chapters. However, disturbances or sudden changes (e.g., faults, sudden load changes or shifts, or sudden changes in network configurations or in supply voltages) can cause these machines to behave quite differently. During such transient periods, it often becomes very important to have a knowledge of machine behavior. Here, the **transient period** is defined as the time period between the beginning of the disturbance and the following steady-state operating conditions. In this chapter a brief review of both the **electrical transient behavior** and the **mechanical transient behavior (dynamic response)** of dc and ac machines is presented. However, an in-depth study of this topic is outside the scope of this book.[1]

10.2 DC MACHINES

In this type of study it is customary to simplify the problem by making various assumptions. For example, it is assumed that the field mmf acts only

---

[1] For an excellent reference, see *Power System Control and Stability*, P. M. Anderson and A. A. Fouad, IEEE Press, New York, 1994.

along the d-axis and the armature mmf acts only along the q-axis. In other words, there is no mutual inductance between the field circuit and the armature circuit. Therefore, the armature reaction has no demagnetizing effect. However, the effects of the armature reaction may be added later as an additional field excitation requirement. It can also be assumed that magnetic saturation does not exist and that the air-gap flux is directly proportional to the field current I_f.

A given dc machine can be represented by two coupled electrical circuits both with resistances and inductances, as shown in Figure 10.1. These circuits, representing the field and the armature of the dc machine, are coupled through the electromagnetic field. Similarly, the electrical system is coupled to the mechanical system through the developed electromagnetic torque T_d and external mechanical torque, which can be either an input torque T_s from a prime mover or a load torque T_L.

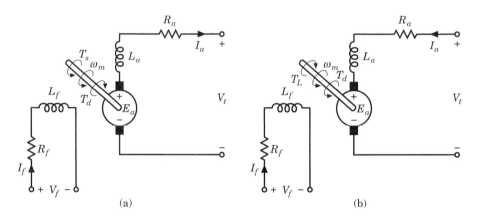

Figure 10.1 Schematic representation of a separately excited dc machine: (a) generator; (b) motor.

10.3 SEPARATELY EXCITED DC GENERATOR

A schematic representation of a separately excited dc generator is shown in Figure 10.1(a). Notice that the inductances of the armature winding and field winding are represented by L_a and L_f, respectively. The developed torque is given by

$$T_d = K_a \Phi_d I_a \tag{10.1}$$

and the internal generated voltage is given by

$$E_a = K_a \Phi_d \omega_m \tag{10.2}$$

Since **magnetic linearity** (i.e., the air-gap flux is directly proportional to the field current I_f) is assumed,

$$T_d = K_f I_f I_a \tag{10.3}$$

and

$$E_a = K_f I_f \omega_m \tag{10.4}$$

where K_f is a constant.

Assume that the generator is driven at a constant speed of ω_m by the prime mover. The voltage equation for the field circuit can then be expressed as

$$V_f = I_f R_f + L_f \frac{dI_f}{dt} \tag{10.5}$$

where I_f, R_f, and L_f are the current, resistance, and self-inductance of the field circuit, respectively. Therefore,

$$\frac{V_f}{R_f} = I_f + \tau_f \frac{dI_f}{dt} \tag{10.6}$$

where $\tau_f = L_f/R_f$ is the time constant of the field circuit. Assuming that the effect of saturation is negligible, the internal generated voltage is

$$E_a = K_f I_f \omega_m = K_g I_f \tag{10.7}$$

where $K_g = K_f \omega_m$ is the slope of the linear section of the magnetization curve. For the armature circuit, the voltage equation can be written in terms of the generator terminal voltage as

$$E_a - V_t = R_a I_a + L_a \frac{dI_a}{dt} \tag{10.8}$$

or

$$\frac{E_a - V_t}{R_a} = I_a \tau_a \frac{dI_a}{dt} \tag{10.9}$$

where $\tau_a = L_a/R_a$ is the time constant of the armature circuit. When the separately excited dc generator is providing an armature current I_a to an electrical load, the developed electromagnetic torque is

$$T_d = \frac{P_d}{\omega_m} = \frac{E_a I_a}{\omega_m} \tag{10.10}$$

The **dynamic equation** of the dc machine is a function of the mechanical torque applied to its shaft by the prime mover. Therefore,

$$T_{shaft} = T_d + J \frac{d\omega_m}{dt} \tag{10.11}$$

or

$$T_{shaft} - T_d = J\frac{d\omega_m}{dt} \qquad (10.12)$$

where J is the moment of inertia of the rotor and the prime mover.

The electrical **transient** behavior of a dc generator involves both the transient behavior of its field circuit and the transient behavior of its armature circuit. It is easier to study this behavior of the machine by first finding the appropriate transfer function and then applying the techniques of Laplace transform theory.

10.3.1 FIELD CIRCUIT TRANSIENT

Assume that the generator is being run by a prime mover at a constant speed of ω_m, with its armature circuit open and its field circuit having just been closed, as shown in Figure 10.1(a). The Laplace transform of the voltage equation (10.5) for the field circuit with zero initial conditions can be given as

$$V_f(s) = I_f(s)R_f + L_f s I_f(s) \qquad (10.13)$$

or

$$V_f(s) = I_f(s)(R_f + sL_f) \qquad (10.14)$$

where $I_f(s)$ is the Laplace transform of the time function I_f, and $V_f(s)$ is the Laplace transform of V_f. Thus, the transfer function relating the field current to the field voltage is given by

$$\frac{I_f(s)}{V_f(s)} = \frac{1}{R_f + sL_f} \qquad (10.15)$$

or

$$\frac{I_f(s)}{V_f(s)} = \frac{1}{R_f(1 + s\tau_f)} \qquad (10.16)$$

where $\tau_f = I_f/R_f$ as before. From equation (10.7), the Laplace transform of the internal generated voltage is

$$E_a(s) = K_g I_f(s) \qquad (10.17)$$

Therefore, the total transfer function relating the internal generated voltage to the field circuit voltage can be expressed as

$$\frac{E_a(s)}{V_f(s)} = \frac{E_a(s)}{I_f(s)}\frac{I_f(s)}{V_f(s)} = \frac{K_g}{R_f(1 + s\tau_f)} \qquad (10.18)$$

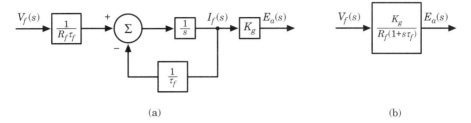

Figure 10.2   Block diagram of a separately excited dc generator with the armature
circuit closed.

The block diagram representation of this equation is shown in Figure 10.2.
The time domain response associated with this total transfer function can be
expressed as

$$e_a(t) = \frac{K_g V_f}{R_f}(1 - e^{-t/\tau_f})\qquad(10.19)$$

or

$$E_a(t) = E_a(1 - e^{-t/\tau_f})\qquad(10.20)$$

Therefore, in the steady-state the internal generated voltage becomes

$$E_a = \frac{K_g V_f}{R_f} = K_g I_f\qquad(10.21)$$

where the steady-state field current is

$$I_f = \frac{V_f}{I_f}\qquad(10.22)$$

10.3.2 ARMATURE CIRCUIT TRANSIENT

Assume that the armature circuit of the generator has just been closed over
an electrical load and that the armature speed and the field current are kept
constant. Therefore, the internal generated voltage of the generator can be expressed as

$$E_a = R_a I_a + L_a \frac{dI_a}{dt} + R_L I_a + L_L \frac{dI_a}{dt}\qquad(10.23)$$

where R_L and L_L are the resistance and the inductance of the electrical load,
respectively. Thus,

$$E_a = (R_a + R_L) + (L_a + L_L)\frac{dI_a}{dt} \qquad (10.24)$$

Its Laplace transform is

$$E_a(s) = I_a(s)(R_a + R_L)(1 + s\tau_{at}) \qquad (10.25)$$

where τ_{at} is the armature circuit time constant determined by

$$\tau_{at} = \frac{L_{at}}{R_{at}} = \frac{L_a + L_L}{R_a + R_L} \qquad (10.26)$$

Therefore, the transfer function is

$$\frac{I_a(s)}{E_a(s)} = \frac{1}{R_{at}(1 + s\tau_{at})} \qquad (10.27)$$

Hence, its time domain response is

$$i_a(t) = \frac{E_a}{R_{at}}(1 - e^{-t/\tau_{at}}) \qquad (10.28)$$

Similarly, the total transfer function relating the armature current to the field voltage can be expressed as

$$\frac{I_a(s)}{V_f(s)} = \frac{I_a(s)}{E_a(s)} \frac{E_a(s)}{V_f(s)} \frac{K_g}{R_f R_{at}(1 + s\tau_f)(1 + s\tau_{at})} \qquad (10.29)$$

Figure 10.3 shows the corresponding block diagram. Since the Laplace transform of the field voltage for a step change of voltage is

$$V_f(s) = \frac{V_f}{s} \qquad (10.30)$$

then

$$I_a(s) = \frac{K_g V_f}{R_f R_{at} s(1 + s\tau_f)(1 + s\tau_{at})} \qquad (10.31)$$

Example 10.1

Assume that a separately excited dc generator has an armature circuit resistance and inductance of 0.2 Ω and 0.2 H, respectively. Its field winding resistance and inductance are 120 Ω and 30 H, respectively. The generator constant K_g is 120 V per field ampere at rated speed. Assume that the generator is driv-

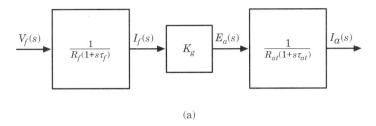

(a)

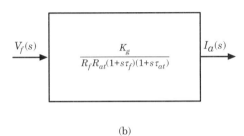

(b)

Figure 10.3   Block diagram of a separately excited dc generator with the armature circuit closed.

en by the prime mover at rated speed and that a 240-V dc supply is suddenly connected to the field winding. Determine the following:

(a) The internal generated voltage of the generator.

(b) The internal generated voltage in the steady state.

(c) The time required for the internal generated voltage to rise to 99 percent of its steady-state value.

SOLUTION:

(a) The time constant of the field circuit is

$$\tau_f = \frac{L_f}{R_f} = \frac{30}{120} = 0.25 \text{ s}$$

From equation (10.19),

$$E_a(t) = \frac{K_g V_f}{R_f}(1 - e^{-t/\tau_f})$$

$$= \frac{(120)(240)}{120}(1 - e^{-t/0.25}) = 240(1 - e^{-4t})$$

(b) The internal generated voltage in the steady state is

$$E_a(\infty) = 240 \text{ V}$$

(c) The time required for the internal generated voltage to rise to 99 percent of its steady-state value is found from

$$0.99(240) = 240(1 - e^{-4t})$$

as

$$t = 0.821 \text{ s}$$

Example 10.2

Assume that the generator given in Example 10.1 is driven by the prime mover at rated speed and is connected to a load made up of a resistance of 3 Ω and inductance of 1.4 H that are connected in series with respect to each other. If a 240 V dc supply is suddenly connected to the field winding, determine the armature current as a function of time.

SOLUTION:
From equation (10.29), the transfer function relating the armature current to the field voltage is

$$\frac{I_a(s)}{V_f(s)} = \frac{K_g}{R_f R_{at}(1 + s\tau_f)(1 + s\tau_{at})}$$

where

$$\tau_f = \frac{L_f}{R_f} = \frac{30}{120} = 0.25 \text{ s}$$

and

$$\tau_{at} = \frac{L_{at}}{R_{at}} = \frac{L_a + L_L}{R_a + R_L} = \frac{0.2 + 1.4}{0.2 + 3} = \frac{0.2 + 1.4}{0.2 + 3} = 0.5 \text{ s}$$

Therefore,

$$\frac{I_a(s)}{V_f(s)} = \frac{120}{(120)(3.2)(1 + 0.25s)(1 + 0.5s)}$$

The Laplace transform of the field voltage for a step change of 240 V is

$$V_f(s) = \frac{240}{s}$$

Thus

$$I_a(s) = \frac{240}{s} \frac{120}{384(1+0.25s)(1+0.5s)}$$

$$= \frac{750}{s(s+4)(s+2)}$$

$$= \frac{A_0}{s} + \frac{A_1}{s+4} + \frac{A_2}{s+2}$$

where

$$A_0 = \frac{750}{(s+4)(s+2)|_{s=0}} = 93.75$$

$$A_1 = \frac{750}{s(s+2)|_{s=-4}} = 93.75$$

$$A_2 = \frac{750}{s(s+4)|_{s=-2}} = -187.5$$

Hence

$$I_a(s) = \frac{93.75}{s} + \frac{93.75}{s+4} - \frac{187.5}{s+2}$$

Therefore, taking the inverse Laplace transform, the armature as a function of time can be found as

$$I_a(t) = 93.75 + 93.75e^{-4t} - 187.5e^{-2t}$$

10.4 SEPARATELY EXCITED DC MOTOR

Assume that a separately excited dc motor is operating at a constant field current I_f, as shown in Figure 10.1(b), and that its speed is controlled by changing its terminal voltage V_t. Ignoring the effects of any saturation, the developed torque T_t and the internal generated voltage E_a can be expressed as

$$T_d = K_f I_f I_a = K_m I_a \tag{10.32}$$

$$E_a = K_f I_f \omega_m = K_m \omega_m \tag{10.33}$$

where $K_m = K_f I_f$ is a constant. This motor constant K_m can also be found from the magnetization curve as

$$K_m = \frac{E_m}{\omega_m} \tag{10.34}$$

The Laplace transforms of equations (10.32) and (10.33) are

$$T_d(s) = K_m I_a(s) \tag{10.35}$$

$$E_a(s) = K_m \omega_m(s) \tag{10.36}$$

When the armature circuit is just energized at $t = 0$, the terminal voltage of the motor is

$$V_t = E_a + R_a I_a + L_a \frac{dI_a}{dt} \tag{10.37}$$

By substituting equation (10.33) into this equation,

$$V_t = K_m \omega_m + R_a I_a + L_a \frac{dI_a}{dt} \tag{10.38}$$

Its Laplace transform then is

$$V_t(s) = K_m \omega_m(s) + R_a I_a(s) + L_a s I_a(s) \tag{10.39}$$

or

$$V_t(s) = K_m \omega_m(s) + I_a(s) R_a (1 + s\tau_a) \tag{10.40}$$

where $\tau_a = L_a / R_a$ is the electrical time constant of the armature.

The developed torque of the motor must be equal to the sum of all opposing torques. Thus, the **dynamic equation** of the motor can be expressed as

$$T_d = K_m I_a = J \frac{d\omega_m}{dt} \tag{10.41}$$

where J is the moment of inertia including the load, B is the equivalent viscous friction constant of the motor including the load, and T_L is the mechanical load torque. Here, $B\omega_m$ is the rotational loss torque of the system. The Laplace transform of equation (10.41) is

$$T_d(s) = K_m I_a(s) = Js\omega_m(s) + B\omega_m(s) + T_L(s) \tag{10.42}$$

so that

$$\omega_m(s) = \frac{T_d(s) - T_L(s)}{B(1 + sJ/B)} \tag{10.43}$$

or

$$\omega_m(s) = \frac{K_m I_a(s) - T_L(s)}{B(1 + s\tau_m)} \quad (10.44)$$

where $\tau_m = J/B$ is the mechanical time constant of the system. Therefore, the Laplace transform of the armature current can be found from equations (10.36) and (10.40) as

$$I_a(s) = \frac{V_t(s) - E_a(s)}{R_a(1 + s\tau_a)} \quad (10.45)$$

or

$$I_a(s) = \frac{V_t(s) - K_m \omega_m(s)}{R_a(1 + s\tau_a)} \quad (10.46)$$

The corresponding block diagram is shown in Figure 10.4.

If mechanical damping B is ignored, the overall transfer function can be found as

$$\frac{\omega_m(s)}{V_t(s)/K_m} = \frac{1}{\tau_i s(1 + s\tau_a) + 1} \quad (10.47)$$

or

$$\frac{\omega_m(s)}{V_t(s)/K_m} = \frac{1/(\tau_a \tau_i)}{s(1/\tau_a + s) + 1/(\tau_a \tau_i)} \quad (10.48)$$

where $\tau_i = JR_a/K_m^2$ is the inertial time constant. The corresponding block diagram is shown in Figure 10.5. From the overall transfer function, the characteristic equation of the speed response to the voltage input is determined as

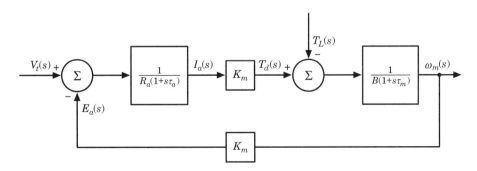

Figure 10.4   Block diagram of a separately excited dc motor with the armature circuit closed.

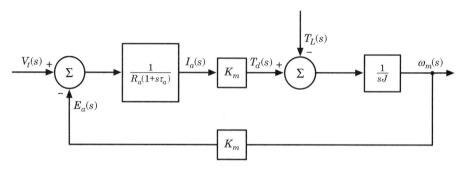

Figure 10.5   Block diagram of a separately excited dc motor with the mechanical damping neglected.

$$s\left(\frac{1}{\tau_a} + s\right) + \frac{1}{\tau_a \tau_i} = s^2 + s\left(\frac{1}{\tau_a} + s\right) + \frac{1}{\tau_a \tau_i} = 0 \tag{10.49}$$

The standard form of the characteristic equation of a second-order system is

$$s^2 + 2\alpha s + \omega_n^2 = 0 \tag{10.50}$$

By comparing equations (10.49) and (10.50), the undamped natural frequency ω_n can be expressed as

$$\omega_n = \left(\frac{1}{\tau_a \tau_i}\right)^{1/2} \tag{10.51}$$

and the damping factor α is

$$\alpha = \frac{1}{2\tau_a} \tag{10.52}$$

Therefore, the damping ratio is

$$\zeta = \frac{\alpha}{\omega_n} = \frac{1}{2}\left(\frac{\tau_i}{\tau_a}\right)^{1/2} \tag{10.53}$$

Note that to study the response of the motor to load changes, the block diagram shown in Figure 10.5 can be modified to the one shown in Figure 10.6.

The overall transfer function relating the speed response to a change in load torque with $V_t = 0$ can be determined from Figure 10.5 by eliminating the feedback path as

$$\frac{\omega_m(s)}{T_L(s)} = -\frac{1 + s\tau_a}{Js(1 + s\tau_a) + (K_m^2/R_a)} \tag{10.54}$$

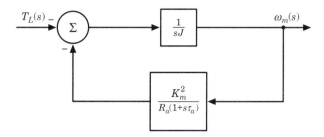

Figure 10.6   Simplified block diagram of a dc motor.

or

$$\frac{\omega_m(s)}{T_L(s)/J} = -\frac{1/\tau_a + s}{s(1/\tau_a + s) + 1/(\tau_a \tau_i)} \tag{10.55}$$

Here, the negative sign indicates that the additional load torque causes a reduction in speed. Also notice that its characteristic equation has the same undamped natural frequency and damping factor given in equations (10.51) and (10.52).

For a given mechanical load torque T_L, the required armature current can be found from the dynamic equation of the motor. Therefore, dividing equation (10.41) by K_m and substituting $\omega_m = E_a/K_m$, the armature current is found as

$$I_a = \frac{J}{K_m^2}\frac{dE_a}{dt} + \frac{BE_a}{K_m^2} + \frac{T_L}{K_m} \tag{10.56}$$

Example 10.3

Assume that a 250-V separately excited dc motor has an armature circuit resistance and inductance of 0.10 Ω and 14 mH, respectively. Its moment of inertia J is 20 kg/m^2 and its motor constant K_m is 2.5 N·m/A. The motor is supplied by a 250-V constant-voltage source. Assume that the motor is initially operating at steady-state without a load. Its no-load armature current is 5 A. Ignore the effects of saturation and armature reaction. If a constant load torque T_L of 1000 N·m is suddenly connected to the shaft of the motor, determine the following:
(a) The undamped natural frequency of the speed response.
(b) The damping factor and damping ratio.
(c) The initial speed in rpm.
(d) The initial acceleration.
(e) The ultimate speed drop.

SOLUTION:
(a) The armature time constant of the motor is

$$\tau_a = \frac{L_a}{R_a} = \frac{0.014}{0.1} = 0.14 \text{ s}$$

and its inertial time constant is

$$\tau_i = \frac{JR_a}{K_m^2} = \frac{(20)(0.1)}{2.5^2} = 0.32 \text{ s}$$

Thus, the undamped natural frequency of the motor is

$$\omega_n = \left(\frac{1}{\tau_a \tau_i}\right)^{1/2} = \left(\frac{1}{(0.14)(0.32)}\right)^{1/2} = 4.72 \text{ rad/s}$$

(b) The damping factor of the system is

$$\alpha = \frac{1}{2\tau_a} = \frac{1}{(2)(0.14)} = 0.5$$

and its damping ratio is

$$\zeta = \frac{\alpha}{\omega_n} = \frac{0.5}{4.72} = 0.11$$

(c) When the mechanical load is suddenly connected at $t = 0$, the internal generated voltage is

$$E_a = V_t - I_a R_a = 250 - (5)(0.1) = 249.5 \text{ V}$$

Hence, the corresponding initial speed of the motor is

$$\omega_m = \frac{E_a}{K_m} = \frac{249.5}{2.5} = 99.8 \text{ rad/s}$$

or

$$\omega_m = (99.8)\left(\frac{60}{2\pi}\right) = 953 \text{ rpm}$$

(d) Assuming that the losses are small enough to ignore, the initial acceleration of the motor can be found from equation (10.41) as

$$\alpha = \frac{d\omega_m}{dt} = \frac{K_m I_a - T_L}{J}$$

$$= \frac{(2.5)(5) - 1000}{20} = -49.38 \text{ rad/s}^2$$

(e) By applying the final-value theorem of Laplace transforms to equation (10.55), the ultimate drop in speed of the motor can be found as

$$\Delta \omega_m = \lim_{s \to \infty} \left\{ \frac{s\left[-(1/J)(1/\tau_a + s)\right]}{s(1/\tau_a + s) + 1/(\tau_a \tau_i)} \frac{\Delta T_L}{s} \right\}$$

$$= -\frac{\tau_i \Delta T_L}{J} = -\frac{(0.32)(1000)}{20} = -16 \text{ rad/s}$$

$$= -16\left(\frac{60}{2\pi}\right) = -152.8 \text{ rpm}$$

10.5 SYNCHRONOUS GENERATOR TRANSIENTS

Whenever there is a sudden change in the shaft torque applied to a synchronous generator or in its output load, there is always a transient that lasts for a very short time before the generator resumes its steady-state operation. Such operation of a generator is defined as **transient operation** and can be electrical or mechanical in nature. For example, a sudden three-phase short-circuit at the stator terminals is an **electrical transient** and a sudden load change may result in a **mechanical transient**.

10.6 SHORT-CIRCUIT TRANSIENTS

The most severe transient in a synchronous generator takes place when its three stator terminals are suddenly shorted out while the generator is operating at synchronous speed with constant excitation under no-load conditions. Such a short on a power system is called a **fault**.

Figure 10.7 represents a typical short-circuit oscillogram[2] that shows the three-phase armature current waves as well as the field current. Notice that the traces of the armature-phase currents are not symmetrical at the zero-current axis, and clearly exhibit the dc components responsible for the offset waves. In other words, each phase current can be represented by a dc transient component of current added on top of a symmetrical ac component. The symmetrical ac component of current by itself is shown in Figure 10.8. As shown in Figure 10.7, the dc component of the armature current is different in each phase and depends on the point of the voltage wave at which the fault takes

---

[2] These oscillograms of short-circuit currents can be used to determine the values of some of the reactances and time constants.

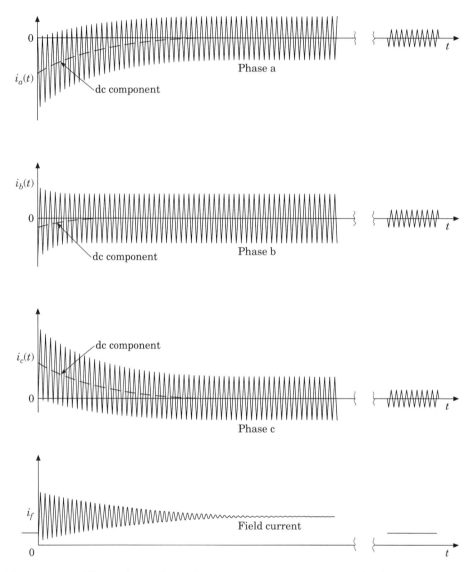

Figure 10.7   Three-phase short-circuit armature currents and field current waves during a three-phase fault at the terminals of a synchronous generator as a function of time.

place. The initial value of the dc component[3] of the fault current can be as large as the magnitude of the steady-state current. The decay rate of this dc component is found from the resistance and equivalent inductance seen at the stator terminals. The dc component plays a role here because the synchronous

---

[3] Also note that the presence of the dc component in the stator phase coil induces an additional current component in the field winding which assumes a damped sinusoidal character.

generator is essentially inductive and in an inductive circuit a current cannot vary instantaneously.

As previously stated, Figure 10.8 shows the ac symmetrical component of the short-circuit armature current. This symmetrical trace can be found oscillographically if the short circuit takes place at the instant when the prefault flux linkage of the phase is zero. The envelope of the short-circuit current represents three different periods: the **subtransient period**, the **transient period**, and the **steady-state period**. For the purpose of short-circuit current calculations, the variable reactance of a synchronous machine can be represented by the following reactance values:

X_s'' = **subtransient reactance** determines the short-circuit current during the first cycle or so after the short circuit occurs. In about $0.05 - 0.1$ s this reactance increases to:

X_s' = **transient reactance** determines the short-circuit current after several cycles at 60 Hz. In about $0.2 - 2$ s this reactance increases to:

X_s = **synchronous reactance** determines the short-circuit current after a steady-state condition is reached.

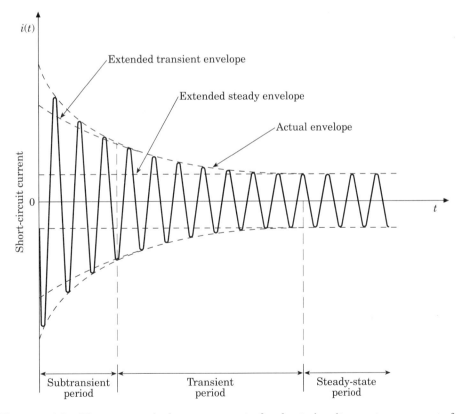

Figure 10.8   The symmetrical ac component of a short-circuit armature current of a synchronous generator.

This representation of the synchronous machine reactance by three different reactances is due to the fact that the flux across the air gap of the machine is much greater at the instant the short circuit takes place than it is a few cycles later. When a short circuit occurs at the terminals of a synchronous machine, it takes time for the flux to decrease across the air gap. As the flux lessens, the armature current decreases since the voltage generated by the air gap flux regulates the current. The subtransient reactance X''_s includes the leakage reactances of the stator and rotor windings of the generator and the influence of the damper windings.[4] Therefore, the subtransient reactance can be expressed as

$$X''_s = X_a + \frac{X_{ar} X_f X_d}{X_f X_d + X_{ar} X_d + X_{ar} X_f} \qquad (10.57)$$

where X_a = leakage reactance of the stator winding per phase
   $\equiv \omega L_a$
   X_f = leakage reactance of the field winding
   $\equiv \omega L_f$
   X_{ar} = reactance associated with the armature mmf acting in the mutual flux path
   $\equiv \omega L_{ar}$
   L_d = leakage reactance of the damper winding

The transient reactance X'_d includes the leakage reactances of the stator and excitation windings of the generator. It is usually larger than the subtransient reactance. The transient reactance can be expressed as

$$X'_s = X_a + \frac{X_f X_{ar}}{X_f + X_{ar}} \qquad (10.58)$$

Alternatively, the three reactance values can be determined from

$$X''_s = \frac{E_a}{I''_a} \qquad (10.59)$$

$$X'_s = \frac{E_a}{I'_a} \qquad (10.60)$$

$$X_s = \frac{E_a}{I_a} \qquad (10.61)$$

where E_a = the internal generated voltage
   I''_a = the subtransient current

---

[4] They are located in the pole faces of generator and are used to reduce the effects of hunting.

I_a' = the transient current
I_a = the steady-state current

It is interesting to observe in Figure 10.8 that the ac component of the short-circuit current consists of the steady-state value and the two components that decay with time constants T_s' and T_s''. Thus, the rms magnitude of the ac current[5] at any time after a three-phase short circuit at the generator terminals can be determined from

$$I_{ac}(t) = I_a + (I_a' - I_a)\,exp\left(-\frac{t}{T_s'}\right) + (I_a'' - I_a')\,exp\left(-\frac{t}{T_s''}\right) \qquad (10.62)$$

where all quantities are in rms values and are equal but displaced 120 electrical degrees in the three phases. However, the instantaneous value of the ac short-circuit current is

$$i_{ac}(t) = \sqrt{2}\left[I_a + (I_a' - I_a)\,exp\left(-\frac{t}{T_s'}\right) + (I_a'' - I_a')\,exp\left(-\frac{t}{T_s''}\right)\right]sin\ \omega t \qquad (10.63)$$

The dc (i.e., unidirectional) component of the short-circuit current is different in each phase and its maximum value is

$$I_{dc,max} = \sqrt{2}I_a'' \qquad (10.64)$$

Since it decays with the armature time constant[6] T_a, it can be expressed as

$$I_{dc} = \sqrt{2}I_a''\,exp\left(-\frac{t}{T_a}\right) \qquad (10.65)$$

Also, the maximum value of the dc component depends upon the point in the voltage cycle where the short circuit takes place. Thus, the dc component of the short-circuit current can be determined from

$$I_{dc} = \sqrt{2}I_a''(cos\ \alpha)\,exp\left(-\frac{t}{T_a}\right) \qquad (10.66)$$

where α is the switching angle. Therefore, the total short-circuit current with the dc offset can be expressed as

---

[5] If such a short circuit does not include all three phases, then the fault currents are determined by using **symmetrical components** methods, which are beyond the scope of this book. For further information, see Gönen [35].
[6] Typically, it varies between 0.1 s and 0.2 s.

$$i_{tot} = \sqrt{2} \left[I_a + (I'_a - I_a) \exp\left(-\frac{t}{T'_s} \right) + (I''_a - I'_a) \exp\left(-\frac{t}{T''_s} \right) \right] \sin \omega t$$

$$(10.67)$$

$$+ \sqrt{2} I''_a (\cos \alpha) \exp\left(-\frac{t}{T_a} \right)$$

Since dc components of current decay very fast, as a rule of thumb, the value of the ac component of current should be multiplied by 1.6 to find the total initial current.

Now assume that the synchronous generator is supplying power to a bus before such a short circuit takes place. Since the reactance of the generator changes from X_s to X''_s, the corresponding internal generated voltages must also vary to maintain the initial condition of flux linkage constancy. Therefore, the internal generated voltages can be expressed as

$$E''_a = V_t + jI_a X''_s \qquad (10.68)$$

$$E'_a = V_t + jI_a X'_s \qquad (10.69)$$

$$E_a = V_t + jI_a X_s \qquad (10.70)$$

where I_a is the prefault load current. Thus, the total short-circuit current with the dc offset can be determined from

$$i_{tot} = \sqrt{2} \left[I_a + (I'_a - I_a) \exp\left(-\frac{t}{T'_d} \right) + (I''_a - I'_a) \exp\left(-\frac{t}{T''_d} \right) \right]$$

$$(10.71)$$

$$+ \sqrt{2} I''_a (\cos \alpha) \exp\left(-\frac{t}{T_a} \right)$$

where

$$T'_d \cong \frac{X'_s}{X_s} T'_s \qquad (10.72)$$

$$T''_d \cong \frac{X''_s}{X_s} T''_s \qquad (10.73)$$

Example 10.4

Assume that a three-phase 200-MVA 13.2-kV 60-Hz wye-connected synchronous generator has the following parameters:

$X_s = 1.0$ pu	$X'_s = 0.23$ pu	$X''_s = 0.12$ pu
$T'_s = 1.1$ s	$T''_s = 0.035$ s	$T_a = 0.16$ s

Assume that the machine is operating at no load when a three-phase short circuit takes place at its terminals. If the initial dc component of the short-circuit is about 60 percent of the initial ac component of the short-circuit current, determine the following:

(a) The base current of the generator.
(b) The subtransient current in per unit and amps.
(c) The transient current in per unit and amps.
(d) The steady-state current in per unit and amps.
(e) The initial value of the ac short-circuit current in amps.
(f) The initial value of the total short-circuit current in amps.
(g) The value of the ac short-circuit current after two cycles.
(h) The values of the ac short-circuit current after 4, 6, and 8 seconds, respectively.

SOLUTION:

(a) From equation (2.99), the base current of the generator is

$$I_B = \frac{S_{3\phi,base}}{\sqrt{3}V_{L,base}} = \frac{200 \times 10^6}{\sqrt{3}(13,200)} = 8,748 \text{ A}$$

(b) The subtransient current is

$$I_a'' = \frac{E_a}{X_s''} = \frac{1.0}{0.12} = 8.333 \text{ pu}$$

or

$$I_a'' = (8.333 \text{ pu})(8748 \text{ A}) = 72,897 \text{ A}$$

(c) The transient current is

$$I_a' = \frac{E_a}{X_s'} = \frac{1.0}{0.23} = 4.348 \text{ pu}$$

or

$$I_a' = (4.348 \text{ pu})(8748 \text{ A}) = 38,036 \text{ A}$$

(d) The steady-state current is

$$I_a = \frac{E_a}{X_s} = \frac{1.0}{1.0} = 1.0 \text{ pu}$$

or

$$I_a = (1.0 \text{ pu})(8,748 \text{ A}) = 8,748 \text{ A}$$

(e) The initial value of the ac short-circuit current is

$$I''_a = 72{,}897 \text{ A}$$

(f) The initial value of the total short-circuit current is

$$I_{tot} \cong 1.6\, I''_a = 1.6(72{,}897 \text{ A}) = 116{,}635 \text{ A}$$

(g) From equation (10.62), the rms value of the ac short-circuit current as a function of time is

$$I_{ac}(t) = I_a + (I'_a - I_a)\exp\left(-\frac{t}{T'_s}\right) + (I''_a - I'_a)\exp\left(-\frac{t}{T''_s}\right)$$

$$= 8{,}748 + 29{,}288\exp\left(-\frac{t}{1.1}\right) + 64{,}149\exp\left(-\frac{t}{0.035}\right)$$

Therefore, the value of the ac current after two cycles (i.e., $t = 1/30$ s) is

$$I_{ac}(t) = 8{,}748 + 29{,}288\exp\left(-\frac{1/30}{1.1}\right) + 64{,}149\exp\left(-\frac{1/30}{0.035}\right) = 61{,}912 \text{ A}$$

(h) The value of the ac current after 4 seconds is

$$I_{ac}(t) = 8{,}748 + 772 + 0 = 9{,}520 \text{ A}$$

After 6 seconds, it is

$$I_{ac}(t) = 8{,}748 + 125 + 0 = 8{,}873 \text{ A}$$

After 8 seconds, it is

$$I_{ac}(t) = 8{,}748 + 20 + 0 = 8{,}768 \text{ A}$$

10.7 TRANSIENT STABILITY

Power system stability can be defined as the *ability of the numerous synchronous machines of a given power system to remain in synchronism (i.e., in step) with each other following a disturbance.* In a stable power system, if synchronous machines are disturbed they will return to their original operating state if there is no change of power, or they will attain a new operating state without a loss of synchronism. The disturbance often causes a transient that is oscillatory in nature, but if the system is stable the oscillations will be damped. In a synchronous generator the most severe disturbance is caused by a short circuit across its terminals. In a synchronous motor, a disturbance may be caused by a sudden application of load torque to the shaft.

Stability can be classified as either transient or dynamic stability[7]. The definition of **transient stability** includes stability after a sudden large disturbance such as a fault, loss of a generator, a sudden load change, or a switching operation. Transient stability is a short-term problem. Today, **dynamic stability** is defined as the *ability of various machines to regain and maintain synchronism after a small and slow disturbance*, such as a gradual change in load. Dynamic stability is a long-term problem with time constants running into minutes. In this study, the effects of regulators, governors, and modern exciters, as well as other factors that affect stability may be included.

Assume that the power system is made up of two synchronous machines and that one of them is operating as a generator and the other as a motor. Also assume that the two-machine system is operating at steady state at point 1 on the power-angle curve given in Figure 10.9(a). Suppose that the generator is supplying electrical power P_1 at an angle δ_1 to the motor and that the motor is driving a mechanical load connected to its shaft. Consider the following specific cases of the operation of this system:

Case 1. Assume that the shaft load of the motor is slowly increasing. The resulting net torque tends to slow down the motor and decreases its speed, causing an increase in the power angle δ. This, in turn, causes the input power to increase until an equilibrium is achieved at a new operating point 2, which is higher than 1.

Case 2. Suppose that the load of the motor is increased suddenly by connecting a large load. The resultant shortage in input power will be temporarily brought about by the decrease in kinetic energy. Consequently, the speed of the motor will decrease, causing the power angle δ and the input power to increase. As long as the new load is less than $P_{\max}$, the power angle δ will increase to a new value so that the input power of the motor is equal to its load. At this point the motor may still be running slowly, causing the power angle to increase beyond its proper

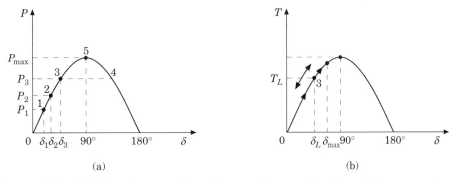

Figure 10.9   (a) Electric power input to a motor as a function of torque angle δ; (b) torque versus the torque angle δ characteristic.

[7] Today, the IEEE does not recognize **steady-state stability** as a separate class of stability. In present practice, it is included in the definition of dynamic stability. However, some authors still refer to the dynamic stability by using the ambiguous name of "steady-state stability".

value. Also an accelerating torque that increases the speed of the motor develops. However, it is possible that when the motor regains its normal speed, the power angle δ may have gone beyond point 4 and the motor input will be less than the load. This causes the motor to pull-out.

Case 3. Assume that the load of the motor is increased slowly until point 5 of maximum power is reached. Any additional load will increase the power δ beyond 90°, causing the input power to decrease further and the net retarding torque to increase again. This torque slows the motor down even more until it pulls out of step.

Case 4. Suppose that the load of the motor is increased suddenly, but the additional load is not too large. The motor will regain its normal speed before the power angle δ becomes too large. The situation is illustrated in Figure 10.9(b) in terms of the resultant torque and δ. When the torque angle δ reaches the value δ_L, the load torque is the same as the torque developed by the motor. However, due to inertia, δ will increase beyond δ_L and the motor will develop more torque than the load needs. The deceleration will decrease, causing the angle δ to reach a maximum value δ_{max} and then swing back. The angle δ will oscillate around δ_L. Such oscillations will later die out because of damping in the system, and the motor will settle down to stable operating conditions at point 3.

The **transient stability limit** can be defined as the *upper limit to the sudden increment in load that the rotor can have without pulling out of step.* This limit is always less than the steady-state limit discussed in Section 7.8. Furthermore, the transient stability limit may have different values depending on the nature and magnitude of the disturbance involved.

10.8 THE SWING EQUATION

Without ignoring the torque that is due to friction, windage, and core losses, the net accelerating torque of a synchronous machine can be expressed, based on Newton's law of rotation, as

$$T_a = T_m - T_d = J\alpha \tag{10.74}$$

or, in terms of the angular position as

$$J\frac{d^2\theta_m}{dt^2} = T_a = T_m - T_d \tag{10.75}$$

where T_a = the net accelerating torque

T_m = the shaft torque corrected for rotational losses including friction, windage, and core losses

T_d = the developed electromagnetic torque

J = the moment of inertia of the rotor

α = the angular acceleration expressed in terms of the angular position θ of the rotor

$$= \frac{d^2\theta}{dt^2}$$

It is customary to use the values of T_m and T_d as positive for generator action and negative for motor action. It is convenient to measure angular position and angular velocity with respect to a reference axis rotating at synchronous speed. Hence, the rotor position can be expressed as

$$\theta_m = \omega_{sm}t + \delta_m \qquad (10.76)$$

Taking the derivatives of θ_m with respect to t,

$$\frac{d\theta_m}{dt} = \omega_{sm} + \frac{d\delta_m}{dt} \qquad (10.77)$$

$$\frac{d^2\theta_m}{dt^2} = \frac{d^2\delta_m^2}{dt^2} \qquad (10.78)$$

By substituting equation (10.78) into equation (10.75),

$$J\frac{d^2\delta_m}{dt^2} = T_a = T_m - T_d \qquad (10.79)$$

and by multiplying both sides of this equation by the angular velocity

$$J\omega_m\frac{d^2\delta_m}{dt^2} = \omega_m T_a = \omega_m T_m - \omega_m T_d \qquad (10.80)$$

Thus, the swing equation can be obtained as

$$M\frac{d^2\delta_m}{dt^2} = P_a = P_m - P_d \qquad (10.81)$$

where $M = J\omega_m$ = inertia constant
$P_a = P_m - P_d$ = net accelerating power
$P_m = \omega_m T_m$ = shaft power input corrected for rotational losses
$P_d = \omega_m T_d$ = electrical power output corrected for electrical losses

The swing equation describes how the machine rotor moves (swings) with respect to the synchronously rotating reference frame in a given disturbance (i.e., when the net accelerating power is not zero). The inertia constant for the synchronous machine is expressed as

$$H = \frac{Kinetic\ energy\ of\ all\ rotating\ parts\ at\ synchronous\ speed}{S_{rated}} \qquad (10.82a)$$

or

$$H = \frac{\frac{1}{2}J\omega_{sm}^2}{S_{rated}} \qquad (10.82b)$$

or

$$H = \frac{1}{2}\frac{M\omega_{sm}}{S_{rated}} \qquad (10.82c)$$

Equation (10.82) can be expressed in terms of per-unit quantities with re-spect to the rated (3ϕ) power of the synchronous generator as

$$\frac{2H}{\omega_{sm}}\frac{d^2\delta_m}{dt^2} = \frac{P_a}{S_{rated}} = \frac{P_m - P_d}{S_{rated}} \qquad (10.83)$$

where the angle δ_m and angular velocity ω_m are in mechanical radians and mechanical radians per second, respectively. For a synchronous generator with p poles, the electrical power angle and radian frequency are associated with the corresponding mechanical variables as

$$\delta(t) = \frac{p}{2}\delta_m(t) \qquad (10.84)$$

$$\omega(t) = \frac{p}{2}\omega_m(t) \qquad (10.85)$$

Also the synchronous electrical radian frequency is related to the synchronous angular velocity as

$$\omega_s = \frac{p}{2}\omega_m \qquad (10.86)$$

Thus, the rated per-unit swing equation (10.83) can be expressed in electrical units as

$$\frac{2H}{\omega_s}\frac{d^2\delta}{dt^2} = P_a = P_m - P_d \qquad (10.87)$$

when δ is in electrical radians

$$\frac{2H}{\omega_s}\frac{d^2\delta}{dt^2} = P_a = P_m - P_d \qquad (10.88)$$

when δ is in electrical degrees

$$\frac{H}{180f}\frac{d^2\delta}{dt^2} = P_a = P_m - P_d \qquad (10.89)$$

The solution of this swing equation is called the **swing curve** $\delta(t)$.

If the synchronous machine is connected to an infinite bus through an external reactance, the electrical power output of the synchronous generator can be expressed as

$$P_d = P_{max}sin\ \delta \qquad (10.90)$$

By substituting this equation into equation (10.87), the swing equation can be found as

$$P_m = \frac{2H}{\omega_s}\frac{d^2\delta}{dt^2} + P_{max}sin\ \delta \qquad (10.91)$$

Since the resulting equation is nonlinear, it is often necessary to use a numerical technique to solve it.

Example 10.5

Assume that a three-phase 250-MVA 15-kV 60-Hz six-pole generator is connected to an infinite bus through a purely reactive network. Also assume that the inertia constant of the generator is 6 MJ/MVA and that it is supplying power of 1.0 per unit to the infinite bus at the steady state. The maximum power that can be supplied is 2.4 per unit. If the output power of the generator becomes zero at a three-phase fault, determine the following:
(a) The angular acceleration of the generator.
(b) The shaft speed of the generator at the end of 12 cycles.
(c) The change in the power angle δ at the end of 12 cycles.

SOLUTION:
(a) Since the generator is operating at steady state before the fault

$$P_m = P_d = 1.0 \text{ pu}$$

Hence, from equation (10.89),

$$\frac{H}{180f}\frac{d^2\delta}{dt^2} = P_m - P_d$$

from which the accelerating torque can be found as

$$\alpha = \frac{d^2\delta}{dt^2} = \frac{180f}{H}(P_m - P_d)$$

$$= \frac{(180)(60)}{6}(1.0 - 0) = 1800 \text{ elec. degrees/s}^2$$

Since the machine has six poles,

$$\alpha = 300\left(\frac{60 \text{ s/min}}{360°\text{/rev}}\right) = 50 \text{ rpm/s}$$

(b) The synchronous speed of the machine is

$$\omega_{sm} = \frac{120f}{p} = \frac{120(60)}{6} = 1200 \text{ rpm}$$

and a 12-cycle interval is

$$t = \frac{12}{60} = 0.2 \text{ s}$$

thus

$$\omega_m = \omega_{sm} + \alpha t$$

$$= 1200 + (50)(0.2) = 1210 \text{ rpm}$$

(c) Since the generator is initially operating at the power angle δ from

$$P_0 = P_{max}\sin \delta_0$$

the initial power angle can be found as

$$\delta_0 = \sin^{-1}\left(\frac{P_0}{P_{max}}\right)$$

$$= \sin^{-1}\left(\frac{P_0}{P_{max}}\right) = 24.62°$$

Therefore,

$$\delta = \delta_0 + \frac{1}{2}\alpha t^2$$

$$= 24.62° + \frac{1}{2}(1800)(0.2)^2 = 60.62°$$

PROBLEMS

10.1   Assume that a 250-V separately excited dc generator has an armature circuit resistance and inductance of 0.10 Ω and 1.2 mH, respectively. Its field winding resistance and inductance are 10 Ω and 52 H, respectively. The generator constant K_g is 104 V per field ampere at 1200 rpm. Assume that the field and armature circuits are initially open and that the prime mover is driving the machine at a constant speed of 1200 rpm. Derive a mathematical expression for the armature terminal voltage as a function of time in terms of the unit step function, if the field circuit is connected to a constant voltage source of 260 V at time $t = 0$.

10.2   Consider Problem 10.1 and assume that after the field circuit has reached a steady state, the armature circuit of the generator is suddenly connected to a load made up of a resistance and inductance that are connected in series with respect to each other. If the values of the resistance and inductance are 1.2 Ω and 1.4 mH, respectively, determine the following in terms of unit step functions:
   (a)   The armature current.
   (b)   The terminal voltage.
   (c)   The developed electromagnetic torque.

10.3   Assume that a 250-V separately excited dc motor has an armature circuit resistance and inductance of 0.15 Ω and 15 mH, respectively. Its moment of inertia J is 18 kg/m^2 and its motor constant K_m is 2.0 N·m/A. The motor is supplied by a constant voltage source of 250 V. Assume that the motor is initially operating at steady state without load. Its no-load armature current is 10 A. Ignore the effects of saturation and armature reaction. If a constant load torque of 500 N·m is suddenly connected to the shaft of the motor, determine the following:
   (a)   The undamped natural frequency of the speed response.
   (b)   The damping factor and damping ratio.
   (c)   The initial speed in rpm.
   (d)   The initial acceleration.
   (e)   The ultimate speed drop.

10.4   A separately excited dc generator has an armature circuit resistance and inductance of 1 Ω and 3 H, respectively. Its field winding resistance and inductance are 20 Ω and 100 Ω, respectively. The generator constant K_g is 100 V per field ampere at rated speed. Assume that the generator is driven by the prime mover at rated speed and that a 250-V dc supply is suddenly connected to the field winding. Determine the following:
   (a)   The internal generated voltage of the generator.

(b) The internal generated voltage in the steady state.
(c) The time required for the internal generated voltage to rise to 99 percent of its steady-state value.

10.5 Assume that the generator given in Problem 10.4 is driven by the prime mover at rated speed and is connected to a load made up of a resistance of 15 Ω and inductance of 5 H that are connected in series with respect to each other. If a 250-V dc supply is suddenly connected to the field winding, determine the armature current as a function of time.

10.6 Consider the synchronous generator given in Example 10.4 and determine the following:
(a) The instantaneous values of the ac short-circuit current for any given time t.
(b) The maximum value of the dc component of the short-circuit current.
(c) The dc component of the short-circuit current at $t = 0.1$ seconds, if the switching angle α is given as 45°.
(d) The total short-circuit current with the dc offset at $t = 0.1$ seconds, if the switching angle α is zero.
(e) The maximum rms value of the total short-circuit current.

A BRIEF REVIEW OF PHASORS

A.1 INTRODUCTION

The instantaneous value of a sinusoidally varying voltage can be expressed as

$$v(t) = V_m \sin \omega t \qquad (A.1)$$

where V_m = amplitude (or maximum value) of the voltage
ωt = argument
ω = radian frequency (or angular frequency)

Since the sine wave is periodic, the function repeats itself every 2π radians. Its period T is 2π radians and its frequency f is $1/T$ hertz. Thus, the radian frequency can be expressed as $\omega = 2\pi f$. The voltage given by equation (A.1) can be expressed as a cosine wave as

$$v(t) = V_m \cos (\omega t - 90°) \qquad (A.2)$$

A more general form of the sinusoid is

$$v(t) = V_m \cos (\omega t + \phi) \qquad (A.3)$$

where ϕ = phase angle

Euler's identity states that

$$e^{j\phi} = \cos \phi + j\sin \phi \qquad (A.4)$$

Therefore, equation (A.3) can be expressed as

$$v(t) = \Re[V_m \cos(\omega t + \phi) + j V_m \sin(\omega t + \phi)]$$

$$= \Re[V_m e^{j(\omega t + \phi)}]$$

$$= \Re[V_m e^{j\phi} e^{j\omega t}] \qquad \text{(A.5)}$$

$$= \sqrt{2}\,\Re[V e^{j\omega t}]$$

Also, by using the definition of the effective (i.e., rms) value of the voltage,

$$V = \frac{V_m}{\sqrt{2}} e^{j\omega t} \qquad \text{(A.6)}$$

Therefore,

$$v(t) = \sqrt{2}\,\Re[V e^{j\omega t}] \qquad \text{(A.7)}$$

Thus, the complex amplitude of the sinusoid given by equation (A.3) can be expressed in exponential form as

$$\boldsymbol{V} = |\boldsymbol{V}|\, e^{j\phi} \qquad \text{(A.8)}$$

or in polar form as

$$\boldsymbol{V} = |\boldsymbol{V}|\, \angle\phi \qquad \text{(A.9)}$$

or

$$\boldsymbol{V} = V\angle\phi \qquad \text{(A.10)}$$

or in rectangular form as

$$\boldsymbol{V} = V(\cos\phi + j\sin\phi) \qquad \text{(A.11)}$$

The complex quantity, given by equation (1.10), is also called a **phasor** (or sometimes, erroneously called a **vector**).[1] Anderson [5] defines a phasor as "a complex number which is related to the time domain sinusoidal quantity by the expression

$$a(t) = \Re(\sqrt{2}\,\boldsymbol{A} e^{j\omega t}) \qquad \text{(A.12)}$$

If we express $\boldsymbol{A}$ in terms of its magnitude $|\boldsymbol{A}|$ and angle α, we have

$$\boldsymbol{A} = |\boldsymbol{A}| e^{j\alpha}$$

---

[1] This subject has been explained in greater depth by Skilling [89] and Nilsson [73].

and

$$a(t) = \Re e(\sqrt{2}\,|\boldsymbol{A}|e^{j(\omega t + \alpha)} = \sqrt{2}\,|\boldsymbol{A}|\cos\,(\omega t + \alpha) \qquad (A.13)$$

Thus, equations (A.12) and (A.13) convert the rms phasor (complex) quantity to the actual time domain variable". For example, assume that a sinusoidal voltage of

$$v(t) = V_m e^{j(\omega t + \phi)} \qquad (A.14)$$

is applied to a circuit with an impedance

$$\boldsymbol{Z} = |\boldsymbol{Z}|e^{j\theta} \qquad (A.15)$$

Therefore, the current can be expressed as

$$i(t) = \frac{V_m e^{j(\omega t + \phi)}}{|\boldsymbol{Z}|e^{j\theta}}$$

$$= \left(\frac{V_m}{|\boldsymbol{Z}|}\right)e^{j(\omega t + \phi - \theta)}$$

$$= I_m e^{j(\omega t + \phi - \theta)}$$

or

$$I_m e^{j(\omega t + \phi - \theta)} = \frac{V_m e^{j(\omega t + \phi)}}{|\boldsymbol{Z}|e^{j\theta}} \qquad (A.16)$$

Since time appears in both the current and voltage expressions, the equation is given in time domain. If the equality is multiplied by $e^{-j\omega t}$ to suppress $e^{j\omega t}$ and multiplied again by $1/\sqrt{2}$ to give the effective current and voltage values, then

$$\frac{e^{-j\omega t}}{\sqrt{2}}I_m e^{j(\omega t + \phi - \theta)} = \frac{e^{-j\omega t}}{\sqrt{2}}\frac{V_m e^{j(\omega t + \phi)}}{|\boldsymbol{Z}|e^{j\theta}} \qquad (A.17)$$

which gives

$$\frac{I_m}{\sqrt{2}}e^{j(\phi - \theta)} = \frac{V_m}{\sqrt{2}}\frac{e^{j\phi}}{|\boldsymbol{Z}|e^{j\theta}} \qquad (A.18)$$

which becomes

$$|\boldsymbol{I}|\angle\phi - \theta = \frac{|\boldsymbol{V}|\angle\phi}{|\boldsymbol{Z}|\angle\theta} \qquad (A.19)$$

TABLE A.1 A Comparison of the Current and Voltage Relationships Between the Time Domain and Frequency Domain for R, L, and C

Time Domain		Frequency Domain (Phasors)	
$i \longrightarrow$ ⊕ $+$ v R $-$	$v = iR$	$I \longrightarrow$ ⊕ $+$ V R $-$	$\mathbf{V} = R\mathbf{I}$
	$i = \dfrac{v}{R}$		$\mathbf{I} = \dfrac{\mathbf{V}}{R}$
$i \longrightarrow$ $+$ v L $-$	$v = L\dfrac{di}{dt}$	$I \longrightarrow$ $+$ V $j\omega L$ $-$	$\mathbf{V} = j\omega L\mathbf{I}$
	$i = \dfrac{1}{L}\displaystyle\int_{-\infty}^{t} v\,dt$		$\mathbf{I} = -j\dfrac{\mathbf{V}}{\omega L}$
$i \longrightarrow$ $+$ v C $-$	$v = \dfrac{1}{C}\displaystyle\int_{-\infty}^{t} i\,dt$	$I \longrightarrow$ $+$ V $-j\dfrac{1}{\omega c}$ $-$	$\mathbf{V} = -j\dfrac{1}{\omega C}\mathbf{I}$
	$i = C\dfrac{dv}{dt}$		$\mathbf{I} = j\omega C\mathbf{V}$

or

$$I = \frac{V}{Z} \tag{A.20}$$

Equations (A.18) to (A.20) are in frequency domain. Equation (A.18) is a **transformed** equation. I and V values without subscripts in equation (A.19) represent the effective current and voltage values. Thus, the I, V, and Z in equation (A.20) are complex quantities. Table A.1 gives a comparison and summary of the relationships between V and I in the time domain and V and I in the frequency domain for the three basic passive ideal circuit elements, R, L, and C. Figure A.1(a) shows the voltage and current functions in the time domain, while Figure A.1(b) shows the voltage and current phasors in the frequency domain.

An impedance that has a resistance R in series with a reactance X can be represented by the impedance operator

$$Z = R + jX \tag{A.21}$$

which is a complex quantity. A sinusoidal current in an impedance Z can be represented by a current phasor I, as shown in Figure A.1(b). Therefore, the voltage drop across the impedance can be expressed as

$$V = IZ$$

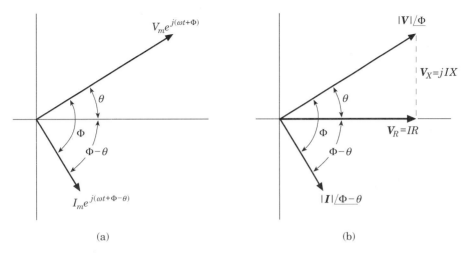

Figure A.1 Voltage and current functions given in: (a) the time domain; (b) the frequency domain.

$$= IR + jIX \qquad (A.22)$$

where IR = horizontal component of the phasor V
     IX = vertical component of the phasor V
      θ = phase angle
       = $\tan^{-1}\dfrac{X}{R}$

Neglecting the corresponding phase angles, the numerical values of the voltage drop and current can be expressed as

$$|V| = |IZ| = \sqrt{R^2 + X^2} \qquad (A.23)$$

and

$$|I| = \frac{|V|}{|Z|} = \frac{|V|}{\sqrt{R^2 + X^2}} \qquad (A.24)$$

If the admittance of a network is already determined, the corresponding impedance can be expressed as

$$Z = 1/Y \qquad (A.25)$$

where Y and Z are complex quantities. Therefore,

$$Z = R + jX$$

$$= \frac{1}{G - jB} \tag{A.26}$$

or

$$\mathbf{Z} = \frac{G}{G^2 + B^2} + j\frac{B}{G^2 + B^2} \tag{A.27}$$

On the other hand, an admittance that has a conductance G in parallel with a susceptance B can be represented by the admittance operator

$$\mathbf{Y} = G - jB \tag{A.28}$$

When inductive and capacitive susceptances are in parallel,

$$B = \frac{1}{\omega L} - \omega C \tag{A.29}$$

When the voltage is represented by a phasor $\mathbf{V}$, the total current can be expressed as

$$\mathbf{I} = \mathbf{VY}$$

$$= VG - jVB \tag{A.30}$$

When two branches of a network are connected in parallel, the complex quantities representing the admittances are added as

$$G - jB = G_1 - jB_1 = G_2 - jB_2 \tag{A.31}$$

Neglecting the corresponding phase angles, the numerical values of the current and voltage drop can be expressed as

$$|\mathbf{I}| = |\mathbf{VY}| = |\mathbf{V}| \sqrt{G^2 + B^2} \tag{A.32}$$

and

$$|\mathbf{V}| = \frac{|\mathbf{I}|}{|\mathbf{Y}|} = \frac{|\mathbf{I}|}{\sqrt{G^2 + B^2}} \tag{A.33}$$

If the impedance is known, the corresponding admittance[2] is found as

---

[2] However, as shown in equations (A.35) and (A.36), G is **not** equal to $1/R$ and B is **not** equal to $1/X$, as is the case when there is only a single circuit element R or X for which $G = 1/R$ and $B = 1/X$, respectively.

$$\boldsymbol{Y} = 1/\boldsymbol{Z} \tag{A.34}$$

Therefore,

$$\boldsymbol{Y} = G - jB$$

$$= \frac{1}{R + jX} \tag{A.35}$$

or

$$\boldsymbol{Y} = \frac{R}{R^2 + X^2} - j\frac{X}{R^2 + X^2} \tag{A.36}$$

However, it makes more sense to define the impedances due to resistor, inductance, and capacitance as

$$\boldsymbol{Z}_R = R \tag{A.37}$$

$$\boldsymbol{Z}_L = j\omega L = jX_L \tag{A.38}$$

$$\boldsymbol{Z}_C = -j\frac{1}{\omega C} = -jX_c \tag{A.39}$$

and by using reduction techniques apply the rules of series and parallel combinations to impedances to determine the equivalent impedance value. Therefore, the equivalent phasor impedance can be defined as the ratio of the phasor voltage to phasor current, as

$$\boldsymbol{Z} = \boldsymbol{V}/\boldsymbol{I} \tag{A.40}$$

The equivalent resistance and reactance values can be found as

$$R = \mathscr{R}e|\boldsymbol{Z}| \tag{A.41}$$

and

$$X = \mathscr{I}e|\boldsymbol{Z}| \tag{A.42}$$

where R and X are determined by taking only the real and imaginary portions of the impedance $\boldsymbol{Z}$, respectively. As given in Table A.1, in a *purely resistive* circuit, current $\boldsymbol{I}$ is in phase with its voltage $\boldsymbol{V}$. However, in a *purely inductive* circuit, current $\boldsymbol{I}$ lags its voltage $\boldsymbol{V}$ by 90°, but in a *purely capacitive* circuit, current $\boldsymbol{I}$ leads its voltage $\boldsymbol{V}$ by 90°.

PROBLEMS

A.1   The root-mean-square (rms) value of a sinusoidal ac current can be expressed as

$$I_{rms} = \sqrt{\frac{1}{T}\int_0^T i^2(t)\,dt}$$

where $i(t) = I_m \cos(\omega t - \theta)$

$$\omega = \frac{2\pi}{T}$$

Verify that the rms value of the current is

$$I_{rms} = \frac{I_m}{\sqrt{2}}$$

A.2   The voltage and current values are given in the time domain as

$$v(t) = 162.6346\cos(10t - 50°)$$

and

$$i(t) = 14.142\sin(5t + 120°)$$

Convert them to the corresponding phasor values expressed in polar forms.

A.3   A voltage value is given in the frequency domain as

$$\mathbf{V} = 70.711\angle - 60°\ \text{kV}$$

Express this in a sine wave form.

A.4   Assume that only three currents $i_1(t)$, $i_2(t)$, and $i_3(t)$ enter a node and that the $i_1(t)$ and $i_2(t)$ are given as $14.14\cos(600t + 30°)$ A and $28.28\sin(600t - 50°)$ A, respectively. Determine the $\mathbf{I}_3$ current in polar form.

A.5   Assume that only three currents enter a node and that the $\mathbf{I}_1$ and $\mathbf{I}_3$ are given as $19\angle - 70°$ A and $24\angle 45°$ A, respectively. Determine the $\mathbf{I}_2$ current in the time domain.

THE PER-UNIT SYSTEM

B.1 INTRODUCTION

Because of various advantages, it is customary in power system analysis calculations to use impedances, currents, voltages and powers in per-unit values (which are *scaled* or *normalized* values) rather than the physical values such as ohms, amperes, kilovolts, and megavoltamperes (or megawatts, or megavars). A per-unit system is an easy way to express and compare quantities. The per-unit value of any quantity is defined as its ratio to an *arbitrarily* chosen base (i.e., *reference*) value with the same dimensions. Therefore, the *per-unit* value of any quantity can be defined as

$$\text{Quantity in per-unit} = \frac{physical\ quantity}{base\ value\ of\ quantity} \qquad (B.1)$$

where the *physical quantity* refers to the given value in ohms, amperes, volts, or other units. The base value is also called the *unit value* since in the per-unit system it has a value of 1, or unity. A *base current* is also referred to as a *unit current*. Since both the physical quantity and base quantity have the same dimensions, the resulting per-unit value, expressed as a decimal, has no dimension and is simply denoted by the subscript pu. The base quantity is indicated by the subscript B. The symbol for per-unit is *pu*. The percent system is obtained by multiplying the per-unit value by 100. Therefore,

$$\text{Quantity in percent} = \frac{physical\ quantity}{base\ value\ of\ quantity} \qquad (B.2)$$

However, the percent system is somewhat more difficult to work with and more subject to error since one must always remember that the quantities have been multiplied by 100. Therefore, the factor 100 has to be continually inserted or removed for reasons which may not be obvious at the time. Thus, the per-unit system is generally preferred in power system calculations.

In applying the per-unit system, base quantities (usually, base apparent power and base voltage) are selected first, and by using these selections and existing electrical laws the other base values are determined. In a *single-phase system*, the following relationships can be defined:

$$I_B = \frac{S_B}{V_B} = \frac{VA_{1\phi,\,base}}{V_{LN,\,base}} = \frac{kVA_{1\phi,\,base}}{kV_{LN,\,base}} \tag{B.3}$$

$$S_B = P_B = Q_B = V_B I_B = V_{LN,\,base} \times I_{base} \tag{B.4}$$

$$Z_B = R_B = X_B = \frac{V_B}{I_B} = \frac{V_{LN,\,base}}{I_{base}} = \frac{kV_{LN,\,base}}{I_{base} \times 1000} \tag{B.5}$$

or

$$Z_B = \frac{V_B}{S_B/V_B} = \frac{V_B^2}{S_B} = \frac{(V_{LN,\,base})^2}{VA_{1\phi,\,base}} = \frac{(kV_{LN,\,base})^2}{MVA_{1\phi,\,base}} \tag{B.6}$$

$$Z_B = \frac{(kV_{LN,\,base})^2 1000}{kVA_{1\phi,\,base}} \tag{B.7}$$

Also,

$$Y_B = G_B = B_B = \frac{I_B}{V_B} = \frac{I_{base}}{V_{LN,\,base}} \tag{B.8}$$

In these equations the subscripts *LN* and ϕ denote *line to neutral* and *per phase*, respectively. The value of S_{base} has to be the same throughout the entire system being studied. For example, the V_{base} values for a given transformer are different on each side but their ratio must be the same as the turns ratio of the transformer. In general, the *rated* or *nominal* voltages of each side of the transformer are selected as the respective base voltages.

If there is only one power equipment under study, its own ratings are usually used as the bases for per-unit calculations. However, if the equipment is used in a system that has its own bases, it is necessary to refer all of the *given* (i.e., *old*) per-unit values to the new system's base values. For example, to convert the per-unit value of an impedance from one (*old*) base to a new base, the following relationship is used:

$$Z_{pu,\,new} = Z_{pu,\,old} \left(\frac{V_{B,\,old}}{V_{B,\,new}} \right)^2 \left(\frac{S_{B,\,new}}{S_{B,\,old}} \right) \tag{B.9}$$

Similarly,

$$(S, P, Q)_{pu,\,new} = (S, P, Q)_{pu,\,old}\left(\frac{S_{B,\,new}}{V_{B,\,new}}\right) \qquad \text{(B.10)}$$

and

$$V_{pu,\,new} = V_{pu,\,old}\left(\frac{V_{B,\,old}}{V_{B,\,new}}\right) \qquad \text{(B.11)}$$

Since power equipment and system data are frequently given as three-phase quantities, the following *three-phase relationships* are used. Notice that the subscript L denotes line-to-line values and 3ϕ denotes three-phase values.

$$(S_{3\phi}, P_{3\phi}, Q_{3\phi})_{base} = 3(S_{1\phi})_{base} \qquad \text{(B.12)}$$

$$V_{L,\,base} = \sqrt{3}\,V_{LN,\,base} \qquad \text{(B.13)}$$

$$I_{base(per\ phase\ delta)} = \frac{I_{base\,(per\ phase\ wye)}}{\sqrt{3}} \qquad \text{(B.14)}$$

$$Z_{base(per\ phase\ delta)} = \frac{I_{base\,(per\ phase\ wye)}}{\sqrt{3}} \qquad \text{(B.15)}$$

Notice the factors $\sqrt{3}$ and 3, which are used to relate delta and wye quantities in volts, amperes, and ohms, are directly taken into account in the per-unit system by the base values. After determining the proper base quantities, the three-phase problems can be treated in a per-unit system as if they were single-phase problems, regardless of the types of transformer connections involved. Hence,

$$I_B = \frac{S_{3\phi,\,base}}{\sqrt{3}V_{L,\,base}} = \frac{kVA_{3\phi,\,base}}{\sqrt{3}kV_{L,\,base}} \qquad \text{(B.16)}$$

and

$$Z_B = \frac{\left(\dfrac{kV_{L,\,base}}{\sqrt{3}}\right)^2 1000}{\left(\dfrac{kVA_{3\phi,\,base}}{\sqrt{3}}\right)} = \frac{\left(kV_{L,\,base}\right)^2 1000}{kVA_{3\phi,\,base}} \qquad \text{(B.17)}$$

or

$$Z_B = \frac{(kV_{L,\,base})^2}{MVA_{3\phi,\,base}}$$

(B.18)

Example B.1

A 240/120 V single-phase transformer rated 5 kVA has a high-voltage winding impedance of 0.3603 Ω. Use 240 V and 5 kVA as the base quantities and determine the following:
(a) The high-voltage side base current.
(b) The high-voltage side base impedance in ohms.
(c) The transformer impedance referred to the high-voltage side in per unit.
(d) The transformer impedance referred to the high-voltage side in percent.
(e) The turns ratio of the transformer windings.
(f) The low-voltage side base current.
(g) The low-voltage side base impedance.
(h) The transformer impedance referred to the low-voltage side in per unit.

SOLUTION:
(a) The high-voltage side base current is

$$I_{B(HV)} = \frac{S_B}{V_{B(HV)}} = \frac{5,000 \text{ VA}}{240 \text{ V}} = 20.8333 \text{ A}$$

(b) The high-voltage side base impedance is

$$Z_{B(HV)} = \frac{V_{B(HV)}}{I_{B(HV)}} = \frac{240 \text{ V}}{20.8333 \text{ A}} = 11.52 \text{ }\Omega$$

(c) The transformer impedance referred to the high-voltage side is

$$Z_{pu(HV)} = \frac{Z_{HV}}{Z_{B(HV)}} = \frac{0.3603 \text{ }\Omega}{11.52 \text{ }\Omega} = 0.0313 \text{ } pu$$

(d) The transformer impedance referred to the high-voltage side is

$$\% Z_{HV} = Z_{pu(HV)} \times 100 = (0.0313 \text{ } pu) \times 100 = 3.13 \text{ }\%$$

(e) The turns ratio of the transformer windings is

$$n = \frac{V_{HV}}{V_{LV}} = \frac{240 \text{ V}}{120 \text{ V}} = 2$$

(f) The low-voltage side base current is

$$I_{B(LV)} = \frac{S_B}{V_{B(LV)}} = \frac{5000 \text{ VA}}{120 \text{ V}} = 41.6667 \text{ A}$$

or

$$I_{B(LV)} = n I_{B(HV)} = 2(20.8333 \text{ A}) = 41.6667 \text{ A}$$

(g) The low-voltage side base impedance is

$$Z_{B(LV)} = \frac{V_{B(LV)}}{I_{B(LV)}} = \frac{120 \text{ V}}{41.6667 \text{ A}} = 2.88 \text{ }\Omega$$

or

$$Z_{B(LV)} = \frac{Z_{B(HV)}}{n^2} = \frac{11.52 \text{ }\Omega}{2^2} = 2.88 \text{ }\Omega$$

(h) The transformer impedance referred to the low-voltage side is

$$Z_{LV} = \frac{Z_{HV}}{n^2} = \frac{0.3606 \text{ }\Omega}{2^2} = 0.0902 \text{ }\Omega$$

therefore

$$Z_{pu(LV)} = \frac{Z_{LV}}{Z_{B(LV)}} = \frac{0.0902 \text{ }\Omega}{2.88 \text{ }\Omega} = 0.0313 \text{ } pu$$

or

$$Z_{pu(LV)} = Z_{pu(HV)} = 0.0313 \text{ } pu$$

Notice that in terms of per units the impedance of the transformer is the same whether it is referred to the high-voltage side or the low-voltage side.

Example B.2

Consider Example B.1 and select 300/150 V as the base voltages for the high-voltage and the low-voltage windings, respectively. Use a new base power of 10 kVA and determine the new per-unit, base, and physical impedances of the transformer referred to the high-voltage side.

SOLUTION:
By using equation (2.92), the new per-unit impedance can be found as

$$Z_{pu,\,new} = Z_{pu,\,old} \left(\frac{V_{B,\,old}}{V_{B,\,new}} \right)^2 \left(\frac{S_{B,\,new}}{S_{B,\,old}} \right)$$

$$= (0.0313 \; pu) \left(\frac{240 \text{ V}}{300 \text{ V}} \right)^2 \left(\frac{10{,}000 \text{ VA}}{5{,}000 \text{ VA}} \right) = 0.0401 \; pu$$

The new current base is

$$I_{B(HV)new} = \frac{S_B}{V_{B(HV)new}} = \frac{10{,}000 \text{ VA}}{300 \text{ V}} = 33.3333 \text{ A}$$

Thus,

$$Z_{B(HV)new} = \frac{V_{B(HV)new}}{I_{B(HV)new}} = \frac{300 \text{ V}}{33.3333 \text{ A}} = 9 \; \Omega$$

Therefore, the physical impedance of the transformer is still

$$Z_{HV} = Z_{pu,\,new} \times Z_{B(HV)new} = (0.0401 \; pu)(9 \; \Omega) = 0.3609 \; \Omega$$

PROBLEMS

B.1   Solve Example B.1 for a transformer rated 100 kVA and 2,400/240 V that has a high-voltage winding impedance of 0.9 Ω.

B.2   Consider the results of Problem B.1 and use 3,000/300 V as new base voltages for the high-voltage and low-voltage windings, respectively. Use a new base power of 200 kVA and determine the new per-unit, base, and physical impedances of the transformer referred to the high-voltage side.

B.3   A 240/120 V single-phase transformer rated 25 kVA has a high-voltage winding impedance of 0.65 Ω. If 240 V and 25 kVA are used as the base quantities, determine the following:
(a)   The high-voltage side base current.
(b)   The high-voltage side base impedance in Ω.
(c)   The transformer impedance referred to the high-voltage side in per unit.
(d)   The transformer impedance referred to the high-voltage side in percent.
(e)   The turns ratio of the transformer windings.
(f)   The low-voltage side base current.
(g)   The low-voltage side base impedance.

(h) The transformer impedance referred to the low-voltage side in per unit.

B.4 A 240/120 V single-phase transformer is rated 25 kVA and has a high-voltage winding impedance referred to its high-voltage side that is 0.2821 pu based on 240 V and 25 kVA. Select 230/115 V as the base voltages for the high-voltage and low-voltage windings, respectively. Use a new base power of 50 kVA and determine the new per-unit base, and physical impedances of the transformer referred to the high-voltage side.

REFERENCES

1. Acarnley, P. P., *Stepping Motors*, 2nd Ed., London: Peter Peregrinus Ltd., 1984.
2. "AIEE Standards Committee Report," *Electrical Engineering*, Vol. 65, No. 11, 1946, pp. 512-516.
3. Alger, P., *Induction Machines*, 2nd Ed., New York: Gordon and Breach, 1970.
4. Anderson, L. R., *Electric Machines and Transformers*, Reston, Virginia: Reston Publishing Company, 1981.
5. Anderson, P. M., *Analysis of Faulted Power Systems*, New York: IEEE Press, 1993.
6. Bergseth, F. R., and Venkata, S. S., *Introduction to Electric Energy Devices*, Englewood Cliffs, New Jersey: Prentice-Hall, Inc., 1987.
7. Bewley, L. V., *Alternating Current Machinery*, New York: Macmillan Publishing Company, 1949.
8. Blume, l. F., et al., *Transformer Engineering: A Treatise on the Theory, Operation, and Application of Transformers*, 2nd Ed., New York: John Wiley & Sons, Inc., 1951.
9. Bose, B. K., *Power Electronics and AC Drives*, Englewood Cliffs, New Jersey: Prentice-Hall Inc., 1951.
10. Brown, D. and Hamilton III, E. P., *Electromechanical Energy Conversion*, New York: Macmillian Publishing Company, 1984.
11. Bumby, J. R., *Superconducting Rotating Electrical Machines*, Oxford: Clarendon Press, 1983.
12. Carr, L., *The Testing of Electric Machines*, London: MacDonald Company Ltd., 1960.
13. Carry, C. C., *Electric Machinery: A Coordinated Presentation of AC and DC Machines*, New York: John Wiley & Sons, Inc., 1958.
14. Chapman, S. J., *Electric Machinery Fundamentals*, New York: McGraw-Hill Book Company, 1985.
15. Concordia, C. D., *Synchronous Machines Theory and Performance*, Schenectady, N. Y.: General Electric Company, 1951.
16. Crosno, C. D., *Fundamentals of Electromechanical Conversion*, New York: Harcourt, Brace & World, Inc., 1968.
17. Daniels, A. R., *The Performance of Electrical Machines*, London: McGraw-Hill Publishing Company, Limited, 1968.
18. Del Toro, V., *Electric Machinery and Power Systems*, Englewood Cliffs, N.J.: Prentice-Hall, Inc., 1985.
19. Del Toro, V., *Electromechanical Devices for Energy Conversion and Control Systems*, Englewood Cliffs, New Jersey: Prentice-Hall, Inc., 1968.
20. Dudley, A. M., and Henderson, S. F., *Connecting Induction Motors: Operation and Practice,* 4th Ed., New York: McGraw-Hill Book Company, 1960.
21. Electro-Craft Corporation, *DC Motors, Speed Controls, Servo Systems*, 3rd Ed., Oxford: Pergamon Press Ltd., 1977.
22. Elgerd, O., *Basic Electric Power Engineering*, Reading, Mass.: Addison Wesley Publishing Company, 1977.
23. Elgerd, O., *Electric Energy System Theory*, 2nd Ed., New York: McGraw-Hill Book Company, 1982.
24. El-Hawary, M. E., *Principles of Electric Machines and Power Electronic Applications*, Reston, Virginia: Reston Publishing Company, Inc., 1966.
25. Emanuel, P., *Motors, Generators, Transformers, and Energy*, Englewood Cliffs, N.J., Prentice-Hall, Inc., 1985.
26. Ergeneli, A., *Elektroteknik*, Vol. 1, Istanbul: Teknik Okulu Yayinlari, 1962.
27. Ergeneli, A., *Elektroteknik*, Vol. 2, Istanbul: Teknik Okulu Yayinlari, 1962.
28. Feinberg, R., *Modern Power Transformer Practice*, New York: John Wiley & Sons, Inc., 1979.
29. Fitzgerald, A. E., Kingsley, Jr., C., and Umans, S. D., *Electric Machinery*, 4th Ed., New York: McGraw-Hill Book Company, 1983.
30. Garik, M. L. and Whipple, C. C., *Alternating-Current Machines*, 2nd Ed., New York: John Wiley & Sons, Inc., 1986.
31. Gehmlich, D. K., and Hammond, S. B., *Electromechanical Systems*, New York: McGraw-Hill Book Company, 1967.

32. Gönen, T., *Engineering Economy for Engineering Managers*, New York: John Wiley & Sons Inc., 1990.

33. Gönen, T., *Electric Power Distribution System Engineering*, New York: McGraw-Hill Book Company, 1986.

34. Gönen, T., *Electric Power Transmission System Engineering: Analysis and Design* , New York: John Wiley & Sons, Inc., 1988.

35. Gönen, T., *Modern Power System Analysis*, New York: John Wiley & Sons, Inc., 1988.

36. Gönen, T., and Haj-mohamadi, M. S., "Electromagnetic Unbalances of Six-Phase Transmission Lines," *Electrical Power & Energy Systems*, Vol. 11, No. 2, 1989, pp. 78-84.

37. Guru, B. S. and Hiziroglu, H. R., *Electric Machinery and Transformers*, Orlando, Florida: Harcourt Brace Jovanocich, Inc., 1988.

38. Hancock, N. N., *Electric Power Utilization*, London: Sir Isaac Pitman & Sons, Ltd, 1967.

39. Hancock, N. N., *Matrix Analysis of Electrical Machinery*, 2nd Ed., Oxford: Pergamon Press Ltd., 1974.

40. Hindmarsh, J., *Electrical Machines*, Oxford: Pergamon Press Ltd., 1965.

41. Hubert, C. I., *Preventive Maintanance of Electrical Equipment*, 2nd Ed., New York: McGraw-Hill Book Company, 1969.

42. IEEE, Std, 112-1978, "IEEE Standard Test Procedure for Polyphase Induction Motors and Generators," New York: IEEE, Inc., 1984.

43. Kenjo, T., *Stepping Motors and Their Microprocessor Controls*, Oxford: Clarendon Press, 1984.

44. Kosow, I. L., *Control of Electric Motors, Englewood Cliffs*, New Jersey: Prentice-Hall, Inc., 1972.

45. Kosow, I. L., *Electric Machinery and Transformers*, Englewood Cliffs, New Jersey: Prentice-Hall, Inc., 1972.

46. Kostenko, M., and Piotrovsky, L., *Electrical Machines*, Vol. 1, Moscow: Mir Publishers, 1974.

47. Kostenko, M., and Piotrovsky, L., *Electrical Machines*, Vol. 2, Moscow: Mir Publishers, 1974.

48. Krause, P. C. and Wasynczuk, O., *Electromechanical Motion Devices*, New York: McGraw-Hill Book Company, 1989.

49. Kuhlmann, J. H., *Design of Electric Apparatus*, 2nd Ed., New York: John Wiley & Sons, Inc., 1040.

50. Kuo, B. C., *Step Motors*, St. Paul, Minnesota: West Publishing Company, 1974.

51. Lawrence, R. R. and Richards, H. E., *Principlesof Alternating Current Machinery*, 4th Ed., New York: McGraw-Hill Book Company, 1953.

52. Leonhard, W., *Control of Electric Drives*, Berlin: Springer-Verlag, 1985.

53. Lindsay, J. F. and Rashid, M. H., *Electromechanics and Electrical Machinery*, Englewood Cliffs, New Jersey: Prentice-Hall, Inc., 1986.

54. Loew, E. A. and Bergseth, F. R., *Direct and Alternating Currents: Theory and Machinery*, 4th Ed., New York: McGraw-Hill Book Company, 1954.

55. Mablekos, V. E., *Electric Machine Theory for Power Engineers*, New York: Harper & Row Publishers, 1980.

56. Majmudar, H., *Electromechanical Energy Converters*, Boston: Allyn and Bacon, Inc., 1965.

57. Matsch, L. W., and Morgan, J. D., *Electromagnetic and Electromechanical Machines*, 3rd Ed., New York: Harper & Row Publishers, 1986.

58. McIntyre, R. L., *Electric Motor Control Fundamentals*, 3rd Ed., New York: McGraw-Hill Book Company, 1974.

59. McLaren, P. G., *Elementary Electric Power and Machines*, Chichester, West Sussex, England: Ellis Horwood Ltd., 1984.

60. McLyman, Wm. T., *Transformer and Inductor Design Handbook*, New York: Marcel Dekker, Inc., 1978.

61. McPherson, G., *An Introduction to Electrical Machines and Transformers*, New York: John Wiley & Sons, Inc., 1981.

62. Meisel, J., *Principles of Electromechanical Energy Conversion*, New York: McGraw-Hill Book Company, 1966.

63. Millermaster, R., *Harwood's Control of Electric Motors*, 4th Ed., New York: John Wiley & Sons, Inc., 1970.

64. MIT Staff, *Magnetic Circuits and Transformers*, New York: John Wiley & Sons, Inc., 1943.

65. Molloy, E., (Ed.), *Small Motors and Transformers: Design and Construction*, London: George Newnes Ltd., 1953.

66. Nasar, S. A., and Boldea, I., *Linear Motion Electric Machines*, New York; John Wiley & Sons, Inc., 1976.

67. Nasar, S. A., (Ed.), *Handbook of Electric Machines*, New York: McGraw-Hill Book Company, 1987.

68. Nasar, S. A., *Electric Energy Conversion and Transmission*, New York: Macmillan Publishing Company, 1985.

69. Nasar, S. A., *Electric Machines and Electromechanics*, Outline Series in Engineering, New York: McGraw-Hill Book Company, 1981.

70. Nasar, S. A., *Electric Machines and Transformers*, New York: Macmillan Publishing Company, 1984.

71. Nasar, S. A., *Electromechanics and Electric Machines*, New York: John Wiley & Sons, Inc., 1979.

72. National Electric Manufacturing Association, Publication No. MGI-1972: *Motors and Generators,* New York: 1972.

73. Nilsson, J. W., *Introduction to Circuits, Instruments, and Electronics*, New York: Harcourt, Brace & World, Inc., 1968.

74. Patrick, D. R., and Fardo, S. W., *Rotating Electrical Machines and Power Systems*, Englewood Cliffs, N.J.: Prentice-Hall, Inc., 1985.

75. Pearman, R. A., *Power Electronics: Solid State Motor Control*, Reston, Virginia: Reston Publishing, 1980.

76. Puschstein, A. F., Lloyd, T. C ., and Conrad, A. G., *Alternating Current Machines*, 3rd Ed., New York: John Wiley & Sons, Inc., 1954.

77. Ramshaw, R. and Van Heeswijk, R. G., *Energy Conversion: Electric Motors and Generators*, Orlando, Florida: Saunders College Publishing, 1990.

78. Richardson, D. V., *Handbook of Rotating Electric Machinery*, Reston, Virginia: Reston Publishing Company, Inc., 1980.

79. Robertson, B. L., and Black, L. J., *Electic Circuits and Machines*, New York: D. Van Nostrand Company, inc., 1949.

80. Sarma, M. S., *Electric Machines: Steady-State Theory and Dynamic Performance*, Dubuque, Iowa: Wm. C. Brown Publishers, 1985.

81. Say, M. G., *Electric Engineering Design Manual*, 3rd Ed., London: Chapman and Hall Ltd., 1962.

82. Say, M. G., *Introduction to the Unified Theory of Electromagnetic Machines*, London: Pitman Publishing, 1971.

83. Seely, S., *Electromechanical Energy Conversion*, New York: McGraw-Hill Book Company, 1962.

84. Sen, P. C., *Principles of Electric Machines and Power Electronics*, New York: John Wiley & Sons, Inc., 1989.

85. Shultz, R. D., and Smith, R. A., *Introduction to Electric Power Engineering*, New York: Harper & Row Publishers, 1985.

86. Siskind, C. S., *Direct Current Machinery*, New York: McGraw-Hill Book Company, 1952.

87. Siskind, C. S., *Electrical Control Systems in Industry*, New York: McGraw-Hill Book Company, 1963.

88. Siskind, C. S., *Electrical Machines, Direct and Alternating Current*, 2nd Ed., New York: McGraw-Hill Book Company, 1959.

89. Skilling, H. H., *Electrical Engineering Circuits*, 2nd Ed., New York: John Wiley & Sons, Inc., 1966.

90. Skilling, H. H., *Electromechanics*, New York: John Wiley & Sons, Inc., 1962.

91. Skrotzki, B. G. A., and Vopat, W. A., *Power Station Engineering and Economy*, New York: McGraw-Hill Book Company, 1960.

92. Slemon, G. R., and Straughen, A., *Electric Machines*, Reading, Mass.: Addison-Wesley Publishing Company, 1980.

93. Smeatson, R. W., (Ed.), *Motor Application and Maintenance Handbook*, New York: McGraw-Hill Book Company, 1969.

94. Smith, R. T., *Analysis of Electric Machines*, New York: Pergamon Press, 1982.

95. Stein, R., and Hunt, W. T., *Electric Power System Components: Transformers and Rotating Machines*, New York: Van Nostrand Reinhold Company, 1979.

96. Stevenson, W. D., *Elements of Power System Analysis*, 4th Ed., New York: McGraw-Hill Book Company, 1981.

97. Stigant, S. A. and Franklin, A. C., *J&P Transformer Book*, 10th Ed., London: Newnes-Butterworth, 1973.

98. Thaler, G. J., and Wilcox, M. L., *Electric Machines: Dynamics and Steady State*, New York: John Wiley & Sons, Inc., 1966.

99. Veinott, C. G., and Martin, J. E., *Fractional- and Subfractional-Horsepower Electric Motors*, 4th Ed., New York: McGraw-Hill Book Company, 1986.

100. Veinott, C. G., *Theory and Design of Small Induction Motors*, New York: McGraw-Hill Book Company, 1959.

101. Wenick, E. H., (Ed.), *Electric Motor Handbook*, London: McGraw-Hill Book Company, 1978.

102. Westinghouse Staff, *Electrical Transmission and Distribution Reference Book*, East Pittsburgh, Pa.: Westinghouse Electric Corporation, 1964.

103. Zorbas, D., *Electric Machines: Principles, Applications, and Control Schematics*, St. Paul, Minnesota: West Publishing Company, 1989.

ANSWERS TO SELECTED PROBLEMS

Chapter 1

1.1 $P_{tot} = 18$ kW

1.2 $C = 378.93 \ \mu$F

1.4   **(a)** $S_{tot} = 309{,}232.9\angle - 14.04°$ A;   **(b)** $I_{tot} = 64.4\angle - 14.04°$ A;   **(c)** $PF = 0.97$ leading

Chapter 2

2.2   **(a)** $I_L = 75.4$ A;   **(b)** $\theta = 31.8°$;   **(c)** $V_{R(\phi)} = 66{,}474$ V;   **(d)** $I_L Z = 8{,}474.4\angle 31.6°$ V;
(e) $V_{s(L)} = 127{,}692.9$ V;   **(f)** $P_{3\phi} = 852.77$ kW

2.5   **(a)** $S_{12} = 146{,}715.5 - j138{,}741.9$ VA;   **(b)** $P_{12} = 146{,}715.5$ W;   **(c)** $Q_{12} = - 138{,}741.9$ var

2.8   **(a)** $P_1 = 8{,}860.5$ W, $P_2 = 4{,}897.5$ W;   **(b)** $P_{tot} = 13{,}796.5$ W

2.10 $I_a = 19.9\angle - 35°$ A, $I_b = 16.98\angle 9.6°$ A

2.12   **(a)** $Z = 2.2412\angle 51.66°$ Ω;   **(b)** $I_a = 53.5427\angle - 51.66°$A;   **(c)** $S_{tot} = 19{,}275.4\angle 51.66°$ VA

2.14   **(a)** $I_L = 73.83$ A;   **(b)** $C = 0.608 \ \mu$F

2.16   **(a)** $I_a = 566.1\angle 45°$ A, $I_b = 566.1\angle - 75°$ A, $I_c = 566.1\angle 165°$ A;   **(b)** $P_{tot} \cong 2{,}884.3$ MW

2.18   **(a)** $I_a = 1{,}044.14\angle 54.9°$ A, $I_b = 626.3\angle - 26.6°$ A, $I_c = 1{,}0294.55\angle - 153.69°$ A;
(b) $V_{DE} = 1944.3\angle - 36.6°$ V;   **(c)** $P_{tot} = 6.656$ MW

2.21   **(a)** $S_{tot} = 4 + j3$ MVA;   **(b)** $Z_\phi = 13.84\angle 36.87°$ Ω

2.23   **(a)** $I_a = 100\angle - 30°$ A, $I_b = 100\angle - 150°$ A, $I_c = 100\angle 90°$ A;   **(b)** $I_n = 0$

2.26   **(a)** $Z_{eq} = 2.5473\angle 55.84°$ Ω;   **(b)** $I_a = 108.79\angle - 55.84°$ A;   **(c)** $S_{tot} = 90{,}449\angle 55.84°$ VA

2.30   **(a)** $V_1 = 479.9513\angle 30°$ V, $V_2 = 479.9513\angle 150°$ V, $V_3 = 479.9513\angle - 90°$ V;   **(b)** $I_1 = $
$17.3205\angle 70°$ A, $I_2 = 17.3205\angle 190°$ A, $I_3 = 17.3205\angle - 50°$ A;   **(c)** $I_{L1} = 30\angle 40°$ A,
$I_{L2} = 30\angle - 80°$ A, $I_{L3} = 30\angle 160°$ A

2.33   **(a)** $P_1 = 8{,}860.53$ W, $P_2 = 4{,}897.51$ W;   **(b)** $P_{tot} = 13{,}758.04$ W

Chapter 3

3.1   **(a)** $l = 0.4084$ m, $A = 0.0033$ m^2;   **(b)** $B = 1.06$ T;   **(c)** $\mathscr{F} = 408.4$ A·turns;   **(d)** $I = 1.06336$ A

3.3 $\mathscr{F}_{coil} = 5260.986$ A·turns

3.5   **(a)** $\Phi = 0.0018$ Wb;   **(b)** $\Phi = 0.0009$ Wb

3.7   **(a)** $I = 5.6$ A;   **(b)** $\mu_r = 426.5$;   **(c)** $\mathscr{R}_c = 93{,}333$ A·turns/Wb

3.10   **(a)** $I = 1.2031$ A;   **(b)** $\mu_r = 723.4$;   **(c)** $\mathscr{R}_c = 76{,}998.4$ A·turns/Wb

3.13   **(a)** $I = 7.2$ A;   **(b)** $\mu_r = 221$;   **(c)** $\mathscr{R}_c = 144{,}000$ A·turns/Wb

3.17 $\mathscr{F}_{coil} = 5{,}092.6$ A·turns

3.20 $\Phi_{tot} = 0.26$ Wb

3.23 $\Phi_{tot} = 7.9 \times 10^{-2}$ Wb

3.26 $B = 0.634$ Wb/m^2

3.30 $B = 2.44$ Wb/m^2

3.34 $\Phi_{tot} = 2.0196 \times 10^{-3}$ Wb

Chapter 4

4.1   (a) $a = 5.75$;   (b) $I_1 = 7.2464$ A, $I_2 = 41.6667$ A;   (c) $Z_L = 1{,}904.4$ W;   (d) $Z_L = 57.6$ W

4.4   (a) $I_1 = 20.8333$ A, $I_2 = 208.3333$ A;   (b) $a = 10$, $N_1 = 500$ turns;   (c) $\Phi_m = 0.018$ Wb

4.7   (a) $a = 6.32$;   (b) $I_1 = 0.8$ mA, $I_2 = 0.0047$ A, $V_2 = 0.47$ V, $P_2 = 2.2$ mW

4.11 $I_{load} = 20.2973\angle - 73.05°$ Ω, $V_{load} = 211.91\angle 0.25°$ V, $P_{line\ loss} = 65.92$ W

4.15 $I_{load} = 19.7815\angle - 73.38°$ Ω, $V_{load} = 159.5\angle - 13.13°$ V, $P_{line\ loss} = 1.96$ W

4.21   (a) $R_c = 88.617$ kΩ, $X_m = 3.696$ kΩ;   (b) $Z_{eq,\ 1} = 0.69 + j2.05$ Ω   (c) $R_{eq,\ 1} = 0.69$ Ω,

        $X_{eq,\ 1} = 2.05$ Ω

4.25 $Z_{th} = 0.002 + j0.009$ Ω

4.31   (a) $VReg = 5.15$ %;   (b) $VReg = 1.9\%$;   (c) $VReg = -1.73\%$

4.35 $kVA_{rated} = 125$ kVA, $I_{se} = 52.1$ A, $I_L = 62.5$ A, $I_c = 10.4$ A

4.38   (a) $\eta = 97.1\%$;   (b) $\eta = 97.38\%$;   (c) $\eta = 97.1\%$

Chapter 5

5.1 $\omega = 377$ rad/s

5.4   (a) $P = 75{,}398.2$ W;   (b) $P = 101.07$ hp

5.8   (a) $F = 0.1125$ N;   (b) to the left;   (c) to the right

Chapter 6

6.2   (a) $n_s = 2$ poles;   (b) $s_{FL} = 5\%$;   (c) 3 Hz;   (d) $n_{rr} = 180$ rpm;   (e) $n_{rs} = 3600$ rpm;

        (f) $n_{rsf} = 0$ rpm

6.5   (a) $n_s = 3600$ rpm;   (b) $n = 3420$ rpm;   (c) $f_2 = 3$ Hz; $T_{shaft} = 52.07$ N·m or $T_{shaft} = 38.39$ lb.ft

6.8   (a) $P_g = 34{,}334$ W;   (b) $P_d = 33{,}834$ W;   (c) $P_{shaft} = 32{,}834$ W;   (d) $\eta = 92.9\%$

6.16   (a) $n = 1740$ rpm;   (b) $I_1 = 116.33\angle - 26.05°$ A; $PF = 0.9$

6.21   (a) $s = 5.5\%$;   (b) $T_d = 308.76$ lb.ft;   (c) $n_m = 1602$ rpm;   (d) $140{,}516.4$ W

6.25   (a) $s_{max} = 0.1858$;   (b) $n_m = 2442.6$ rpm;   (c) $T_{max} = 470.3$ N·m;   (d) $T_{shaft} = $

        191.55 N·m;   (e) $s_{max} = 1885.15$ rpm;   (f) $R_2 = 0.14$ Ω;   (g) $T_{shaft} = 328.1$ N·m

6.30   (a) $s = 0.001107$;   (b) $I_1 = 38.426\angle - 83.7°$ A;   (c) $P_{in} = 3221$ W

Chapter 7

7.2   (a) $\omega = 188.5$ rad/s;   (b) $E_{max} = 391.723$ V;   (c) 277 V;   (d) 480 V

7.6 $E_\phi = 2{,}731.3\angle 8.59°$ V

7.9   (a) $P_{in} = 20{,}271.74$ W;   (b) $S_{in} = 25{,}339.67$ VA

7.13   (a) $E_a = 227.36\angle - 40.52°$ V;   (b) $\delta = 40.52°$;   (c) $\eta = 90.23\%$;   (d) $P_s = 76{,}692$ W,

        $T_s = 407$ N·m

7.18   (a) $n_s = 3600$ rpm;   (b) $I_f = 4.7$ A;   (c) $I_f = 5.9$ A;   (d) $\eta = 92.73\%$;   (e) $E_a = 534.77$ V;

        (f) $I_f = 4.25$ A

7.22   (a) at 0.8 lagging PF: $V_t = 423$V, at 1.0 PF: $V_t = 472$ V, at 0.8 leading PF: $V_t = 527$ V;

        (b) $\eta = 94.7\%$;   (c) $T_{app} = 328.5$ N·m $T_{ind} = 311$ N·m;   (d) at 0.8 lagging PF:

$VReg = 13.5\%$, at 1.0 PF: $VReg = 1.7\%$, at 0.8 leading PF: $VReg = -8.9\%$

7.27 (a) $I_a = 29.6\angle36.87°$ A, $E_a = 360.7 - 15.21°$ V; (b) $P_{out} = 55.95$ kW; (c) $I_L = 129.6$ A, $PF = 0.976$ lagging

7.30 $S_{tot} = 665.87\angle33.39°$ kVA

Chapter 8

8.2 (a) $Z = 500\ \Omega$; (b) $A_p = 0.0495$ m^2, $\Phi_d = 0.0346$ Wb; (c) $a = 6$ lap, $K_a = 79.58$; (d) $E_a = 259.5$ V; (e) $I_{coil} = 33.33$ A

8.4 (a) $E_a = 264.9$ V; (b) $E_a = 235.1$ V

8.7 $T_{d(125)} = 304.8$ N·m

8.11 (a) $R_a = 0.03\ \Omega$; (b) $I_a = 403$ W; (c) total losses $= 11{,}928$ W; (d) $\eta = 89.3\%$

8.14 (a) $E_a = 262.3$ V; (b) $T_d = 171.16$ N·m; (c) $\eta = 88.9\%$

8.17 $N_{se} = 7$ turns

8.21 (a) $n_{100} = 1{,}061$ rpm; (b) $T_{d(100)} = 206.9$ N·m

8.26 (a) $R_{a,\,B} = 0.3556\ \Omega$; (b) $R_a = 0.0178\ \Omega$; (c) $I_{a,\,start} = 13{,}500$ A

8.30 $R_a = 01202\ \Omega$, $R_f = 78.13\ \Omega$

Chapter 9

9.2 (a) $Z_f = 27.893 + j36.65\ \Omega$, $Z_b = 1.83 + j1.9\ \Omega$; (b) $I_1 = 4.35\angle51.05°$ A; (c) $P_{in} = 328.2$ W; (d) $P_g = 246.6$ W; (e) $T_d = 1.31$ N·m; (f) $P_d = 234.3$ W; (g) $P_{out} = 190.3$ W; (h) $T_{out} = 1.06$ N·m; $\eta = 58\%$

9.6 (a) $I_1 = 7.533\angle -51.05°$ A, $P_g = 739.5$ W, $T_d = 3.92$ N·m; (b) $I_1 = 8.69\angle -51.05°$ A, $P_g = 984.2$ W, $T_d = 5.22$ N·m

9.9 $C = 187.9\ \mu$F

9.12 (a) $n_m = 476.6$ rpm; (b) $T_d = 23.57$ N·m

Chapter 10

10.1 $V_t = E_a = 260(1.0 - \exp -2t)u(t)$

10.4 (a) $E_a(t) = 250(1 - \exp -5t)$; (b) $E_a(\infty) = 250$ V; (c) $t = 0.92$ s

10.6 (b) $i_{ac}(t) = I_{dc,\,max} = 103{,}092$ A; (c) $I_{dc} = 39{,}019$ A; (d) $I_{dc} = 55{,}181$ A

Appendix A

A.2 $I = 10\angle30°$ A

A.4 $I_1 = 10\angle30°$ A, $I_2 = 20\angle -140\infty$ A, $I_2 = 10.30\angle49.7°$ A

Appendix B

B.2 $Z_{pu,\,new} = 0.02$ pu, $I_{B(HV)new} = 666.6667$ A, $Z_{B(HV)new} = 0.45\ \Omega$

INDEX

Find % VR + η

20 kVA   6 Hz   2400/240 V

0...04   -0.0279   0. 204   0.0279   P.F. = 0.8 lagging

I_1 I_2 +

g=0.0021 V $j \cdot 0.04414 V$ V_2 ← Thevenin eq

j 0.0558   0.408

$10 I_1 = I_1'$ g $-j$ I_2 $V_2 = 240 \angle 0$

$\frac{V_1}{10} = V_1'$

$I_2 = \frac{S}{V_2}$ $-36.87 = \frac{20000}{240} \angle -36.87 = 83.33 \angle -36.87$

$V_1' = 240 \angle 0 + (83.33 \angle -3.87)(0.0408 + j 0.0558) = 245.5 \angle 0.39$

$\% \, VR = \frac{245.5 - 240.0}{240.0} \times 100 = 2.29 \%$

$P_{cu} = (83.33)^2 (0.0408) = 283.3$

$P_{core} = \frac{V_1'^2}{R_c} = V_1'^2 g = (245.5)^2 (0.002198) = 132.5$

$P_\ell = 283.3 + 132.5 = 415.8$

$\eta = \frac{P_{out}}{P_{in}} = \frac{16000}{16415.8} = 97.5 \%$

$V_1' = 240 \angle 0$

$I_1 = \frac{20000}{240} \angle -36.87 = 83.33 \angle -36.87$

$\bar{I}_2 = I_1' - (240 \angle 0)(0.002198)(240 \angle 0)(0.04414 \angle -90 = 76.98 \angle -30.8$

$V_2 = 240 \angle 0 - (76.98 \angle -30.8)(0.0408 + j 0.0558) = 235.1 \angle -0.51$

$\% \, VR = \frac{240 - 235.1}{235.1} = 2.08 \%$

$I_a = \left(\frac{-V t}{\varepsilon R} + I_L \right) \left(1 + \frac{N_s}{N_f} \right)^{-1}$ $I_{feq} = I_f + \frac{N_s}{N_f} I_a$

$\bar{I}_a = \bar{I}_L - I_{feq}$

$E_a = V_t - I_a R_a - 2$

$w = \frac{E_a}{mag \, curve}$ * rating of mag curve